The ACS Style Guide

A Manual for Authors and Editors

2ND EDITION

Janet S. Dodd, Editor

American Chemical Society
Washington, DC

Library of Congress Cataloging-in-Publication Data

The ACS style guide: a manual for authors and editors / Janet S. Dodd, editor—2nd ed.

p. cm.

Includes bibliographical references and index.

ISBN 0–8412–3461–2.—ISBN 0–8412–3462–0 (pbk.)

1. Chemical literature—Authorship—Handbooks, manuals, etc. 2. English language—Style—Handbooks, manuals, etc.

I. Dodd, Janet S., date

QD8.5.A25 1997
808′.06654—dc21 96–49413
CIP

The paper used in this publication meets the minimum requirements of American National Standard for Information Sciences—Permanence of Paper for Printed Library Materials, ANSI Z39.48-1984.

PRINTED IN THE UNITED STATES OF AMERICA

Contributors

◆ Janet S. Dodd
Chemical & Engineering News
American Chemical Society
1155 16th Street, NW
Washington, DC 20036

◆ Marvin Coyner
Office of Public Outreach
American Chemical Society
1155 16th Street, NW
Washington, DC 20036

◆ Madeleine Jacobs
Chemical & Engineering News
American Chemical Society
1155 16th Street, NW
Washington, DC 20036

◆ Barbara Friedman Polansky
Copyright Administrator
American Chemical Society
1155 16th Street, NW
Washington, DC 20036

◆ K. Barbara Schowen
University of Kansas
Deparment of Chemistry
Lawrence, KS 66045-0046

◆ David Weisgerber
Chemical Abstracts Service
2540 Olentangy River Road
P.O. Box 3012
Columbus, OH 43210-0012

◆ Larry James Winn
Western Kentucky University
Department of Communications
Bowling Green, KY 42101

Contents

CHAPTER *9*

Illustrations and Tables *281*

CHAPTER *10*

Peer Review *305*

Preface

Publishing has always involved numerous editorial decisions, and now decisions about technology have been added. Computers have changed everything. It is hard to think of one aspect of life that has not been touched by computers in some way. Certainly the publishing industry has been greatly affected by computer technology, from the authors writing their papers in their offices to the editors carrying out the production process in publishing offices. The printing process itself has undergone enormous changes, and the final version of a paper may be an electronic product and not a printed product. Authors and editors are performing functions now that they never imagined 10 years ago. What's more, it seems that as soon as they get accustomed to one level of technology, it changes significantly, and they need to be learning constantly.

In the midst of all this change, the comforting thought is that one goal of authors and editors has not changed: to communicate information in the most understandable and expedient fashion in publications of the highest quality. To accomplish that goal, we need guidelines. This book is intended to guide and answer questions for authors and editors, to save them time, and to ensure clarity and consistency. In any publication, quality cannot exist without consistency, and consistency helps readers focus on content as opposed to style. By lessening the burden of one type of decision, this style guide will make the publishing process faster and more efficient. The book is addressed to both authors and editors because, essentially, authors start the process and editors finish it.

Editors and authors need each other, they need to cooperate, and they need clear guidelines. At ACS, we would like to achieve good author–editor relationships so that we will publish readable and well-read scientific literature. We hope that this book will help us to accomplish that goal.

Acknowledgments

I am grateful for the suggestions of all of my colleagues who took time to review the chapters or provide information: Pamela D. Angulo, Jay Cherniak, Zeki Erim (who also did the index), Elaine Firestone of NASA, Susan Fisher, Amie Jackowski, Alan Kahan, Betsy Kulamer, Donna Lucas, Judith Marcus, Beth Mitchell, Jennie Reinhardt, Anna Tecson, Louise Voress, Mary Warner, Elizabeth Wood, and Celia McFarland of the Columbus Editorial Office. I give special thanks to my good friend and long-time coworker, Paula M. Bérard, who not only reviewed the chapters and made excellent suggestions, but also was the production editor for the book.

JANET S. DODD
American Chemical Society
Washington, DC

CHAPTER *1*

Writing a Scientific Paper

This chapter is a general guide to writing a scientific paper. Specific guidelines for text length, preparation of figures and tables, and instructions on how to submit your paper differ from journal to journal and publisher to publisher. For ACS journals and special publications, read the Guide, Notes, Notice, or Instructions for Authors that appear in each publication's first issue of the year and on the World Wide Web at http://pubs.acs.org. For ACS books, consult the brochure "How To Prepare Your Manuscript for the ACS Symposium Series" or "Instructions for Authors", available from the Books Department or on the World Wide Web at the same address.

Getting Started

Although there is no fixed set of "writing rules" to be followed like a cookbook recipe or an experimental procedure, some guidelines can be helpful. Start by answering some questions:

- What is the function or purpose of this paper? Are you describing original and significant research results? Are you reviewing the literature? Are you providing an overview of the topic? Something else?
- How is your work different from that described in other reports on the same subject? (Unless you are writing a review, be sure that your paper will make an original contribution. Some publishers, including ACS, do not publish previously published material.)

- What is the best place for this paper to be published—in a journal or as part of a book? If a journal, which journal is most appropriate? (Appendix I, "ACS Publications", describes ACS journals and books.)
- Who is the audience? What will you need to tell them to help them understand your work?

Answering these questions will clarify your goals and thus make it easier for you to write the paper with the proper amount of detail. It will also make it easier for editors to determine the paper's suitability for their publications. Writing is like so many other things: if you clarify your overall goal, the details fall into place.

Advice from the Authorities

The Elements of Style
by William Strunk, Jr., and E. B. White

Omit needless words. Vigorous writing is concise. A sentence should contain no unnecessary words, a paragraph no unnecessary sentences, for the same reason that a drawing should have no unnecessary lines and a machine no unnecessary parts. This requires not that the writer make all his sentences short, or that he avoid all detail and treat his subjects only in outline, but that every word tell....

Avoid fancy words. Avoid the elaborate, the pretentious, the coy, and the cute. Do not be tempted by a twenty-dollar word when there is a ten-center handy, ready, and able.... All [words] are good, but some are better than others.

Errors in English and Ways To Correct Them
by Harry Shaw

No standards can be absolute. Our language is constantly changing. Also, diction, like fashions in dress and food, is influenced by changes in taste. Again, what is acceptable in daily speech and conversation may not be suitable in written form. The use of this or that word cannot be justified by saying that it is often heard or seen in print. Advertisements, newspapers, magazines, and even some "good" books may exhibit faulty diction.

Once you know the function of your paper and have identified its audience, review your material for completeness or excess. Then, organize your material into the standard format: introduction, experimental details or theoretical basis, results, discussion, and conclusions. This format has become standard because it is suitable for most reports of original research, it is basically logical, and it is easy to use. The reason it accommodates most reports of original research is that it parallels the scientific method of deductive reasoning: define the problem, create a hypothesis, devise an experiment to test the hypothesis, conduct the experiment, and draw conclusions. Furthermore, this format enables the reader to understand quickly what is being presented and to find specific information easily. This ability is crucial now more than ever because scientists, if not all professionals, must read much more material than their time seems to allow.

Even if your results are more suited to one of the shorter types of presentation, the logic of the standard format applies, although you might omit the standard headings or one or more entire sections. As you write, you can modify, delete, or add sections and subsections as appropriate.

An extremely important step is to check the specific requirements of the publication you have targeted and follow them. Most publications require revisions of manuscripts that are not in their requested format. Thus, not following a publication's requirements can delay publication and make more work for you. Finally, your paper will be peer-reviewed, so a good idea is to pay attention to the aspects that the reviewers will be considering. Chapter 10 presents the opinions of many reviewers.

Writing Style and Word Usage

Short declarative sentences are the easiest to write and the easiest to read, and they are usually clear. However, too many short sentences in a row can sound abrupt or monotonous. To add sentence variety, it is better to start with simple declarative sentences and then combine some of them than to start with long rambling sentences and then try to shorten them.

You and your colleagues probably have been discussing the project for months, so the words seem familiar, common, and clear to you. However, the readers will not have been part of these discussions. That is where copy editors can help. Their job is to make sure that readers understand the material you are presenting.

By all means, write in your own personal style, but keep in mind that scientific writing is not literary writing. Scientific writing serves a purpose completely different from that of literary writing, and it must therefore be precise and unambiguous.

If English is not your first language, ask an English-speaking colleague—if possible, a native English speaker—for help with grammar and diction.

Choosing the Correct Word or Phrase

◆ Use words in their primary meanings; do not use a word to express a thought if such usage is uncommon, informal, or primarily literary. Examples are using "since" when you mean "because", and "while" when you mean "although". Many words are clear when you are speaking because you can amplify your meaning with gestures, expressions, and vocal inflections—but when these same words are written, they may be clear only to you.

◆ Use appropriate verb tenses.

- Simple past tense is correct for stating what was done, either by others or by you: "The solutions were heated to boiling." "The spectra were recorded." "Jones reviewed the literature and gathered much of this information." "We recently found that relativistic effects enhance the bond strength." "The structures were determined by neutron diffraction methods."

- Present tense is correct for statements of fact: "Absolute rate constants for a wide variety of reactions are available." "Hyperbranched compounds are macromolecular compounds that contain a branching point in each structural repeat unit."

- Present and simple past tenses may both be correct for results, discussion, and conclusions: "The characteristics of the voltammetric wave indicate that electron transfer and breaking of the carbon–iodine bond are concerted." "The absence of substitution was confirmed by preparative-scale electrolysis at a potential located at the foot of the voltammetric wave." "IR spectroscopy shows that nitrates are adsorbed and are not removed by washing with distilled water."

◆ Use the active voice when it is less wordy and more direct than the passive.

Poor The fact that such processes are under strict stereoelectronic control is demonstrated by our work in this area.

Better Our work in this area demonstrates that such processes are under strict stereoelectronic control.

◆ Use first person when it helps to keep your meaning clear and to express a purpose or a decision.

> Jones reported xyz, but I (or we) found...
>
> I (or we) present here a detailed study...
>
> My (or our) recent work demonstrated...
>
> To determine the effects of structure on photophysics, I (or we)...

However, avoid phrases such as "we believe", "we feel", "we concluded", and "we can see", as well as personal opinions.

◆ Use an affirmative sentence rather than a double negative.

Instead of	***Consider using***
This reaction is not uncommon	This reaction is common This reaction is not rare This reaction occurs about 40% of the time
This transition was not unexpected	This transition was expected We knew that such transitions were possible
This strategy is not infrequently used	This strategy is frequently used This strategy is occasionally used
This result is not unlikely to occur	This result is likely to occur This result is possible

◆ Watch the placement of the word "only". It has different meanings in different places in the sentence.

> Only the largest group was injected with the test compound. (Meaning: and no other group)
>
> The largest group was only injected with the test compound. (Meaning: and not given the compound in any other way)
>
> The largest group was injected with only the test compound. (Meaning: and no other compounds)
>
> The largest group was injected with the only test compound. (Meaning: there were no other test compounds)

◆ Be sure that the antecedents of the pronouns "this" and "that" are clear. If there is a chance of ambiguity, use a noun to clarify your meaning.

> ***Ambiguous*** The photochemistry of transition-metal carbonyl complexes has been the focus of many investigations. This is due to the central role that metal carbonyl complexes play in various reactions.
>
> ***Unambiguous*** The photochemistry of transition-metal carbonyl complexes has been the focus of many investigations. This interest is due to the central role that metal carbonyl complexes play in various reactions.

◆ Use the proper subordinating conjunctions. "While" and "since" have strong connotations of time. Do not use them where you mean "although", "because", or "whereas".

> ***Poor*** Since solvent reorganization is a potential contributor, the selection of data is very important.
>
> ***Better*** Because solvent reorganization is a potential contributor, the selection of data is very important.
>
> ***Poor*** While the reactions of the anion were solvent-dependent, the corresponding reactions of the substituted derivatives were not.
>
> ***Better*** Although the reactions of the anion were solvent-dependent, the corresponding reactions of the substituted derivatives were not.
>
> ***Also*** The reactions of the anion were solvent-dependent, but (or whereas) the corresponding reactions of the substituted derivatives were not.

◆ Use "respectively" to relate two or more sequences in the same sentence.

> The excitation and emission were measured at 360 and 440 nm, respectively. (That is, the excitation was measured at 360 nm and the emission was measured at 440 nm.)

◆ Use the more accurate terms "greater than" or "more than" rather than the imprecise "over" or "in excess of".

> greater than 50%, *not* in excess of 50%
> more than 100 samples, *not* over 100 samples
> more than 25 mg, *not* in excess of 25 mg, *not* over 25 mg

◆ Use "fewer" to refer to number; use "less" to refer to quantity.

> fewer than 50 animals　　fewer than 100 samples
> less product　　less time　　less work

◆ However, use "less" with number and unit of measure combinations because they are regarded as singular.

> less than 5 mg　　less than 3 days

◆ Use "between" with two named objects; use "among" with three or more named or implied objects.

> Communication between scientists and the public is essential.
>
> Communication among scientists, educators, and the public is essential.
>
> Communication among scientists is essential.

◆ Choose "assure", "ensure", and "insure" depending on your meaning. To assure is to affirm; to ensure is to make certain; to insure is to indemnify for money.

> He assured me that the work had been completed.

The procedure ensures that clear guidelines have been established.

You cannot get a mortgage unless you insure your home.

◆ Choose "affect", "effect", and "impact" depending on your meaning. "Affect" is a verb meaning to influence, modify, or change. "Effect" as a verb means to bring about, but as a noun it means consequence, outcome, or result. "Impact" is a noun meaning a significant effect.

The increased use of pesticides affects agricultural productivity.

The use of polychlorinated benzenes has an effect on the cancer rate.

The effect of the added acid was negligible.

The new procedure effected a 50% increase in yield.

The impact of pesticide use on health is felt throughout the world.

The acid did not have a great impact on the reaction rate.

◆ It is acceptable to use split infinitives to avoid awkwardness or ambiguity.

Awkward The program is designed to assist financially the student who is considering a career in chemistry.

Better The program is designed to financially assist the student who is considering a career in chemistry.

Ambiguous The bonded phases allowed us to investigate fully permanent gases.

Better The bonded phases allowed us to fully investigate permanent gases.

◆ Use "whether" to introduce at least two alternatives, either stated or implied.

I am not sure whether I should repeat the experiment.

I am not sure whether I should repeat the experiment or use a different statistical treatment.

I am going to repeat the experiment whether the results are positive or negative.

◆ Use "whether or not" to mean "regardless of whether".

Incorrect I am not sure whether or not to repeat the experiment.

Correct I am not sure whether to repeat the experiment.

Also correct Whether or not the results are positive, I will repeat the experiment.

Also correct Whether or not I repeat the experiment, I will probably leave the laboratory late tonight.

◆ Use "to comprise" to mean "to contain" or "to consist of"; it is not a synonym for "to compose". The whole *comprises* the parts, or the whole *is com-*

Advice from the Authorities

Writing Successfully in Science
by Maeve O'Connor

When you start writing the draft, or your share of it, you should, ideally, cut yourself off from the outside world. Try to find a time when you can remain undisturbed for several hours and a place where no one will interrupt you. Write at the time of day when you feel freshest and most alert....

If you find it difficult to start writing on the blank page or screen in front of you, leave the introduction for later and start with any section you have already drafted or made detailed notes about. The materials and methods section is often the easiest place to begin, and the results section the next easiest. Once you get going, write as quickly as you can. If the article is short, try to finish it in one sitting, to give it as much unity as possible....

Long words and complicated sentences are not essential features of good scientific writing, although they are often thought to be so. The best writing in science, as elsewhere, is simple, clear, precise, and vigorous. Decide what you want to say and say it as simply, informatively, and directly as possible.

On Writing Well
by William Zinsser

Clutter is the disease of American writing. We are a society strangling in unnecessary words, circular constructions, pompous frills, and meaningless jargon.... But the secret of good writing is to strip every sentence to its cleanest components. Every word that serves no function, every long word that could be a short word, every adverb which carries the same meaning that is already in the verb, every passive construction that leaves the reader unsure of who is doing what—these are the thousand and one adulterants that weaken the strength of a sentence. And they usually occur, ironically, in proportion to education and rank.

posed of the parts, but the whole is not comprised of the parts. Never use "is comprised of".

> ***Incorrect*** A book is comprised of chapters.
>
> ***Correct*** A book comprises chapters.
>
> ***Also correct*** A book is composed of chapters.
>
> ***Incorrect*** Our research was comprised of three stages.
>
> ***Correct*** Our research comprised three stages.

Articles

◆ Choose the articles "a" and "an" according to the pronunciation of the words or abbreviations they precede.

> a nuclear magnetic resonance spectrometer
> an NMR spectrometer

◆ Use "a" before an aspirated "h"; use "an" before the vowel sounds of a, e, i, o, "soft" u, and y.

a house	a history, *but* an hour	an honor
a union	a U-^{14}C, *but* an ultimate	
an ylide	an yttrium compound	

◆ Choose the proper article to precede B.A., B.S., M.A., M.S., and Ph.D., according to pronunciation of the first letter.

a B.S. degree	an M.S. degree	a Ph.D.

Comparisons

◆ Introductory phrases that imply comparisons should refer to the subject of the sentence and be followed by a comma.

> ***Incorrect*** Unlike alkali-metal or alkaline-earth-metal cations, hydrolysis of trivalent lanthanides proceeds significantly at this pH.
>
> ***Correct*** Unlike that of alkali-metal or alkaline-earth-metal cations, hydrolysis of trivalent lanthanides proceeds significantly at this pH.
>
> ***Also correct*** Unlike alkali-metal or alkaline-earth-metal cations, trivalent lanthanides hydrolyze significantly at this pH.
>
> ***Incorrect*** In contrast to the bromide anion, there is strong distortion of the free fluoride anion on the vibrational spectroscopy time scale.
>
> ***Correct*** In contrast to the bromide anion, the free fluoride anion is strongly distorted on the vibrational spectroscopy time scale.

◆ Use the verb "compare" followed by the preposition "to" when similarities are being noted. Use "compare" followed by the preposition "with" when differences are being noted. Only things of the same class should be compared.

> Compared to compound **3**, compound **4** shows an NMR spectrum with corresponding peaks.
>
> Compared with compound **3**, compound **4** shows a more complex NMR spectrum.

◆ Do not omit words needed to complete comparisons, and do not use confusing word order. The subordinating conjunction "than" is often used to introduce the second element in a comparison, following an adjective or adverb in the comparative degree.

> ***Incorrect*** The alkyne stretching bands for the complexes are all lower than the uncoordinated alkyne ligands.
>
> ***Correct*** The alkyne stretching bands for the complexes are all lower than those for the uncoordinated alkyne ligands.
>
> ***Also correct*** The alkyne stretching bands are all lower for the complexes than for the uncoordinated alkyne ligands.
>
> ***Incorrect*** The decrease in isomer shift for compound **1** is greater in a given pressure increment than for compound **2**.
>
> ***Correct*** The decrease in isomer shift for compound **1** is greater in a given pressure increment than that for compound **2**.
>
> ***Also correct*** The decrease in isomer shift in a given pressure increment is greater for compound **1** than for compound **2**.

◆ Idioms often used in comparisons are "different from", "similar to", "identical to", and "identical with". Generally these idioms should not be split.

> ***Incorrect*** The complex shows a significantly different NMR resonance from that of compound **1**.
>
> ***Correct*** The complex shows an NMR resonance significantly different from that of compound **1**.
>
> ***Incorrect*** Compound **5** does not catalyze hydrogenation under similar conditions to compound **6**.
>
> ***Correct*** Compound **5** does not catalyze hydrogenation under conditions similar to those for compound **6**.

Exception These idioms can be split if an intervening prepositional phrase modifies the first word in the idiom.

> The single crystals are all similar in structure to the crystals of compound **7**.
>
> Solution A is identical in appearance with solution B.

◆ Phrases such as "relative to", "as compared to", and "as compared with" and words such as "versus" are also used to introduce the second element in a comparison. The things being compared must be in parallel structure (that is, grammatically equal).

> The greater acidity of nitric acid relative to nitrous acid is due to the initial-state charge distribution in the molecules.
>
> The lowering of the vibronic coupling constants for Ni as compared with Cu is due to configuration interaction.
>
> This behavior is analogous to the reduced Wittig-like reactivity in thiolate versus phenoxide complexes.

Parallelism

◆ Use coordinating conjunctions ("and", "but", "or", "nor", "yet", "for", and sometimes "so"), correlative conjunctions ("either, or"; "neither, nor"; "both, and"; "not only, but also"; "not, but"), and correlative constructions ("as ... as"; e.g., "as well as") to connect words or groups of words of equal grammatical rank.

> ***Incorrect*** Compound **12** was prepared analogously and by Lee's method (5).
>
> ***Correct*** Compound **12** was prepared in an analogous manner and by Lee's method (5).
>
> ***Incorrect*** It is best to use alternative methods both because of the condensation reaction and because the amount of water in the solvent increases with time.
>
> ***Correct*** It is best to use alternative methods both because of the condensation reaction and because of the increase in the amount of water in the solvent with time.
>
> ***Incorrect*** The product was washed either with alcohol or acetone.
>
> ***Correct*** The product was washed with either alcohol or acetone.
>
> ***Also correct*** The product was washed either with alcohol or with acetone.
>
> ***Incorrect*** Not only was the NiH functionality active toward the C-donor derivatives but also toward the N donors.
>
> ***Correct*** The NiH functionality was active not only toward the C-donor derivatives but also toward the N donors.
>
> ***Also correct*** The NiH functionality was not only active toward the C-donor derivatives but also active toward the N donors.
>
> ***Also correct*** Not only was the NiH functionality active toward the C-donor derivatives, but it was also active toward the N donors.

◆ Use parallel constructions in series and lists, including section headings and subheadings in text and tables and listings in figure captions.

◆ Do not try to use parallel construction around the word "but" when it is not used as a coordinating conjunction.

> Increasing the number of fluorine atoms on the adjacent boron atom decreases the chemical shift, but only by a small amount.
>
> The reaction proceeded readily, but with some decomposition of the product.

Words and Phrases To Avoid

◆ Avoid slang and jargon.

◆ If you have already presented your results at a symposium or other meeting and are now writing the paper for publication in a book or journal, delete all references to the meeting or symposium such as "Good afternoon, ladies and gentlemen", "This morning we heard", "in this symposium", "at

Advice from the Authorities

The Scientist as Editor
by Maeve O'Connor

Write simply and concisely.... Use short words rather than long ones, and concrete rather than abstract terms; where appropriate, prefer the first person singular or plural to the third person, and the active to the passive voice. Avoid vague statements, jargon and laboratory slang, and words not defined in dictionaries.

Scientific English: A Guide for Scientists and Other Professionals
by Robert A. Day

Scientists (and perhaps scholars in all fields) should learn to use English *simply*. Short, simple words—in short, straightforward sentences—usually convey meaning more clearly than do esoteric words and convoluted sentences. This concept is a bit controversial, because the skilled writer, using that wonderful, massive vocabulary we have available in English, can paint word pictures of overwhelming beauty. On the other hand, clarity and meaning can easily fade into the background.

this meeting", and "I am pleased to be here". Such phrases would be appropriate only if you were asked to provide an exact transcript of a speech.

◆ Be brief. Wordiness obscures your message, annoys the reader, and displeases the publisher because the resulting lengthy paper is more expensive to produce and to print.

- Omit phrases such as

  As already stated
  It has been found that
  It has long been known that
  It is interesting to note that
  It is worth mentioning at this point
  It may be said that
  It was demonstrated that

- Omit excess words.

Instead of	*Use*
It is a procedure that is often used.	This procedure is often used.
There are seven steps that must be completed.	Seven steps must be completed.
This is a problem that is...	This problem is...
These results are preliminary in nature.	These results are preliminary.

- Use single words instead of phrases.

Instead of	*Use*
a number of	many, several
a small number of	a few
are in agreement	agree
are found to be	are
are known to be	are
at present	now
at the present time	now
based on the fact that	because
by means of	by
despite the fact that	although
due to the fact that	because
during that time	while
fewer in number	fewer
for the reason that	because
has been shown to be	is
if it is assumed that	if
in color, e.g., red in color	*just state the color*, e.g., red
in consequence of this fact	therefore, consequently
in length	long
in order to	to

Instead of	*Use*
in shape, e.g., round in shape	*just state the shape*, e.g., round
in size, e.g., small in size	*just state the size,* e.g., small
in spite of the fact that	although
in the case of ...	in ..., for ...
in the near future	soon
in view of the fact that	because
is known to be	is
it appears that	apparently
it is clear that	clearly
it is likely that	likely
it is possible that	possibly
it would appear that	apparently
of great importance	important
on the order of	about
owing to the fact that	because
prior to	before
reported in the literature	reported
subsequent to	after

- Do not use contractions in scientific papers.

  Incorrect The identification wasn't confirmed by mass spectrometry.

  Correct The identification was not confirmed by mass spectrometry.

- Do not use the word "plus" or the plus sign as a synonym for "and".

  Incorrect Two bacterial enzymes were used in a linked-enzyme assay for heroin plus metabolites.

  Correct Two bacterial enzymes were used in a linked-enzyme assay for heroin and its metabolites.

- Do not use "respectively" when you mean "separately" or "independently".

  Incorrect The electrochemical oxidations of chromium and tungsten tricarbonyl complexes, respectively, were studied.

  Correct The electrochemical oxidations of chromium and tungsten tricarbonyl complexes were studied separately.

- Avoid misuse of prepositional phrases introduced by "with".

  Poor Nine deaths from leukemia occurred, with six expected.

  Better Nine deaths from leukemia occurred, and six had been expected.

  Poor Of the 20 compounds tested, 12 gave positive reactions, with three being greater than 75%.

  Better Of the 20 compounds tested, 12 gave positive reactions; three of these were greater than 75%.

Poor Two weeks later, six more animals died, with the total rising to 25.

Better Two weeks later, six more animals died, and the total was then 25.

- Do not use a slash to mean "and" or "or".

Incorrect Hot/cold extremes will damage the samples.

Correct Hot and cold extremes will damage the samples.

- Replace "and/or" with either "and" or "or", depending on your meaning.

Incorrect Our goal was to confirm the presence of the alkaloid in the leaves and/or roots.

Correct Our goal was to confirm the presence of the alkaloid in the leaves and roots.

Also correct Our goal was to confirm the presence of the alkaloid in either the leaves or the roots.

Also correct Our goal was to confirm the presence of the alkaloid in the leaves, the roots, or both.

Advice from the Authorities

"The Development of Research Writing"
by Robert A. Day

A scientific experiment is not complete until the results have been published. Therefore, to *do* science, one must also *write* science. Realizing this, scientists should weigh the words in their manuscripts as carefully as they weigh the reagents in their laboratories....

In scientific writing, there is no room for and no need for ornamentation. The flowery literary embellishments, the metaphors, the similes, and the idiomatic expressions are very likely to cause confusion and should seldom be used in writing research papers. Science is simply too important to be communicated in anything other than words of certain meaning. The meaning should be clear not only to peers of the author, but also to students just embarking on their careers, to scientists reading outside their narrow discipline, and especially to those readers (the majority of readers today) whose native language is other than English.

Gender-Neutral Language

The U.S. government and many publishers have gone to great effort to encourage the use of gender-neutral language in their publications. Gender-neutral language is also a goal of many chemists. Recent style guides and writing guides urge copy editors and writers to choose terms that do not reinforce outdated sex roles. Gender-neutral language can be accurate and unbiased and not necessarily awkward.

The most problematic words are the noun "man" and the pronouns "he" and "his", but there are usually several satisfactory gender-neutral alternatives for these words. Choose an alternative carefully and keep it consistent with the context.

◆ Instead of "man", use "people", "humans", "human beings", or "human species", depending on your meaning.

> ***Outdated*** The effects of compounds **I–X** were studied in rats and man.
>
> ***Gender-neutral*** The effects of compounds **I–X** were studied in rats and humans.
>
> ***Outdated*** Men working in hazardous environments are often unaware of their rights and responsibilities.
>
> ***Gender-neutral*** People working in hazardous environments are often unaware of their rights and responsibilities.
>
> ***Outdated*** Man's search for beauty and truth has resulted in some of his greatest accomplishments.
>
> ***Gender-neutral*** The search for beauty and truth has resulted in some of our greatest accomplishments.

◆ Instead of "manpower", use "workers", "staff", "work force", "labor", "crew", "employees", or "personnel", depending on your meaning.

◆ Instead of "manmade", use "synthetic", "artificial", "built", "constructed", "manufactured", or even "factory-made".

◆ Instead of "he" and "his", change the construction to a plural form ("they" and "theirs") or first person ("we", "us", and "ours"). Alternatively, delete "his" and replace it with "a", "the", or nothing at all. "His or her", if not overused, is not terribly unpleasant.

> ***Outdated*** The principal investigator should place an asterisk after his name.
>
> ***Gender-neutral*** Principal investigators should place asterisks after their names.
>
> ***Gender-neutral*** If you are the principal investigator, place an asterisk after your name.

Gender-neutral The name of the principal investigator should be followed by an asterisk.

However, do not use a plural pronoun with a singular antecedent.

Incorrect The principal investigator should place an asterisk after their name.

- Instead of "wife", use "family" or "spouse" where appropriate.

Outdated The work of professionals such as chemists and doctors is often so time-consuming that their wives are neglected.

Gender-neutral The work of professionals such as chemists and doctors is often so time-consuming that their families are neglected.

Outdated the society member and his wife

Gender-neutral the society member and spouse

Components of a Paper

Use the standard format, which is described next, for reports of original research but not necessarily for literature reviews or theoretical papers. Present all parts of your paper as concisely as possible.

Title

The best time to determine the title is after you have written the text, so that the title will reflect the paper's content and emphasis accurately and clearly. The title must be brief and grammatically correct but accurate and complete enough to stand alone. A two- or three-word title may be too vague, but a 14- or 15-word title is unnecessarily long. Choose terms that are as specific as the text permits: "a vanadium–iron alloy" rather than "a magnetic alloy". Avoid phrases such as "on the", "a study of", "research on", "report on", "regarding", and "use of". In most cases, omit "the" at the beginning of the title. Avoid nonquantitative, meaningless words such as "rapid" and "new".

Spell out all terms in the title, and avoid jargon, symbols, formulas, and abbreviations. Whenever possible, use words rather than expressions containing superscripts, subscripts, or other special notations. Do not cite company names, specific trademarks, or brand names of chemicals, drugs, materials, or instruments.

The title serves two main purposes: (1) to attract the potential audience and (2) to aid retrieval and indexing. Therefore, be sure to include several keywords. The title should provide the maximum information for a computerized title search.

Series titles are of little value. Some publications do not permit them at all. If consecutive papers in a series are published simultaneously, a series title may be relevant, but in a long series, paper 42 probably bears so limited a relationship to paper 1 that they do not warrant a common title. In addition, an editor or reviewer seeing the same title repeatedly may reject it on the grounds that it is only one more publication on a general topic that has already been discussed at length.

If you cannot create a title that is short, consider breaking it into title and subtitle.

Byline and Affiliation

Include in the byline all those, and only those, who made substantial contributions to the work, even if the paper was actually written by only one person. Appendix III, "Ethical Guidelines to Publication of Chemical Research", is more explicit on this topic.

Many ACS publications specifically request at least one full given name for each author, rather than only initials. Use your first name, initial, and surname (e.g., John R. Smith) or your first initial, second name, and surname (e.g., J. Robert Smith). Whatever byline you use, be consistent. Papers by John R. Smith, Jr., J. Smith, J. R. Smith, Jack Smith, and J. R. Smith, Jr., will not be indexed in the same place; the bibliographic citations may be listed in five different locations, and ascribing the work to a single author will therefore be difficult if not impossible.

Do not include professional, religious, or official titles or academic degrees.

The affiliation is the institution (or institutions) at which the work was conducted. If there is more than one author, use an asterisk or superscript (check the specific publication's style) to indicate the author or authors to whom correspondence should be addressed. Clarify all corresponding authors' addresses by accompanying footnotes if they are not apparent from the affiliation line. Telephone numbers, fax numbers, and electronic mail (e-mail) addresses may be included in corresponding author footnotes.

Also provide the corresponding author's e-mail address and fax number, in addition to postal address and telephone number.

Abstract

Most publications require an informative abstract for every paper, even if they do not publish abstracts. For a research paper, briefly state the problem or the purpose of the research, indicate the theoretical or experimental plan used, summarize the principal findings, and point out major conclusions. Include chemical safety information when applicable. Do not supplement or evaluate the conclusions in the text. For a review paper, the abstract describes the topic, the scope, the sources reviewed, and the conclusions. Write the abstract last to be sure that it accurately reflects the content of the paper.

The abstract allows the reader to determine the nature and scope of the paper and helps editors identify key features for indexing and retrieval.

Although an abstract is not a substitute for the article itself, it must be concise, self-contained, and complete enough to appear separately in abstract publications. Often, authors' abstracts are used in *Chemical Abstracts*. Further-

Advice from the Authorities

Writing To Learn
by William Zinsser

I would say this to everyone who feels that his main aptitude is for science or technology, or for any other field that lies outside the humanities, and that therefore he can't write: Learn to use the tools without fear. They are not some kind of secret apparatus owned by the English teacher or any other teacher. They are simple mechanisms for putting your thoughts on paper. Enjoy finding out how they work. Take as much pleasure in what an active verb will do for you as in what a mathematical formula will do, or a computer, or a centrifuge.

The Chemist's English
by Robert Schoenfeld

Don't be frightened of grammar. When you sit down to write your paper or thesis or report, your most dangerous enemy is not the split infinitive—it is ambiguity. A split infinitive is very often acceptable anyway, but where it needs correcting it can be corrected by a copy editor. However, the copy editor, unless he is a mind-reader, cannot correct an ambiguity. So, even if you are not a smooth writer, don't sit there staring at the blank page: get your facts down first and fix up the dangling participles afterwards.

more, abstracts of full papers submitted to ACS journals will be published in *Advance ACS Abstracts* several weeks before the journal is published.

The optimal length is one paragraph, but it could be as short as two sentences. The length of the abstract depends on the subject matter and the length of the paper. Between 80 and 200 words is usually adequate.

Do not cite references, tables, figures, or sections of the paper in the abstract. Do not include equations, schemes, or structures that require display on a line separate from the text.

Use abbreviations and acronyms only when it is necessary to prevent awkward construction or needless repetition. Define abbreviations at first use in the abstract (and again at first use in the text).

Introduction

A good introduction is a clear statement of the problem or project and the reasons that you are studying it. This information should be contained in the first few sentences. Give a concise and appropriate background discussion of the problem and the significance, scope, and limits of your work. Outline what has been done before by citing truly pertinent literature, but do not include a general survey of semirelevant literature. State how your work differs from or is related to work previously published. Demonstrate the continuity from the previous work to yours. The introduction can be one or two paragraphs long. Often, the heading "Introduction" is not used because it is superfluous; opening paragraphs are usually introductory.

Experimental Details or Theoretical Basis

In research reports, this section can also be called "Experimental Methods", "Experimental Section", or "Materials and Methods". Check the specific publication. For experimental work, give sufficient detail about your materials and methods so that other experienced workers can repeat your work and obtain comparable results. When using a standard method, cite the appropriate literature and give only the details needed.

Identify the materials used, and give information on the degree of and criteria for purity, but do not reference standard laboratory reagents. Give the chemical names of all compounds and the chemical formulas of compounds that are new or uncommon. Use meaningful nomenclature; that is, use standard systematic nomenclature where specificity and complexity require, or use trivial nomenclature where it will adequately and unambiguously define a well-established compound.

Describe apparatus only if it is not standard or not commercially available. Giving a company name and model number in parentheses is nondistracting and adequate to identify standard equipment.

Avoid using trademarks and brand names of equipment and reagents. Use generic names; include the trademark in parentheses after the generic name only if the material or product you used is somehow different from others. Remember that trademarks often are recognized and available as such only in the country of origin. In ACS publications, *do not use* trademark (^{TM}) and registered trademark (®) symbols.

Describe the procedures used, unless they are established and standard.

Note and emphasize any hazards, such as explosive or pyrophoric tendencies and toxicity, in a separate paragraph introduced by the word "Caution:". Include precautionary handling procedures, special waste disposal procedures, and any other safety considerations in adequate detail so that workers repeating the experiments can take appropriate safety measures. Some ACS journals also indicate hazards as footnotes on their contents pages.

In theoretical reports, this section is called, for example, "Theoretical Basis" or "Theoretical Calculations" instead of "Experimental Details" and includes sufficient mathematical detail to enable other researchers to reproduce derivations and verify numerical results. Include all background data, equations, and formulas necessary to the arguments, but lengthy derivations are best presented as Supporting Information.

Results

Summarize the data collected and their statistical treatment. Include only relevant data, but give sufficient detail to justify your conclusions. Use equations, figures, and tables only where necessary for clarity and brevity.

Discussion

The purpose of the discussion is to interpret and compare the results. Be objective; point out the features and limitations of the work. Relate your results to current knowledge in the field and to your original purpose in undertaking the project: Have you resolved the problem? What exactly have you contributed? Briefly state the logical implications of your results. Suggest further study or applications if warranted.

Present your results and discussion either as two separate sections or as one combined section if it is more logical to do so. Do not repeat information given elsewhere in the manuscript.

Conclusions

The purpose of the Conclusions section is to put the interpretation into the context of the original problem. Do not repeat discussion points or include irrelevant material. Your conclusions should be based on the evidence presented.

Summary

A summary is unnecessary in most papers. In long papers, a summary of the main points can be helpful, if you stick to the main points only. If the summary itself is too long, its purpose is defeated.

Acknowledgments

Generally, the last paragraph of the paper is the place to acknowledge people, organizations, and financing. As simply as possible, thank those persons, other than coauthors, who added substantially to the work, provided advice or technical assistance, or aided materially by providing equipment or supplies. Do not include their titles. If applicable, state grant numbers and sponsors here, as well as auspices under which the work was done, including permission to publish.

Follow the journal's guidelines on what to include in the Acknowledgments section. Some journals permit financial aid to be mentioned in acknowledgments, but not meeting references. Some journals put financial aid and meeting references together, but not in the Acknowledgments section.

References

In many journals and books, references are placed at the end of the article or chapter; in others, they are treated as footnotes. In any case, place your list of references at the end of the manuscript.

In ACS books and most journals, the style and content of references are standard regardless of where they are located. Follow the reference style presented in Chapter 6.

The accuracy of the references is the author's responsibility. If you copy citations from another source, check the original reference for accuracy and appropriate content.

Special Sections

This discussion on format applies to most manuscripts, but it is not a set of rigid rules and headings. If your paper is well-organized, scientifically sound, and appropriate to the publication for which you are preparing it, you may include other sections and subsections. For example, an appendix contains material that is not critical to understanding the text but provides important background information.

Supporting Information

Material that may be essential to the specialized reader but not require elaboration in the paper itself is published as Supporting Information. Examples are large tables, extensive figures, lengthy experimental procedures, mathematical derivations, analytical and spectral characterization data, biological test data for a series, molecular modeling coordinates, modeling programs, crystallographic information files (CIF), instrument and circuit diagrams, and expanded discussions of peripheral findings. More journals are encouraging this type of publishing to keep printed papers shorter.

For complete instructions on how to prepare this material for publication, check the Guide, Notes, Notice, or Instructions for Authors that appear in each publication's first issue of the year and on the World Wide Web at http://pubs.acs.org.

When you include Supporting Information, place a statement to that effect at the end of the paper using the format specified in the author instructions for the specific journal.

ACS publishes this material in the CD-ROM, microfilm, and microfiche editions of its journals. For many papers published in ACS journals, Supporting Information is also available electronically on the World Wide Web at http://pubs.acs.org.

Types of Presentations

The following are general descriptions; Appendix I discusses each type of presentation with specific reference to ACS publications.

Articles

Articles, also called full papers, are definitive accounts of significant, original studies. They present important new data or provide a fresh approach to an established subject.

The organization and length of an article should be determined by the amount of new information to be presented and by space restrictions within the publication. The standard format is suitable for most papers in this category.

Notes

Notes are concise accounts of original research of a limited scope. They may also be preliminary reports of special significance. The material reported must be definitive and may not be published again later. Appropriate subjects for notes include improved procedures of wide applicability or interest, accounts of novel observations or of compounds of special interest, and development of new techniques. Notes are subject to the same editorial appraisal as full-length articles.

Communications

Communications, called "Letters" or "Correspondence" in some publications, are usually preliminary reports of special significance and urgency that are given expedited publication. They are accepted if the editor believes that their rapid publication will be a service to the scientific community. Communications are subject to strict length limitations; they must contain specific results to support their conclusions, but they may not contain nonessential experimental details.

The same rigorous standards of acceptance that apply to full-length papers also apply to communications. Communications are submitted to review, and they are not accepted if the editor believes that the principal content has been published elsewhere. In many cases, authors are expected to publish complete details (not necessarily in the same journal) after their communications have been published. Acceptance of a communication, however, does not guarantee acceptance of the detailed manuscript.

Reviews

Reviews integrate, correlate, and evaluate results from published literature on a particular subject. They seldom report new experimental findings. Effective review articles have a well-defined theme, are usually critical, and present novel theoretical interpretations. Ordinarily, they do not give experimental details, but in special cases (as when a technique is of central interest), experimental procedures may be included. An important function of reviews is to serve as a guide to the original literature; for this reason, accuracy and completeness of references cited are essential. Reviews critically analyze the literature.

Advice from the Authorities

Line by Line: How To Improve Your Own Writing
by Claire Kehrwald Cook

You probably should delete all intensive adverbs—*very, really, truly, actually,* and the like. If you've chosen the right word, adding a *very* defeats your purpose. If you haven't got the right word, the *very* offers poor compensation. Readers pay no attention to this overused word. If you want to put a *very* in front of a *large*, you should consider substituting *enormous, huge, gigantic,* or *massive.*

Scientists Must Write: A Guide to Better Writing for Scientists, Engineers, and Students
by Robert Barrass

In science, every statement should be based on evidence and not on unsupported opinion. Speculation cannot take the place of evidence. The scientist should therefore avoid excessive qualification. Words and phrases such as *possible, probably, perhaps, it is likely to,* and *is better referred to perhaps* should cause you to think again. Have I considered the evidence sufficiently? Is there enough evidence for the qualification to be omitted? If not, are further investigations needed before the work is ready for publication?

Book Chapters

In multiauthored books, chapters may be accounts of original research or literature reviews (like journal articles), but they may also be topical overviews. They may be developed and expanded from presentations given at symposia, or they may be written especially for the book in which they will be published. Multiauthored books should contain at least one chapter that reviews the subject thoroughly and also provides an overview to unify the chapters into a coherent treatment of the subject. In a longer book that is divided into sections, each section may need a short overview chapter.

In books entirely written by one author or collaboratively by more than one author, each chapter treats one subdivision of the broader topic, and each is a review and overview.

Bibliography

Barrass, Robert. *Scientists Must Write: A Guide to Better Writing for Scientists, Engineers, and Students;* Chapman & Hall: New York, 1978.

Cook, Claire Kehrwald. *Line by Line: How To Improve Your Own Writing;* Houghton Mifflin: Boston, MA, 1985.

Day, Robert A. The Development of Research Writing. *Scholarly Publishing,* January 1989, pp 107–115.

Day, Robert A. *Scientific English: A Guide for Scientists and Other Professionals,* 2nd ed.; Oryx Press: Phoenix, AZ, 1995.

Eisenberg, Anne. *Writing Well for the Technical Professions;* Harper & Row: New York, 1989.

King, Lester S. *Why Not Say It Clearly: A Guide to Scientific Writing,* 2nd ed.; Little, Brown: Boston, MA, 1991.

O'Connor, Maeve. *Writing Successfully in Science;* Chapman & Hall: New York, 1992.

Rathbone, Robert R. *Communicating Technical Information: A New Guide to Current Uses and Abuses in Scientific and Engineering Writing,* 2nd ed.; Addison-Wesley: Reading, MA, 1985.

Schoenfeld, Robert. *The Chemist's English,* 3rd ed.; VCH: Weinheim, Germany, 1990.

Shaw, Harry. *Errors in English and Ways To Correct Them,* 4th ed.; HarperPerennial: New York, 1993.

Strunk, William, Jr.; White, E. B. *The Elements of Style,* 3rd ed.; Macmillan: New York, 1979.

Zinsser, William. *On Writing Well: An Informal Guide to Writing Nonfiction,* 5th ed.; HarperPerennial: New York, 1994.

Zinsser, William. *Writing To Learn;* Harper & Row: New York, 1988.

CHAPTER 2

Communicating in Other Formats: Posters, Letters to the Editor, and Press Releases

Tips for Effective Poster Presentations

K. Barbara Schowen
University of Kansas

Think back to the last poster session you attended. Which were the memorable posters? Which communicated their science the most successfully, and how? When asked these questions, most people will agree that effective posters clearly state the research problem and the conclusion reached; use a minimum of words and panels, a readable font, and clearly labeled graphs and diagrams; and look simple, neat, and pleasing to the eye. How to prepare such a poster is the subject of this section.

A poster is one of two common methods used at meetings and conferences for communicating the results of recent scientific investigations. As with an oral slide presentation, a poster is expected to contain the information typical of scientific papers: background, purpose, methods, results, interpretation, and conclusions.

In a poster, this information is customarily displayed in concise form with a minimum of text on a wall or easel-like apparatus. The space is usually designated by number in accordance with a printed program, and the area at one's disposal is on average about 5–6 feet wide and 3–4 feet high.

(ACS meetings generally provide space 6 feet wide and 4 feet high; 2 m × 1 m is common at international meetings.)

The poster is commonly scheduled for viewing during a specified time period, and the presenter is expected to be available for questions and discussion during a designated part of that time. The advantage and pleasure of poster presentation for many individuals is this opportunity for meaningful one-on-one dialogue with viewers of common scientific interests.

How Does a Poster Differ from a Slide Talk?

A poster uses a very different mode of communication, has a different kind of audience, and induces a different type of discussion from those of a slide talk, and a poster can be appreciated long after its official use.

A poster is alone in the world of scientific reporting in its minimal use of words, relying mainly on nonverbal visual means of communication. With a poster, any oral component is minimal and less structured and will depend on the nature of a viewer's questions and comments.

The attendees at a talk constitute a "captive" audience. The attendees at a poster session, on the other hand, are more fluid and varied. They may happen on your poster more or less by accident, or they may have specifically sought it out. They may glance at your poster cursorily and move on, or they may pore over each panel and stay for an extended discussion. Most will fall somewhere between these extremes.

Time for questioning at a talk is often limited and necessarily involves (at least passively) the entire group. With a poster the possibility exists for more extensive, lively, and individual contact with genuinely interested people.

A poster, furthermore, because it is tangible, can later be put on display—for example, at the home institution.

When Is a Poster Most Appropriate?

Talks and posters lend themselves to different kinds of scientific communicating. A poster is ideal for reporting a contained body of work, a single experiment (or related set of experiments), or something with a straightforward question posed and a clear and clean-cut conclusion. An oral presentation, on the other hand, is much better suited for instruction, persuasion, development of arguments, and evaluation (of conflicting data or theories, for example), for critical literature review, or for reporting on a number of loosely related experiments. None of these activities works well in a poster format. When given a choice, bear in mind what posters can and cannot do well, and choose accordingly.

Designing the Poster: Concept, Style, and Tone

Before buying poster paper and going to the word processor, think back on poster sessions you have attended. If possible, attend one or two again, look at samples that may be on display at your institution, and evaluate what works. Your goal in this survey is not to critique the scientific content or the superficial impact of the presentation, but to see (1) which posters allow you to get the most information plus a notable message in the most efficient and straightforward manner and (2) which ones catch your eye and make you stop to read more closely. These—the simultaneously effective and pleasing posters—are your models.

Creativity Versus Communication

There are, of course, many possible physical designs for an effective poster. Devising one (if you do not have the services of a professional graphic arts facility) can be a pleasurable creative activity and a welcome change in routine, especially for those who have an eye for color and design as well as the artistic ability and time to execute their visions. Most of us, however, have neither the time nor the talent for such indulgence and should repress over-ambitious creative urges in favor of simplicity. Strike a balance, keep things simple, and let the message through.

What you want is a good-looking poster—one that looks as though you care about presenting your work and one that will entice a viewer to stop and look—but above all one that promulgates your scientific message. The message should leap to the eye and be remembered, not the fact that you have pasted your information on purple poster board at an unusual angle.

Know Your Audience

Poster sessions fall into two main types: those at large meetings and those at small special-interest conferences. The poster-session audience at a large meeting will not necessarily be as well versed in the area of your poster as the audience at a smaller meeting. Give some thought to this difference, therefore, when planning the introduction, background, and significance sections of your poster. For example, it may be more important to include a detailed review of the reaction catalyzed by a particular enzyme and its generally accepted mechanism in a general session of a large meeting than it would be at a conference devoted to enzyme mechanisms and attended only by specialists in that area. Nevertheless, a common error is to overestimate the viewers' familiarity with your subject. It is never a mistake to introduce your material assuming no background knowledge.

Overall Considerations

Four things need to be considered—often simultaneously—in designing a successful poster:

1. the physical aspects (mechanics of display and general appearance);
2. the scientific message and how best to communicate it;
3. the details of the presentation (arrangement, layout, and size) of this message (be it textual, graphical, or pictorial) on individual poster panels; and
4. the arrangement of the individual panels with respect to one another and the space provided.

Designing the Physical Display: Concept and Mechanics

The first step always is to find out the dimensions of the space allotted to you.

Basic Recipe

Here is a simple recipe for a bare-bones, no-frills, simple, attractive, effective, easy-to-put-together, easy-to-transport, and easy-to-hang generic poster. Create a title–author panel as described in the next section. Then, print what you want to say (or show) in large format on letter-size paper, number each sheet, and then affix each sheet to a slightly larger sheet of colored construction paper. Arrange these panels in logical, sequential order at eye level in the space provided. Construction paper has the advantage over poster board because it weighs less and is easier to push thumbtacks through; the rigidity of the heavier poster board, however, results in a sturdier, more permanent finished product. (ACS meeting guidelines expressly advise against using heavy stock because of mounting difficulty.) The choice of color is certainly up to you, but some things work better than others. In general, using one color throughout gives a cleaner and more unified look to the display, and darker colors often look better than pale or "fluorescent" tones. Cut the paper or board and attach the printed sheets so that everything is straight. It sometimes is nice to have varying sizes of paper, but keep in mind that anything larger than letter size may be harder to transport and may get damaged.

Something To Catch the Eye

The basic recipe is too bland and boring? In that case, you can allow a bit of creative free rein and think of ways to personalize or "spice up" the poster. Some effective touches are a molecular model glued onto a colored piece of paper, a photo of your apparatus or of computer-generated models or graphics, an actual piece of equipment, or a vial of a brightly colored or crystalline compound. One of these accessories per poster is probably sufficient.

The Title–Author Panel

This panel, of course, is a must. It is customary to make a banner that fits across the top of your assigned space. The heading should contain the title, authors, and affiliations, usually one line for each, for example:

> THE STEREOCHEMISTRY OF THIOLATES
> Michael F. Houk, Dale A. Johnson, and Andrew R. Marcos
> Department of Chemistry, University of Rochester, Rochester, NY 14627

Be sure that all agrees with the printed program. The lettering should be large enough to be read from across a room. Letters for the title should be at least 1.5 in. (3.8 cm) high boldface type; letters for the authors and affiliations can be somewhat smaller.

Banners are generally the most variable and creative elements in posters, but they can be time-consuming to prepare. Many organizations have good software available for landscape printing onto continuous sheets of computer paper; sometimes splicing of smaller sheets of printed paper works; and, as a last resort, so does hand lettering on a long strip of paper. Do not hesitate to consult experienced poster presenters at your institution for tips. Also, many copy centers are well set up and willing to help with such projects, by blowing up to specified size a smaller strip of paper containing the desired text, for example. You may wish to incorporate your institution's logo or colors.

The Scientific Message: Getting the Story Out

Operate on the assumption that, given the usual number of posters and the time allotted per session, a person will seldom have more than five minutes to devote to your poster. Then design the reading component of your poster such that, on average, the gist of each panel can be understood and appreciated in less than a minute. This step takes some care and forethought; a poster will not work if you simply write a report, print it, tape it up, and hope that someone will take the time to go through it. Here, then, are some general considerations when planning for each component of a typical poster with sections for background, purpose, experimental or theoretical methods, results, interpretation, and conclusion.

Background and Introduction

As with any scientific communication, providing the context for the results to be reported is of paramount importance and may determine whether the reader will appreciate the rest of the report. It will be most effective if the introduction can be summarized in one paragraph or less and printed with a large font on no more than a single sheet of paper. Leading references can be included.

Purpose or Statement of Problem

In general, the objective of a particular investigation of the sort presented in a poster can be expressed in one or two succinct sentences. Again, these can be printed in large type on a sheet of paper. Ideally, and if possible, express this "statement" as a single question—for example, "Given that the ^{13}C-isotope effect for nonenzymic decarboxylation of $CH_3CO–COOH$ is 5% (1.05), does decarboxylation, product release, or both limit the rate of pyruvate decarboxylase action with saturating pyruvate at 25 °C and pH 6?"

Experimental or Theoretical Approach

Here there is a great deal of variability, and your own situation will determine how much is needed and what detail is warranted. The experimental method may be a major, or even the main, point of the poster and require some elaboration of procedure, schematics of the apparatus, and so forth. Absolutely to be avoided, however, is a report of the type that would go in the experimental section of a formal paper, dissertation, or report. Use bulleted lists of procedures; sketches, figures, diagrams, or photos of equipment; and a listing of conditions. Essentials should be given, but not detail.

Data and Results

The cardinal rule for effective presentation of experimental data in posters does not really differ from that for preparing slides or overhead transparencies for a talk: keep the information per panel to a minimum and make it easy to absorb at a glance. For example, graphical presentation of numerical data is generally more effective than tabulation (be sure the axes are clearly labeled); give structures of compounds rather than names wherever possible; and label key spectral and chromatographic peaks. Be sure that each figure, graph, or spectrum is completely labeled so that it can stand on its own and not require you to interpret it.

Interpretation

Summarize as much as possible. Avoid complicated prose and long paragraphs. Use short numbered or bulleted phrases whenever possible.

Conclusion and "Take-Home Message"

This is probably the most important part of your poster, so be sure it is short, pithy, and attractive to the eye. Two sentences is the recommended maximum. Ideally, the conclusion will be the answer to the question posed at the outset. For example, "The observed ^{13}C-isotope effect is 2.5%. Decarboxylation is thus 50% rate-limiting, and product release is also 50% rate-limiting, both steps occurring at equal rates." The conclusion should not be a simple summary.

Further Plans

Consider including next steps if you plan to continue work in the area. These plans may be outlined in a short numbered list or, if succinct, be made a part of the conclusion: "Compound 8 was successfully synthesized in five steps in 35% overall yield. Its activity as a cholinesterase inhibitor is under investigation, and samples have been submitted to the National Cancer Center for screening."

Acknowledgments

It is proper to acknowledge funding sources, individuals, facilities, and so on as appropriate. A small, separate panel near the end is fine for this purpose.

Designing the Individual Panels: Simple, Legible, and Complete

The overall challenge is to convey all information concisely. I have seen very effective posters that consisted of only four panels: introduction and purpose, experimental method, results, and conclusion. The style of lettering, spacing, placement on the page, and so forth are a matter of individual discretion. An essential consideration, however, is legibility. The most important rule is to choose fonts and construct graphs and other art such that the entire poster can be read from a distance of at least 3 ft (1 m), preferably 4 ft.

Type and figures should be balanced over the entire poster as much as possible to avoid distraction; that is, font size and style should be uniform, and figures should be of similar size. Type with letters that are 3/8 in. high (e.g., Times New Roman 50-point boldface font) is recommended. The size of Times New Roman 24-point boldface should be considered an absolute minimum for the text. Headings should be larger type than the text, boldface, on lines by themselves, and have blank space above and below them.

Avoid, under all circumstances, panels that consist of a sheet of paper filled with normal-size (10- or 12-point) typed text and material that reads as if taken from a formal paper, progress report, or grant application. No one will have time to read and do justice to such documents in the time available. A poster is not a journal paper.

Designing the Overall Layout: Clean and Logical

While you are deciding on the scientific content and how it will appear on the individual panels, give thought to the overall appearance and layout of your poster, and the placement of the individual panels with respect to each other and within the assigned space. The first consideration is the logical flow of your panels. Most poster readers will look for the beginning at the top left-hand corner. You will need to guide them from there on, even though each sheet should be numbered. Are the panels to be read in horizontal rows from left to right or in vertical columns from top to bottom? The decision is dictated by the shape and number of your individual panels and by the dimensions of your space. Poster areas that are wider than they are high lend themselves to the horizontal layout and vice versa. The viewer's eye can be directed by means of numbered panels (preferable), arrows, red thread, or appropriate headings, but in all cases the direction should be very clear. Most often, the conclusion is somewhere on the right.

Another consideration is placement of the panels with respect to the floor. Keep in mind that it is very difficult for people to comfortably read material that is much above eye level or below hip level. The best way to solve this problem is to keep the number of panels to a minimum and to place them at a convenient average height. This generally means that the top of the highest panel (not counting title banner) should be no more than 6 ft (1.8 m) from the floor.

Putting the Poster Together: Cutting and Pasting

When you have printed your material and have any photos or objects for display, you are ready to put everything together for the final presentation format. Have at hand two or three large sheets of poster board or heavy construction paper (available at art supply outlets or university bookstores) of the color you have chosen. This colored board or paper may then be cut so as to be somewhat larger than the sheets of paper you have printed. Uniform margins of approximately ½ in. or 1 cm generally look best. For 8½ × 11 in. paper, the colored poster sheets could be cut to approximately 9½ × 12 in., for example. The sheets of paper may then be attached to the poster sheets with glue or double-sided sticky tape. I prefer to use regular clear tape rolled into a 1-in. diameter ring (sticky side out) as the "glue", pressing the tape onto the back corners of the paper panels and then pressing the taped panels onto the colored poster paper. Doing the top corners first usually ensures that the sheets are hung straight. The advantage of this kind of tape over the double-sided variety or glue is that it is quite easy to lift the tape and start over if things do not look right. If you have not already printed numbers on your white sheets (or planned on other devices to guide the viewer from panel to panel), now is the time to number your poster panels. Not only will it be easier for your readers to follow, but it will also allow you to hang the poster more quickly. I like to use round, colored labels found in stationery stores and to stick them on the top right-hand corner of each panel.

Using a large table or the floor, practice laying out your completed poster panels within the area you will have at your disposal at the meeting. Keep in mind again that most of the panels need to be at a comfortable viewing level for the average person, that the sequence (right to left or up and down) needs to be clear, and that the panels need to "flow" from one to the other. After you have laid out the panels, look for overall symmetry and design. Is the arrangement pleasing? Does it catch the eye? How could it be improved?

When you are satisfied with the layout, collect and stack the panels in numerical order. If none of your panels involve paper larger than letter size, your entire poster should now easily fit into a briefcase or folder for safe transport to the meeting, ready to be hung with a minimum of last-minute worry about order, fit, or design.

Further Preparations

Be ready with a short oral summary of the main points of your poster. A brief synopsis of the purpose of your experiments, the results you achieved, and the conclusions you draw is very useful, particularly for those viewers who say, "Can you tell me in a few words what this is all about?"

Also prepare brief oral explanations of the important features of each panel, particularly those containing tables, figures, spectra, and so forth.

This groundwork will allow you to "walk through" the poster with anyone who seems particularly interested. Preparing an oral summary and explanations may sound easy (after all, you prepared the panels and must know what they contain), but it takes some thought and planning, particularly if it is to be done clearly and quickly.

It also can be handy to have several copies of a one-page (or half-page) handout with your name, poster title, affiliation, meeting date, and addresses (e-mail, etc.) along with a summary of your poster, structure of the compound you prepared, or whatever is most appropriate. These handouts can be given to those who seem particularly interested and with whom you may later wish to correspond. It may be difficult to estimate demand for these—you do not want to hand them out like advertising leaflets—but don't be too modest.

Hanging the Poster

At ACS meetings, your poster should be mounted before the opening of the poster session and left in place until the closing of the session. Generally you will have been given instructions as to when to hang your poster or told the time that it is expected to be available for viewing.

Well-organized meetings, such as ACS national meetings, will have your assigned space clearly marked with the poster number as it appears in the meeting program, along with push pins for your use. A very good idea, however, is to bring along some of your own favorite thumbtacks or colored pins just in case there are not enough provided or you don't like the style. Remember that most ordinary thumbtacks are not long enough to go through heavy poster stock and into the boards.

It is assumed that you know the dimensions of your allotted space, that you designed your poster with this space in mind, and that, indeed, the space you have been given closely approximates what you expected. It is then a matter of scanning the area and pinning your prepared panels neatly, straight, and in sequence according to your predetermined plan. Put up the title panel first, particularly if the title is long. Again, try to have most of your important information at eye level. If you have a prepared summary handout, place copies on a convenient surface or chair (most often not available) or pin them to the poster board with a small note saying "Take one."

Monitoring the Poster

Very often posters are expected to be up and available for viewing for some specified time, with you expected to be there in person for some shorter but clearly designated time interval. At ACS meetings, you are encouraged to be present for the entire session. Assume that some people will really want to see your poster and may therefore wish to speak to you. Stand next to the poster, but do not obstruct the view. Avoid chatting with your neighboring poster presenter or with your research-group fan club. If you are already speaking with someone, you tend to discourage dialogue with viewers who do not want to interrupt but may otherwise be quite interested in your work. In other words, be professional and "at the ready". Be sure your name tag is visible so that a visitor can identify you and address questions to you. If at all possible, plan to stay for more than the minimum time assigned.

You will discover that people who visit your poster have their own individual styles, and it is well to be prepared for them. Some will just glance at the title and possibly the conclusion and move on. That's fine and is to be expected; you have probably done the same thing when the subject is not one that happens to be in your area; if you are in a hurry; or if the poster is unattractive, carelessly presented, or does not look readable.

Some people will come by and say, "Tell me about your poster." These are the folks for whom you have prepared your short synopsis. Remember to keep it brief and to the point, emphasizing the purpose of your study and the main conclusions, so that you will have time for other people.

Some viewers will quietly read each word and leave without comment. It is perfectly acceptable to stand by without saying anything yourself while someone reads your poster. It is also acceptable, especially if there is not much other activity, and if the person seems to be interested, to offer to go through the poster (your panel-by-panel explanation) and ask whether he or she has any comments or questions.

Still other people will scan the poster carefully and then ask questions. Occasionally these questions are quite simple: "What temperature did you run this reaction at?" But other questions, especially if they are from someone working in a related area, can be based on a real understanding of your problem or system. These questions may be answered at more length and in more detail. Often they prove very helpful, in that they may inform you of related work or references of which you were unaware, suggest further lines of experimentation, or postulate alternative conclusions. This type of one-on-one scientific dialogue is, of course, what makes poster sessions so special. Real discussion can take place, and contacts, friendships, and even future collaborations can be established. When you realize that your visitor is working in a related area, give out your one-page summary handout.

You may receive questions for which you do not have a ready answer. If you do not know the answer to a question, admit it. You could say something like, "I'm not certain on that point; I'll need to check on it," or "That is an excellent point; I'll need to take that into consideration." The best response may be to take the name and address of the questioner and send the answer when you know it. In any case, do not try to cover up, change the subject, or answer a different question. Your questioner will certainly notice, and you will not rise in his or her estimation by being evasive.

In all these situations, the usual rules of conversational discourse apply: Make eye contact, speak clearly but at low volume and with enthusiasm, listen to your visitors' comments and questions, and express appreciation for their interest.

Finally, you may provide sign-up sheets to record the names and addresses of attendees who might want you to send them more information.

Removing and Saving the Poster

In many cases, other poster sessions follow yours or other activities are planned for the poster room, so it is essential (and prudent, if you don't want to lose the poster!) to take it down promptly at the designated time. At ACS national meetings, it should be removed immediately after the close of the session. "Used" posters are nice to save for display at the home institution—outside your office or lab, for example, to inform (impress) visitors, colleagues, and future co-workers about your productivity and accomplishments.

Letters to the Editor

Madeleine Jacobs
Editor, *Chemical & Engineering News*

Letters to the Editor is one of the best-read sections of nearly every magazine and journal. At *Chemical & Engineering News*, for example, 75% of the readers always or sometimes read the section. A well-edited Letters section provides an opportunity for readers to comment on topics covered in a particular issue, a forum for dissenting views, and a place to stimulate further dialogue on topics of interest. The Letters section of a publication can help involve readers in every issue.

If you want to write a letter to the editor, do so as quickly as possible after the article stimulating your interest has appeared. Your letter may provide additional information about an article, or it may provide a correction or a clarification. You may also write about other letters, as well as topics that are of general interest to the particular audience of the publication (for example, comments on employment trends are always appropriate for *Chemical & Engineering News,* as this is an ongoing area of coverage).

In writing a letter, be concise—no more than 400 words if possible. Also, try to keep a neutral tone in your letter. Inflammatory letters are unlikely to be published, as are letters that insult previous letter writers and/or authors of articles. Passion is acceptable, but bear in mind that there is a fine line between passion and outrage. After writing a passionate letter, let it sit for a few hours or even a day. This is especially true if you are sending it by e-mail. Frequently, you'll find that you wish to revise a letter after thinking about it for a while.

Most editors limit contributions from individual writers in order to give as many readers as possible an opportunity to be heard. At *Chemical & Engineering News*, the policy is normally one letter every six months. Writers can send letters more frequently, but they should understand that their letters will not be published more frequently except in extenuating circumstances. Some publications do not acknowledge incoming mail; *Chemical & Engineering News* acknowledges all letters to the editor, but it may sometimes take several weeks for a decision to be made, and writers should be patient.

Letters to the Editor are almost always edited slightly for style, grammar, and syntax, and occasionally for length. Many editors will send an edited letter back to the writer for approval; others will not.

Most publication editors receive dozens of letters each week. Although policies vary for every publication, usually an editor reads every incoming letter. Sometimes, an editor will have an experienced staff member screen the most promising letters. At *Chemical & Engineering News*, the editor reads every letter. Letters arrive at *Chemical & Engineering News* by mail, fax, and e-mail. Letters are normally selected on the basis of a combination of relevance, editorial clarity, and the opportunity for it to shed further light on a topic and stimulate further discussion.

Press Releases for the Lay Media

Marvin Coyner
Office of Public Outreach

This section introduces the process of preparing and formatting press releases and the basic elements of news. An expanded version can be found on the World Wide Web at http://www.acs.org/pafgen/relgde.htm.

"Just the Facts, Ma'am..."

You have just heard on the radio that there has been an explosion and fire in the downtown area. Details are sketchy. You wonder **who** might have more information. You remember a friend who lives near the area. You call your friend.

"I just heard about the explosion. **What** happened?" you ask.

"A propane tank exploded and blew a big hole in the ground next to Main Street downtown," your friend says.

"Was anyone hurt?" you inquire.

"I don't know. But I've heard that several cars burned."

"**When** did the explosion happen?"

"About 4 o'clock this morning," your friend replies.

"Any idea **why** the tank exploded?"

"No. But I was told a construction crew was working overnight in the area," your friend says.

"**Where** on Main Street was the accident?"

"Between Elm and Maple."

"That's right in the middle of the business district," you recall.

"Yeah. I guess a lot of people won't get much business done today," your friend speculates.

The Five W's

Even if you have never encountered the incident just described, you probably have gone through the same fact-finding process at some point. The questions most people ask when trying to get information are the same basic ones a reporter asks when trying to report a news story: who, what, where, when, why—the five W's of journalism.

If you ever need to write a press release, answering the five W's in your release will generally ensure that you have included the essential information that others need to know about your story or event. Of course, you can and, in most cases, should include more details. Usually, though, the most important information will be covered by the five W's.

Know Your Audience

Other than the five W's, another extremely important question to ask before writing your press release is, "Who is the audience I want to reach?" If, for example, your release is related to chemistry or the ACS, you need to determine whether your intended audience is primarily chemists or nonchemists.

Assume you have been asked to prepare a press release inviting the general public to hear a speaker at your next ACS local section meeting. The first thing to do is gather the five W's and any interesting and pertinent details. Then decide to whom you want to deliver the message. In this case, the audience is the general public because those are the people you are trying to attract to the meeting. Therefore, write your press release in plain English, and leave out as much of the jargon and technical terms as possible.

Now that you know the audience for whom you are writing, ask yourself where they are most likely to see or hear your message. Usually, it is a community newspaper, a local radio station, or a local television station—the lay media.

Keep in mind that your chemist colleagues will no doubt also see your message in the local media. So, be sure you are accurate when you convert science information into lay language. If your message is targeted specifically to other chemists, you can stand to be a little more technical in your presentation, but it is still better to keep it as simple as possible. Save the jargon for publications with a primarily technical audience.

(Most ACS local sections have public relations chairs who have been given training in preparing press releases. Find out whether your section has a public relations chair before writing a press release about your local section for the lay media.)

Elements of News

Just as the chemical elements are the building blocks for everything around us, so too are there news elements that determine the structure of news stories. Knowing these news elements can help you understand how reporters and editors determine what gets into print and on the air. The following list is by no means complete, but it contains the news elements that you are most likely to encounter.

- **Immediacy** or timeliness. Something that has just occurred or is about to occur contains the element of immediacy. Reporters usually are not interested in something that took place last month. If your press release is about an upcoming event, check with your local media to see how soon in advance they would like to have the information.

- **Proximity** or localization. The closer to home the story is, the more important it becomes to people in that area.

- **Consequence** or relevancy. The more people are affected or potentially affected by something, the better chance it has of being reported.

- **Prominence.** Names make news, and not just people's names, but places or events as well. If the mayor or governor is participating in an ACS-sponsored activity, your local media might very well cover it because of that person's involvement. If the activity is taking place at a local historic site or as part of a well-known event, reporters will take notice.
- **Oddity.** Unusual and out-of-the-ordinary occurrences are newsworthy.
- **Emotion.** Highlighting the emotional angle of a story can elicit reactions ranging from mild amusement to sympathy to outright anger.
- **Controversy.** People love reading about controversy, and reporters love writing about it. That's why opposing viewpoints are given so much space in a newspaper or broadcast.

Constructing a Press Release

News Peg

The most dominant news element becomes your news peg. It is called that because it is the peg on which you hang the rest of your story. Once you determine the news peg, put it right up front in your press release.

The right news peg presented early in the release will make it easier for reporters and editors to decide whether they want to use your release. The wrong news peg or one that is buried in the copy will almost certainly guarantee that your release will not be used; it may find its way into the infamous "circular file".

Lead

Next, develop your lead (pronounced "leed"), the opening sentence(s) or paragraph. Often, the safest lead includes the most important information right up front: the five-W lead, the who, what, where, when, and why of the story. It also is called a "summary" lead because it summarizes the most important points of a story.

News Hook

Another method of leading off your press release is to quickly "hook" the audience, that is, entice them to read, listen to, or watch your story. A news hook sets the stage for your five W's. News hooks should be kept short.

Tie-Ins

Your press release will be more appealing to reporters, editors, and the general public if you can tie your story to something already in the news, and if you can localize it. Piggybacking on current news improves the chances that your article will be used, because you are building on something in which the public is already interested. For example, explaining the chemistry involved in a current space shuttle experiment is an excellent news tie-in. If a chemist in your geographic area helped design the experiment, then you have an excellent opportunity for a local tie-in as well.

Order of Descending Importance

After you write the lead, you need to expand on the five W's and present other relevant information. Editors and writers in the media will look at your press releases the same way people look at newspapers. They typically glance at the headlines on a page, decide what looks as if it might interest them, and then read the first few sentences of the story before determining whether they want to read further. Write your story in order of descending importance. Put the most important facts at the beginning and work down toward the least important. That sequence allows readers (including reporters and editors) to quickly get the gist of a story, even if they read only the first few sentences. This is called the "inverted pyramid" style of writing. It has an added benefit for editors; they can shorten the length of stories—without losing the essential facts—by cutting from the bottom.

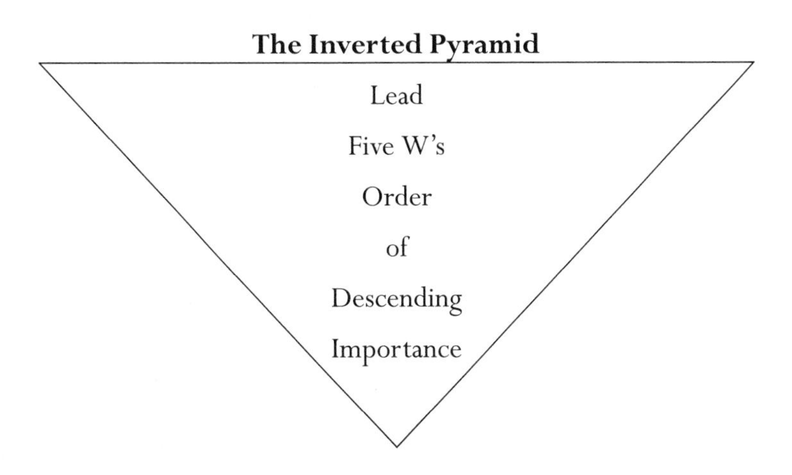

Honesty and Accuracy

The people to whom you mail your press releases expect them to be truthful and accurate. Never mislead or attempt to cover up. Always be truthful. Double-check all your facts and spellings, especially dates, times, and names of people and places. Publications and broadcast stations will judge your organization by the reliability of the information you put in your press release. A shoddy release can cause damage that could take years to repair.

Types of Press Releases

Three basic types of press releases are hard news, feature news, and announcements.

Often, hard news stories dealing with chemistry come from papers presented at ACS meetings or appearing in ACS journals. Press releases based on these papers are typically prepared by the ACS News Service. Discuss ideas you may have for such releases with the News Service.

Ideas for feature releases can come from anywhere. You are limited only by your imagination. Of course, keep in mind that your audience has to judge your topics to be of interest to them.

Announcements are by far the most common type of press release sent to the lay media by community organizations. Local newspapers frequently use releases about meetings, awards, and elections from such groups. The receptiveness of the media to these types of stories will vary from area to area. In general, the smaller the community, the better the chances for getting routine feature and announcement stories published or broadcast. However, even in large cities, suburban newspapers and radio stations featuring local news programs might welcome this sort of material. Also, innumerable specialized publications—alumni letters, company magazines, trade journals—make heavy use of people-oriented material when there is a good local tie-in.

A reminder: If you want to do a press release about an ACS event in your area, check first with the ACS local section to see whether someone is already preparing something.

Format Guidelines

Full Releases or Summaries?

You can write a full press release in the journalistic style you find in mass media publications, or you can give an abbreviated summary containing the most important of the five W's but little else. This summary is called by various names: Fact Sheet, Tip Sheet, News Summary, Media Alert, and so forth.

Hard news and feature stories generally benefit when presented as full releases. For a meeting announcement, a fact sheet might be sufficient. Of course, if you have an exciting and interesting speaker whom you would like the public to hear, then you may wish to expand beyond just the bare facts. Either way, remember to include the five W's.

Line Spacing

Use double spacing for full press releases. This makes it much easier for reporters to read and make notes. Use single spacing for summaries, provided you keep them brief, limited to one page, and in a logical layout that is easy to follow.

Length

Follow the K.I.S.S. rule: Keep it short and simple. Always keep summaries to one page, and try to keep full press releases to one page whenever possible.

There are exceptions, of course, to the one-page-only guideline; features and hard news sometimes warrant expansion. Remember, though, that the subject matter on the second page needs to be as interesting and relevant to the lay public as on the first.

Immediate Release or Hold for Release?

Most press releases should be "For Immediate Release" and you should indicate that at the top of your release. Only in special cases should you ask a reporter or editor to hold your release for later. If you are going to "embargo" a release, as it is called, you need a valid reason. As an example, the ACS News Service routinely embargoes releases it prepares, sometimes weeks in advance, on papers being published in upcoming ACS journals. The embargoed release date is typically the publication date of the journal. In effect, the News Service is asking correspondents not to preempt the ACS publication. In return, the ACS is willing to give the correspondents advance notice of interesting material so that they have time to conduct interviews and write stories that they can then publish or broadcast at the same time the journal article appears in print. Thus, both parties benefit.

Let reporters and editors know whether they can use your press release as soon as they receive it or whether you want them to hold it for later. If you forget to indicate whether a release is immediate or embargoed, reporters will assume it is immediate.

Contact Information

If reporters are interested in using your press release for a story, they may have some questions. Always include on your release the name and telephone number of at least one person who can answer questions or who can put reporters in touch with the right person. Usually, the contact is the person who wrote the release. If someone else's name is included as a contact, be sure the person agrees in advance to be listed on the release.

Headlines

Include a short headline at the top of the press release that indicates the subject. Headlines are not part of the body of the story. You still need to include all the necessary information in the body of your release even if you have written some of it in the headline.

Datelines

Generally, it is not necessary to include a dateline. If you do, include the city in which the event is taking place or where the announcement is being made. The city name should be in all-capital letters. In most cases, the name of the state, in upper- and lowercase letters, should follow the city. The date is optional, despite the fact that it is called a dateline.

For embargoed releases, use the date that you indicated at the top of your release as the "Hold for Release" date. Use only the month, spelled out in upper- and lowercase letters, and the date in arabic numerals.

Datelines are placed at the very beginning of a release and typically separated from the body of the copy by a dash. For example

COLUMBUS, Ohio, July 25—Chemical Abstracts Service today announced...

Page Numbers

Include page numbers if your press release is longer than one page. Remember, in most cases, it probably should be only one page.

End of Release

Common ways to indicate the end of a press release include

-30- -end- xxxx

One of these designations should be centered just below the last line of your release. Some organizations like to include a single-spaced, brief blurb about themselves at the bottom of the final page, after the end indicator. This blurb is a handy reference for reporters and editors and a nice publicity statement for your organization.

Release Numbers

These numbers are useful to organizations in filing and keeping track of press releases. The release number should be placed on the first page in a position where it does not interfere with the text or detract from the overall appearance of the page. Release numbers should appear in the same location on each press release.

Letterhead

Letterhead stationery indicates to readers that the press releases are official communications from your organization.

"They Didn't Use It!"

You won't be the first person to agonize over writing the perfect press release that nobody had the wisdom to use. It happens to everybody. Trust me!

There is something you can do, however, to improve your chance for success. Call the newspapers and broadcast stations to which you plan to send your releases. Here are some things you should ask:

- What type of information do you typically use from organizations such as mine?
- To whom should I send my releases?
- What are your deadlines for receiving releases?
- Is a fact sheet sufficient, or do you prefer a full release?
- What can you recommend I do that will make my releases useful to you?
- How do you prefer receiving information from me? Mail, fax, e-mail, or phone?

CHAPTER 3

Grammar, Punctuation, and Spelling

This chapter presents grammatical points that cover most situations. It does not attempt to discuss all the rules of grammar; many excellent grammar texts are available for that purpose, such as those given in the bibliography at the end of the chapter. Writing style and word usage are discussed in Chapter 1. Punctuation, spelling, and word usage are also discussed in Chapter 5 with respect to numbers, mathematics, and units of measure and in Chapter 7 with respect to chemical names.

Grammar

Subject–Verb Agreement

Everyone knows that a subject and its verb must agree in number. Nevertheless, errors in subject–verb agreement are quite common. The primary cause is confusion about the number of the subject.

◆ The number of the subject can be obscured when one or more prepositional phrases come between the subject and the verb.

> Application of this technique to studies on the phytoplankton biomass and its environments is described. (The subject is "application", which is singular.)

◆ Two singular subjects joined by "and" require a plural verb.

> Growth and isolation of M13 virus were described.

Exception A subject that is plural in form but singular in effect takes a singular verb. Here a compound subject functions as a single entity.

Research and development is attracting a growing number of young scientists.

Its inventor and chief practitioner is a native son of Boston, Robert Coles.

Much inconsistency and confusion exists with technical documentation.

Bacon and eggs was on the menu.

◆ When two or more subjects are joined by "or", the verb takes the number of the closer or closest subject.

The appropriate metal ion concentration or the rate constants were used.

The rate constants or the appropriate metal ion concentration was used.

◆ Collective nouns take a singular verb when the group as a whole is meant; in that case they are often preceded by the word "the". Collective nouns take a plural verb when individuals of the group are meant; in that case, they are often preceded by the word "a".

contents	majority	range
couple	number	series
dozen	pair	variety
group		

The number of metal amides synthesized was the largest to date. (Refers to the number as a unit.)

A number of metal amides were synthesized. (Refers to each amide.)

The series of compounds was prepared to test the hypothesis. (Refers to the series as a unit.)

A series of compounds were tested. (Refers to each compound.)

The variety of materials tested was sufficient for comparative analysis. (Refers to variety as a unit.)

A variety of materials are being tested for selective removal of ^{90}Sr from nuclear waste solutions. (Refers to the materials individually.)

This group of workers is well aware of its responsibilities. (Refers to the group as a unit.)

This group of workers are willing to sign their names. (Refers to the individuals.)

◆ "Data" can be a singular or plural noun.

After the data is printed and distributed, we can meet to discuss it. (Refers to the whole collection of data as one unit.)

Experimental data that we obtained are compared with previously reported results. (Refers to the data as individual results.)

◆ Units of measure are treated as collective nouns and therefore take a singular verb.

The mixture was stirred, and 5 mL of diluent was added.

Five grams of NaCl was added to the solution.

Three weeks is needed to complete the experiment.

To the mixture was added 5 g of compound B.

Under high pressure, 5 volumes of solution A was added.

◆ Nouns ending in *ics* and denoting a scientific discipline are usually singular.

dynamics	mechanics
kinetics	physics
mathematics	thermodynamics

Mechanics involves the application of Newton's three laws of motion.

The kinetics of electron transfer to and from photogenerated radicals was examined by laser flash photolysis.

The thermodynamics is governed by the positions of the valence and conduction bands.

◆ Compound subjects containing the words "each", "every", and "everybody" take singular verbs.

Each flask and each holder was sterilized before use.

Every rat injected and every rat dosed orally was included.

Everybody in the group and every visitor is assigned a different journal each month.

◆ Sometimes, one of these words is implicit; such cases take a singular verb.

Each name and address is entered into the database.

◆ If both components of the compound subject do not contain, explicitly or implicitly, one of the words "each", "every", or "everybody", the verb must be plural.

Each student and all the professors were invited.

◆ Indefinite pronouns themselves (or adjectives combined with the indefinite pronoun "one") can be the subject of the sentence.

- Those that take a singular verb are "each", "either", "neither", "no one", "every one", "anyone", "someone", "everyone", "anybody", "somebody", and "everybody".

Each was evaluated for its effect on metabolism.

Neither disrupts the cell membrane.

Regarding compounds **1–10**, every one reacts with the control agent.

Someone measures the volume every day.

- Those that take a plural verb are "several", "few", "both", and "many".

Several were evaluated for their effects on metabolism.

Few disrupt the cell membrane.

Regarding compounds **1** and **2**, both react with the control agent.

Many were chosen to be part of the study.

- Those that take either a singular or a plural verb, depending on context, are "some", "any", "none", "all", and "most". The number of the object of the preposition determines the number of the indefinite pronoun related to it.

  Some of the money was stolen.

  Some of the books were lost.

  Not all the disks are here; some were lost.

◆ When a fraction is the subject of the sentence, the number of the attendant object of the preposition determines the number of the subject.

One-third of the precipitate was dissolved.

One-fourth of the electrons are excited.

The remainder of the compounds are yet to be described.

◆ When a subject and its predicate nominative disagree in number, the verb takes the number of the subject.

The preparation and structure determination of these three compounds are the topic of this paper.

The topic of this paper is the preparation and structure determination of these three compounds.

Awkward Omissions of Verbs and Auxiliary Verbs

Each subject in a compound sentence must have the proper verb and auxiliary verb.

Incorrect The eluant was added to the column, and the samples collected in 10-mL increments.

Correct The eluant was added to the column, and the samples were collected in 10-mL increments.

Restrictive and Nonrestrictive Expressions

◆ A phrase or clause is restrictive when it is necessary to the sense of the sentence; that is, the sentence would become pointless without the phrase or clause. Restrictive clauses are best introduced by "that", not "which".

It was necessary to find a blocking group that would react with the amino group but not with the hydroxyl group.

> Comparison will be restricted to acetylene compounds that have the same functional end groups.

If the clauses beginning with "that" were deleted, the sentences would not convey the information intended. Therefore, the clauses are restrictive.

Phrases can also be restrictive.

> Reactions leading to the desired products are shown in Scheme 1.

If the phrase "leading to the desired products" were deleted, the sentence would not convey the information intended.

◆ A phrase or clause is nonrestrictive if it adds information but is not essential; that is, the sentence does not lose its meaning if the phrase or clause is deleted. Nonrestrictive phrases and clauses are set off by commas.

> Squalene, a precursor of cholesterol, is a 30-carbon isoprenoid.
>
> This highly readable book, written in nontechnical language, surveys the field of chemistry by describing the contributions of chemistry to everyday life.
>
> Moore, working at the Rockefeller Institute, developed methods for the quantitative determination of amino acids.
>
> The current–voltage curves, which are shown in Figure 6, clearly demonstrate the reversibility of all four processes.
>
> Several hazardous waste disposal sites are located along the shores of the Niagara River, which is a major water source.
>
> Melvin Calvin, who won the Nobel Prize in 1961, elucidated the biochemical pathways in photosynthesis.

Dangling Modifiers

A dangling modifier is a word or phrase that does not clearly and logically modify another word in the sentence. In scientific writing, the passive voice is often necessary ("the solutions were heated"; "melting points were determined"), but its use can lead to dangling modifiers.

◆ If a modifier precedes the subject of a sentence, it must modify that subject and be separated from it by a comma. Otherwise, it is a dangling modifier.

> ***Incorrect*** Splitting the atom, many new elements were discovered by Seaborg.
>
> ***Correct*** Splitting the atom, Seaborg discovered many new elements.
>
> ***Incorrect*** Upon splitting the atom, many new elements were discovered by Seaborg.
>
> ***Correct*** Upon splitting the atom, Seaborg discovered many new elements.

Incorrect When confronted with these limitations, the experiments were discontinued.

Correct When confronted with these limitations, we discontinued the experiments.

Also correct In light of these limitations, the experiments were discontinued.

Incorrect Understanding the effect of substituents on the parent molecules, the ortho hydrogens could be assigned to the high-frequency peak.

Correct Understanding the effect of substituents on the parent molecules, we could assign the ortho hydrogens to the high-frequency peak.

Incorrect Using the procedure described previously, the partition function can be evaluated.

Correct Using the procedure described previously, we can evaluate the partition function.

◆ Phrases starting with "based on" must modify a noun or pronoun that usually immediately precedes or follows the phrase. Use phrases starting with "on the basis of" to modify a verb.

Incorrect Based on resonance enhancement and frequency shifts, changes in the inter-ring separation were calculated.

Correct On the basis of resonance enhancement and frequency shifts, changes in the inter-ring separation were calculated.

Incorrect Based on extensive study, this genetic deficiency was attributed to the loss of one isozyme. ("Based on extensive study" modifies the noun "deficiency", but this is not the meaning.)

Correct On the basis of extensive study, this genetic deficiency was attributed to the loss of one isozyme. ("On the basis of extensive study" modifies the verb "was attributed".)

Correct Style guidelines based on authoritative sources are included in this book. ("Based on authoritative sources" modifies the noun "guidelines".)

◆ "Due to" means "attributable to"; use it only to modify a noun or pronoun directly preceding it in the sentence or following a form of the verb "to be".

Incorrect Delays resulted due to equipment failure.

Correct Delays due to equipment failure were unavoidable.

Also correct The delays were due to equipment failure.

Incorrect This high value resulted due to the high conversion efficiencies of the enzymatic reactor.

Correct This high value is due to the high conversion efficiencies of the enzymatic reactor.

Also correct This high value resulted from the high conversion efficiencies of the enzymatic reactor.

Incorrect Due to exposure to low levels of lead, children can be at risk for developmental problems.

Correct Because of exposure to low levels of lead, children can be at risk for developmental problems.

Also correct Children can be at risk for developmental problems because of exposure to low levels of lead.

◆ Absolute constructions are words, phrases, or clauses that are grammatically unconnected with the rest of the sentence in which they appear. They are sometimes called "sentence modifiers" because they qualify the rest of the sentence. They may occur anywhere in the sentence, and they are always set off by commas. They are not dangling modifiers.

Contrary to the excited-state situation, metal–metal bonding interactions in the ground states are weak.

The conclusions were premature, considering the lack of available data.

Judging from the spectral changes, exhaustive photolysis of compound **4** had occurred.

The conformations about the Re–Re bond, in addition, are different for all three complexes.

When necessary, the solutions were deaerated by bubbling nitrogen.

Clearly, alternative synthetic methods are possible.

The instructor having made her point, the discussion continued.

Absolute constructions often begin with one of the following words:

considering	provided
given	providing
judging	regarding
concerning	failing

◆ In mathematical papers, absolute phrases beginning with the words "assuming" and "taking" are often used as sentence modifiers.

Assuming that distance d is induced by the norm, M is a symmetrical and positively defined matrix.

Taking this value as an upper limit, the two shortest distances are sometimes too long for incipient hydrogen bonds.

◆ A subordinate or elliptical clause may be used as a sentence modifier.

The compound is stable in air, as we concluded from the experimental evidence.

The Mo 5s orbitals, as expected, interact strongly with the ligands.

◆ An introductory infinitive or infinitive phrase may be a sentence modifier.

To prepare compound **2**, the method of Garner was followed.

Reflexive Pronouns

◆ Use the reflexive pronouns "myself", "yourself", "himself", "herself", "itself", "ourselves", and "themselves" only to refer back to a noun or another pronoun in the same sentence.

> ***Incorrect*** Please send your manuscript to the associate editor or myself.
> ***Correct*** Please send your manuscript to the associate editor or me.
>
> ***Correct*** The associate editor herself will review your manuscript.
>
> ***Incorrect*** My collaborators and myself will evaluate the results.
> ***Correct*** My collaborators and I will evaluate the results.
>
> ***Correct*** I myself will evaluate the results.

Punctuation

Comma

◆ Use a comma before Jr. and Sr., but treat II and III according to the person's preference. Within a sentence, use a comma after Jr. and Sr. and after II and III if they are preceded by a comma.

> William M. Delaney, Jr. Charles J. Smith, III John J. Alden II
> William M. Delaney, Jr., was elected to the governing board.
> Charles J. Smith, III, received a majority of the votes.
> John J. Alden II did not run for office this year.

◆ In dates, use a comma after the day, but not after the month when the day is not given.

> June 15, 1996 June 1996

When giving a complete date within a sentence, use a comma after the year as well.

> On August 18, 1984, an extraordinary person was born.

◆ Use a comma after most introductory words and phrases.

> However, the public is being inundated with stories about cancer-causing chemicals.
>
> Therefore, the type of organic solvent used is an important factor in lipase-catalyzed enzymatic synthesis.
>
> After 3 months, the plants grown under phosphorus-deficient conditions were evaluated.

Thus, their motion is the result of the rotation of ferromagnetic domains.

On cooling, a crystalline phase may develop in coexistence with an amorphous phase.

◆ Use a comma after a subordinate clause that precedes the main clause in a complex sentence.

Although 40 different P450 enzymes have been identified, only six are responsible for the processing of carcinogens.

Since the institute opened, plant breeders have developed three new prototypes.

Because the gene and the molecular marker are so close on the chromosome, they segregate together in the progeny.

◆ Use a comma before, but not after, the subordinating conjunction in a nonrestrictive clause.

Incorrect The bryopyran ring system is a unique requirement for anticancer activity whereas, the ester substituents influence the degree of cytotoxicity.

Correct The bryopyran ring system is a unique requirement for anticancer activity, whereas the ester substituents influence the degree of cytotoxicity.

◆ Use commas to set off nonrestrictive phrases or clauses.

The products, which were produced at high temperatures, were unstable.

◆ Phrases introduced by "such as" or "including" can be restrictive (and thus not set off by commas) or nonrestrictive (and thus set off by commas).

Potassium compounds such as KCl are strong electrolytes; other potassium compounds are weak electrolytes.

Previously, we described a mathematical model including a description of chlorophyll degradation in foods.

Divalent metal ions, such as magnesium(II) and zinc(II), are located in the catalytic active sites of the enzymes.

Hydrogen-bonded complexes, including proton-bound dimers, are well-known species.

In the first two sentences, the phrases are restrictive because the sentences do not make their points without the phrases. In the third and fourth sentences, the phrases are nonrestrictive because the sentences can make their points without the phrases.

◆ An appositive is a noun that follows another noun and identifies or explains the meaning of the first noun.

My wife, Jeanne, is a biochemist at the National Institutes of Health.

My son James plays baseball, and my son John plays soccer.

An appositive is nonrestrictive (and therefore set off by commas) when it names the only possibility. In the first sentence, Jeanne is a nonrestrictive appositive. An appositive is restrictive (and therefore not set off by commas) when it points out one of two or more possibilities. In the second sentence, the names of the two sons are restrictive appositives.

◆ Use a comma before the coordinating conjunction in a series of words, phrases, or clauses of equal rank containing three or more items. (This comma is called the *serial comma*.)

> Water, sodium hydroxide, and ammonia were the solvents.
>
> The red needles were collected, washed with toluene, and dried in a vacuum desiccator.
>
> The compound does not add bromine, undergo polymerization by the Diels–Alder reaction, or react with electrophiles.

◆ Use a comma before (and not after) the coordinating conjunctions "and", "or", "nor", "but", "yet", "for", and "so" connecting two or more main clauses (complete thoughts).

> Toluene and hexane were purified by standard procedures, and benzene was redistilled from calcium hydride.
>
> The role of organic templates in zeolite synthesis has been studied extensively, but no general principles have been delineated.
>
> Supported metals are among the most important industrial catalysts, yet only a few have been studied thoroughly.
>
> No dielectric constants are available for concentrated acids, so it is difficult to give a quantitative explanation for the results.

◆ In compound sentences containing coordinating conjunctions, the clause following the conjunction is punctuated as if it were alone.

> The reaction proceeds smoothly, and by use of appropriate reagents, the yields will be enhanced.

◆ Use commas to set off the words "that is", "namely", and "for example" when they are followed by a word or list of words and not a clause. Also use a comma after the item or items being named. Use a comma after "i.e." and "e.g." in parenthetical expressions.

> The new derivatives obtained with the simpler procedure, that is, reaction with organocuprates, were evaluated for antitumor activity.
>
> Alkali metal derivatives of organic compounds exist as aggregates of ion pairs, namely, dimers, trimers, and tetramers, in solvents of low polarity.
>
> Many antibiotics, for example, penicillins, cephalosporins, and vancomycin, interfere with bacterial peptidoglycan construction.

These oxides are more stable in organic solvents (e.g., ketones, esters, and ethers) than previously believed.

◆ Use commas to separate items in a series that contains another series in parentheses already separated by commas.

The structure was confirmed with spectroscopy (^{1}H NMR, UV, and IR), high-resolution mass spectrometry, and elemental analysis.

◆ Use commas to separate two reference citation numbers, but use an en dash (–) to express a range of three or more in sequence, whether they are superscripts or are on the line in parentheses. When they are superscripts, do not use a space after the comma.

Experimental investigations[10,14,18–25] concerned the relative importance of field and electronegativity effects.

Certain complexes of cobalt were reported (*10, 11*) to have catalytic effects on hydrolysis reactions.

Flash photolysis studies (*3–7*) demonstrated the formation of transient intermediate products such as triplet states.

◆ Use a comma to introduce quotations, but not if the quotation is the subject of the sentence or a phrase that is the object of a sentence.

In the words of Pasteur, "Chance favors the prepared mind."

Pasteur said, "Chance favors the prepared mind."

"Chance favors the prepared mind" is a translation from the French. (The quotation is the subject of the sentence.)

Pasteur said that good fortune will favor "the prepared mind". (The quoted phrase is the object of the sentence.)

◆ Use a comma between two or more adjectives preceding a noun only if you can reverse the order of the adjectives without losing meaning. If you can insert the word "and", the comma is correct.

The intense, broad signals of the two groups confirmed their location.

The broad, intense signals of the two groups confirmed their location.

Sample preparation is a repetitious, labor-intensive task.

Sample preparation is a labor-intensive, repetitious task.

A powerful, versatile tool for particle sizing is quasi-elastic light scattering.

A versatile, powerful tool for particle sizing is quasi-elastic light scattering.

But: Polyethylene is an important industrial polymer.

The rapid intramolecular reaction course leads to ring formation.

The backbone dihedral angles were characterized by *J* couplings.

The local structural environment of the Mn cluster was determined.

◆ Do not use a comma to separate a verb from its subject, its object, or its predicate noun.

> ***Incorrect*** The addition of substituted silanes to carbon–carbon double bonds, has been studied extensively.
>
> ***Correct*** The addition of substituted silanes to carbon–carbon double bonds has been studied extensively.
>
> ***Incorrect*** The disciplines described in the brochure include, materials science, biotechnology, and environmental chemistry.
>
> ***Correct*** The disciplines described in the brochure include materials science, biotechnology, and environmental chemistry.
>
> ***Incorrect*** The solvents used in this study were, cyclohexane, methanol, *n*-pentane, and toluene.
>
> ***Correct*** The solvents used in this study were cyclohexane, methanol, *n*-pentane, and toluene.

◆ Do not use a comma preceding "et al." unless commas are needed for other reasons.

> Saltzman et al   Saltzman, M. J., et al.   Saltzman, Brown, et al.

◆ Do not use a comma before the conjunction joining a compound predicate consisting of only two parts.

> ***Incorrect*** The product distribution results were obtained in sodium hydroxide, and are listed in Table 10.
>
> ***Correct*** The product distribution results were obtained in sodium hydroxide and are listed in Table 10.

Period

◆ Use a period at the end of a declarative sentence, but never in combination with any other punctuation marks, even if they are part of a quote.

> He said, "Watch out!"
>
> She asked, "May I go?"

◆ Do not use periods after most abbreviated units of measure, except when the abbreviation could be confused with a word (in. for inches, at. for atomic, no. for number).

◆ If a sentence ends with an abbreviation that includes a period, do not add another period.

> She will return at 3 a.m.

Semicolon

◆ Use a semicolon to separate independent clauses that are not joined by a conjunction.

> All solvents were distilled from an appropriate drying agent; tetrahydrofuran and diethyl ether were also pretreated with activity I alumina.

◆ Use semicolons between items in a series of words, phrases, or data strings if one or more of the items already contain commas.

> We thank Zachary Axelrod, University of Michigan, for spectral data; Caroline Fleissner, Harvard University, for helpful discussions; and the National Science Foundation for financial support (Grant XYZ 123456).

> The product was dried under vacuum to give compound 2: yield 68%; IR1991 m, 1896, s, sh, 1865, s cm^{-1}; ^{1}H NMR 0.36 ppm; ^{13}C NMR 221.3, 8.1 ppm.

> Figure 1. Cyclic voltammograms in dichloromethane: (a) compound **1**, 23 °C; (b) compound **2**, –40 °C; (c) compound **4**, 23 °C.

> Figure 6. Ru–H stretchings in the IR spectrum of compound **5**: ×, 298 K; +, 90 K.

This rule holds even if the only group containing the commas is the last in the series.

> The compounds studied were methyl ethyl ketone; sodium benzoate; and acetic, benzoic, and cinnamic acids.

◆ Use a semicolon between independent clauses joined by conjunctive adverbs or transitional phrases such as "that is", "however", "therefore", "hence", "indeed", "accordingly", "besides", and "thus".

> The rate at which bleaching occurred was dependent on cluster size; that is, the degradation of the mononuclear cluster was about 5 times faster than that of the tetranuclear cluster.

> Many kinetic models have been investigated; however, the first-order reactions were studied most extensively.

> The proposed intermediate is not easily accessible; therefore, the final product is observed initially.

> The restriction of the rotational motions of the *tert*-butyl group gives rise to large entropy changes for the association reaction; hence, the covalent form is relatively easy to identify.

> The efficiency of the cross-coupling depends on the nature of X in RX; thus, the reaction is performed at room temperature by slow addition of the ester.

◆ Do not use a semicolon between dependent and independent clauses.

> ***Incorrect*** The activity on bromopyruvate was decreased; whereas, the activity on pyruvate was enhanced.

Correct The activity on bromopyruvate was decreased, whereas the activity on pyruvate was enhanced.

Colon

◆ Use a colon to introduce a word, phrase, complete sentence, or several complete sentences that illustrate, clarify, or expand the information that precedes it. Capitalize the first word after a colon only if the colon introduces more than one complete sentence, a quotation, or a formal statement.

> The electron density was studied for the ground state of three groups of molecules: (1) methane–methanol–carbon dioxide, (2) water–hydrogen peroxide, and (3) ferrous oxide–ferric oxide.
>
> We now report a preliminary finding: no chemical shift changes were detected in the concentration range 0.1–10 M.
>
> The following are our conclusions: Large-angle X-ray scattering studies give us an accurate picture of structures up to 9 Å. They do not allow the specification of defects, such as random ruptures of the chains. The structural models defined are strongly supported by magnetic measurements.

◆ In figure captions, use a colon to introduce explanations of symbols or other aspects of the figure.

> Figure 1. Variable-temperature ^{1}H NMR spectra of compound **12**: top, 403 K; middle, 353 K; bottom, 298 K.
>
> Figure 3. Brønsted-type plots for aminolysis in 1 M KCl at 25 °C: ○, 2-nitrophenyl acetate; □, 3-chlorobenzoic acid; ◇, 2,6-dinitrobenzoic acid.

◆ Do not use a colon (or any punctuation) between a verb and its object or complement or between a preposition and its object.

> ***Incorrect*** The rate constants for the reaction in increasing concentrations of sodium hydroxide are: 3.9, 4.1, 4.4, 4.6, and 4.9.
>
> ***Correct*** The rate constants for the reaction in increasing concentrations of sodium hydroxide are 3.9, 4.1, 4.4, 4.6, and 4.9.
>
> ***Incorrect*** The thermal decomposition was investigated with: gas chromatography, BET surface areas, and X-ray powder diffraction.
>
> ***Correct*** The thermal decomposition was investigated with gas chromatography, BET surface areas, and X-ray powder diffraction.
>
> ***Incorrect*** Transition-metal nitrides have many properties that make them suitable for industrial applications, including: high wear resistance, high decomposition temperature, and high microhardness.
>
> ***Correct*** Transition-metal nitrides have many properties that make them suitable for industrial applications, including high wear resistance, high decomposition temperature, and high microhardness.

◆ Use either a colon or a slash to represent a ratio, but not an en dash. Use either a slash or an en dash between components of a mixture, but not a colon.

> dissolved in 5:1 glycerin/water
> dissolved in 5:1 glycerin–water
> the metal/ligand (1:1) reaction mixture
> the metal–ligand (1:1) reaction mixture
> the metal–ligand (1/1) reaction mixture
> the methane/oxygen/argon (1/50/450) matrix
> the methane/oxygen/argon (1:50:450) matrix

Quotation Marks

Location of quotation marks is a style point in which ACS differs from other authorities. In 1978, ACS questioned the traditional practice and recommended a deviation: logical placement. Thus, if the punctuation is part of the quotation, then it should be within the quotation marks; if the punctuation is not part of the quotation, the writer should not mislead the reader by implying that it is.

◆ Place closing quotation marks before all punctuation that is not part of the original quotation. Place them after all punctuation that is part of the quotation.

> The sample solution was stirred briefly with a magnetic “flea”.
> Ralph Waldo Emerson said, “The reward of a thing well done is to have done it.”

◆ Use quotation marks around words used in a new sense or words not used literally, but only the first time they appear in text.

> Plastocyanin is a soluble “blue” copper protein.
> The integrated intensity of each diagonal in the spectrum is proportional to a “mixing coefficient”.
> The “electron-deficient” cations are, in fact, well-established intermediates.

◆ Use quotation marks to enclose the titles of uniquely named parts and sections of a book or a paper.

> A complete description of the oils is given in the section “Flavonoids in Citrus Peel Oils”, and other references are listed in the bibliography.

But: The preface describes the complexity of the problem.

◆ Use quotation marks to enclose short direct quotations (two or three sentences).

> In the book *Megatrends*, Naisbitt concludes, “We are moving from the specialist who is soon obsolete to the generalist who can adapt.”

◆ Use a narrower column width (that is, indented on both sides) for longer quotations (extracts) of 50 words or more. Do not use quotation marks.

> Everything is made of atoms. That is the key hypothesis. The most important hypothesis in all of biology, for example, is that everything that animals do, atoms do. In other words, there is nothing that living things do that cannot be understood from the point of view that they are made of atoms acting according to the laws of physics.
>
> —Richard Phillips Feynman

However, this convention does not apply in an article quoting someone who has been interviewed. In such cases, quoted text need not be differentiated by column width, and quotation marks should be used.

◆ Use single quotation marks only when they are within double quotation marks.

> He said, "You should read the article 'Fullerenes Gain Nobel Stature' in the January 6, 1997, issue of *Chemical & Engineering News*."

Parentheses

Parenthetical expressions contain information that is subsidiary to the point of the sentence. The sentence does not depend on the information within the parentheses.

◆ Use parentheses for parenthetical expressions that clarify, identify, or illustrate and that direct the reader.

> The total amount (10 mg) was recovered by modifying the procedure.
>
> The final step (washing) also was performed under a hood.
>
> The curve (Figure 2) obeys the Beer–Lambert law.
>
> The results (Table 1) were consistently positive.
>
> Only 15 samples (or 20%) were analyzed.

◆ If a parenthetical sentence is within another sentence, do not use a final period within the closing parentheses, and do not start the parenthetical sentence with a capital letter.

> Our results (the spectra are shown in Figure 5) justified our conclusions.
>
> Our results justified our conclusions (the spectra are shown in Figure 5).

◆ If a parenthetical sentence is not within another sentence, use a final period inside the closing parenthesis.

> A mechanism involving loss of a CH radical followed by rearrangement was proposed. (The reactions are shown in Scheme 1.)

◆ Use parentheses to enclose numerals in a list. Always use parentheses in pairs, not singly.

> Three applications of this reaction are possible: (1) isomerization of sterically hindered aryl radicals, (2) enol–keto transformation, and (3) sigmatropic hydrogen shift.

◆ Use parentheses to identify the manufacturer of reagents and equipment.

> cobalt chloride (Mallinckrodt)
>
> a pH meter with a glass electrode (Corning)

◆ Do not use parentheses when citing a reference number in narrative text. In such a case, the reference number is the point of the sentence, not subsidiary information, and thus not parenthetical.

> in ref 12, *not* in ref (12), *not* in (12)

◆ Use parentheses in mathematical expressions as discussed in Chapter 5.

Square Brackets

◆ Use square brackets within quotation marks to indicate material that is not part of a direct quote.

> In the words of Sir William Lawrence Bragg, "The important thing in science is not so much to obtain new facts as to *discover new ways* [italics added] of thinking about them."

◆ Use square brackets to indicate concentration: $[Ca^+]$.

◆ Use square brackets in mathematical expressions as discussed in Chapter 5.

Dashes

The shortest dash is the hyphen (-); the en dash (–) is longer; and the em dash (—) is the longest. Hyphens are discussed in the section on hyphenation in Chapter 4.

En Dash

◆ Use an en dash to mean the equivalent of "and", "to", or "versus" in two-word concepts where both words are of equal weight.

> acid–base titration
> bromine–olefin complex
> carbon–oxygen bond
> cis–trans isomerization
> cost–benefit analysis
> dose–response relationship
> ethanol–ether mixture
> host–guest complexation

log–normal function
metal–ligand complex
metal–metal bonding
nickel–cadmium battery
oxidation–reduction potential
producer–user communication
red–black dichroic crystals
structure–activity relationship
structure–property relationship
temperature–time curve
vapor–liquid equilibrium
winter–fall maxima

Exception Use a hyphen for color combinations such as blue-green. See Chapter 4, p 80.

◆ Use an en dash to mean "to" or "through" with a span of three or more numerals or other types of ranges.

12–20 months
sections 1b–1f
Lyon and co-workers (*23–26*)
Figures 1–4
parts C–E
5–50 kg
compounds **A–I**
Lyon and co-workers[23a–d]

Exception 1 When either one or both numbers are negative or include a symbol that modifies the number, use the word "to" or "through", not the en dash.

–20 to +120 K
10 to >600 mL
–145 to –30 °C
<5 to 15 mg
~50 to 60

Exception 2 Do not use an en dash when the word "from" or "between" is used.

from 500 to 600 mL
between 7 and 10 days

◆ Use an en dash to link the names of two persons of equal importance used as a modifier.

Bednorz–Müller theory
Beer–Lambert law
Bose–Einstein statistics
Debye–Hückel theory
Diels–Alder reaction
Fermi–Dirac statistics
Fischer–Tropsch effect
Fisher–Johns hypothesis
Flory–Huggins interaction
Franck–Condon factor
Friedel–Crafts reaction
Geiger–Müller effect
Henderson–Hasselbalch equation
Jahn–Teller effect
Lineweaver–Burk method
Mark–Houwink plot
Meerwein–Ponndorf theory
Michaelis–Menten kinetics
Stern–Volmer plot
van't Hoff–Le Bel theory
Wolff–Kishner theory
Young–Laplace equation
Ziegler–Natta-type catalyst

Treatment of double surnames is covered in Chapter 4, p 79.

◆ Use an en dash between components of a mixed solvent. (A slash can also be used.)

The melting point was unchanged after three crystallizations from hexane–benzene.

Em Dash

◆ Use em dashes to set off words that would be misunderstood without them.

> ***Incorrect*** All three experimental parameters, temperature, time, and concentration, were strictly followed.
>
> ***Correct*** All three experimental parameters—temperature, time, and concentration—were strictly followed.

◆ Do not use em dashes to separate phrases or nonrestrictive clauses if another form of punctuation can be used.

> ***Incorrect*** Knauth—not Stevens—obtained good correlation of results and calculations.
>
> ***Correct*** Knauth, not Stevens, obtained good correlation of results and calculations.
>
> ***Incorrect*** The singly charged complexes—which constituted bands 1 and 3—liberated maleate anion upon decomposition.
>
> ***Correct*** The singly charged complexes, which constituted bands 1 and 3, liberated maleate anion upon decomposition.

Ellipsis Points

◆ Within a quotation, use three periods (points of ellipsis) to indicate deleted words or phrases. These three periods are in addition to other needed punctuation. Thus, if a period is already there, the result will be four periods.

> No science is immune to the infection of politics and the corruption of power.... The time has come to consider how we might bring about a separation, as complete as possible, between Science and Government in all countries.
>
> —Jacob Bronowski

◆ Do not begin or end a quotation with ellipsis points.

Spelling

Consult a dictionary to resolve spelling questions. *Merriam-Webster's Collegiate Dictionary* and *Webster's New World Dictionary of the American Language* are the desk dictionaries used by the ACS technical editing staff. ACS staff members also use the unabridged *Webster's Third New International Dictionary.* However, whatever your dictionary, choose the first spelling of a word. Use American spellings, except in proper names and direct quotations (including titles).

Recommended Spelling List

Many words in regular usage, as well as many technical terms, have two or more acceptable spellings. The following list gives recommended spellings and capitalizations where appropriate, for some terms not found in easily accessible dictionaries, words often misspelled, common expressions, and words for which the ACS preference may not match your dictionary's.

absorbance
absorbency
absorbent
accommodate
adsorbent
aerobic
aging
aglycon
air-dry (verb)
Alfa Inorganics
ambiguous
amine (RNH_2)
ammine (NH_3 complex)
amphiphile
ampule
AnalaR
analog (computer)
analogue
analyte
analyze
annelation
annulation
antioxidant
appendixes
aqua regia
Arrhenius
artifact
asymmetry
audio frequency
autoxidation
auxiliary
Avogadro
bacitracin
back-bonding
back-donation
back-titrate (verb)
backscatter
backscattering
backward
baker's yeast
band gap
band-pass
bandwidth
baseline
Beckman (instrument company)
Beckmann (thermometer, rearrangement)
Beer's law
Beilstein
BFGoodrich
Bio-Rad
bit
black box
blackbody
blender
Boltzmann
borderline
Bragg scattering
break-seal
break up (verb)
breakup (noun)
bremsstrahlung
brewer's yeast
bridgehead
Brinkmann
broad band (noun)
broad-band (adjective)
Brønsted
Büchner
build up (verb)
buildup (noun)
buret
butanol, 1-butanol (***never*** *n*-butanol)
n-butyl alcohol
tert-butylation
byline
bypass
byproduct
byte
Calbiochem
canceled
canister
cannot
Cartesian
catalog
Celite
chloramine
chloro amine
chlorophyll *a*
Chromosorb
clean up (verb)
cleanup (noun)
clear-cut
co-ion
co-occurrence
co-worker
coauthor
collinear
colorimetric
complexometric
concomitant
condensable
conductometric
conrotatory
constantan
coordination
Coulombic
counter electrode
counteranion
counterion
coverslip
cross-coupling
cross-link
cross over (verb)
cross-react
cross-reaction
cross section (noun)

cross-sectional (adjective)
crossover (noun, adjective)
cuboctahedron
cut off (verb)
cutoff (noun)
cuvette
cytochrome *c*
Darzens
database
deamino (not desamino)
deoxy (not desoxy)
dependent
desamine (amino acid names only)
desiccator
deuterioxide
deuteroporphyrin
Dewar benzene
Dewar flask
dialogue
Diatoport
diffractometer
disc (anatomy, electrophoresis)
discernible
disk
disrotatory
dissymmetric
distill
dry ice
drybox
dyad
ebullioscopic
eigenfunction
eigenvalue
electroless
electron microscope
electronvolt
electrooptic
electropositive
eluant
eluate
eluent
Elvehjem
end point
enzymatic
enzymic
Erlenmeyer (flask)
exchangeable
fall off (verb)
falloff (noun, adjective)
far-infrared
faradic (referring to current, not the person)
fax (noun, verb, adjective)
feedback
fiber-optic (adjective)
fiber optics (noun)
filterable
firebrick
flavin
Florisil
flow sheet
fluoborate
fluoramine
fluoro amine
Fluorolube
focused
follow up (verb)
follow-up (noun, adjective)
forbear (verb, to refrain)
forebear (noun, ancestor)
forego (verb, to go before)
forgo (verb, to do without)
formulas
forward
freeze-dry (verb)
fulfill
γ ray
gauge
Gaussian
gegenion
glovebag
glovebox
Gouy
graduated cylinder
gram
Gram-negative
Gram-positive
gray
Grignard
groundwater
half-ester
half-life
half-width
halfway
Hamiltonian
Hantzsch
Hazmat
heat-treat (verb)
hemoglobin
hemolysate
heterogeneous
Hoffmann degradation
homogeneous
homologue
Hunsdiecker
hydrindan
hydriodic
hydriodide
hydrolysate
hydrolyzed
ice-cold
ice–water bath
inasmuch
indan
indexes
indices (mathematical)
indispensable
inflection
Infracord
infrared
innocuous
inoculate
insofar
inter-ring
intra-ring
iodometric
iodometry
isooctane
isopiestic
isopropyl alcohol (***never*** isopropanol)
isosbestic
Karl Fischer

kayser
Kekulé
Kjeldahl
Kramers
Kugelrohr
labeled
laser
leukocyte
leveling
levorotatory
lifetime
ligancy
ligate
ligated
line shape
line width
liquefy
liter
lumiflavin
luster
lysate
lysed
make up (verb)
makeup (noun)
Markovnikov
matrices (mathematical)
matrixes (media)
megohm
Mendeleev
mesoporphyrin
metalate
metalation
metallization
metallize
metalloenzyme
meter
methyl Cellosolve
methyl orange
micro-Kjeldahl
mid-infrared
midpoint
Millipore
minuscule
mixture melting point
monochromator
Mössbauer
naphthyl
near-ultraviolet
neopentyl
Nernstian
Norit
nuclide
occurred
occurrence
occurring
ortho ester
orthoformate
orthohydrogen
orthopositronium
outgas
outgassing
overall
Parafilm
parametrization
path length
percent
Petri
pharmacopeia
phenolphthalein
phlorin
phosphomonoester
phosphorous (as in phosphorous acid)
phosphorus (element)
phthalic
pipet
pipetted
plaster of Paris
Plexiglas
point source
porphine
porphyrin
portland cement
programmed
Pronase
2-propanol (***never*** iso-propanol)
pseudo-first-order
pyrolysate
quantitation
radio frequency
radioelement
radioiodine
radionuclide
re-form (to form again)
reform (to amend)
repellent
riboflavin
ring-expand (verb)
rotamer
scale up (verb)
scale-up (noun)
scavengeable
Schwarzkopf
seawater
self-consistent
selfsame
Sephadex
set up (verb)
setup (noun)
side arm (noun)
side chain
sideband
siphon
Soxhlet
spin-label (noun)
spin–lattice
spin–orbit
steam bath
steam-distill (verb)
stepwise
stereopair
stereoptically
Student's *t* test
sulfolane
sulfur
superacid
superhigh frequency
supernatant (adjective)
supernate (noun)
Suprasil
syndet
synthase
synthetase
test tube
θ solvent
thiamin
thioacid
thioester
thioether

thioketone
toward
transmetalation
tropin
Ubbelohde
ultrahigh vacuum
un-ionized
uni-univalent
upfield
urethane
van der Waals
VandenHeuvel
van't Hoff–Le Bel
Vigreux
vis-à-vis
voltameter
voltammeter
voltmeter
wastewater
wave function
wavelength
wavenumber
well-known
Whatman
work up (verb)
workup (noun)
X-irradiation
X-ray
ylide
zerovalent
zigzag
zinc blende

Tricky Possessives

◆ Form the possessive of a joint owner by adding an apostrophe and an "s" after the last name only.

Kanter and Marshall's results   Bausch and Lomb's equipment

◆ Form the possessive of plural nouns that do not end in "s" by adding an apostrophe and an "s". Form the possessive of plural nouns that end in "s" by adding an apostrophe only.

people's rights   children's books   compounds' structures

Tricky Plurals

Sometimes, the plural form is so familiar that it is used erroneously instead of the singular, usually with Latin and other non-English words. The following list shows the correct singulars and plurals. The preferred forms are given first.

Singular	***Plural***
alga	algae
apparatus	apparatus, apparatuses
appendix	appendixes, appendices
bacterium	bacteria
basis	bases
criterion	criteria, criterions
erratum	errata
focus	focuses, foci
formula	formulas, formulae
fungus	fungi, funguses
hypothesis	hypotheses
index	indexes

Singular	*Plural*
index	indices (mathematical)
latex	latices, latexes
locus	loci
matrix	matrices (mathematical)
matrix	matrixes (media)
maximum	maximums, maxima
medium	mediums, media
minimum	minimums, minima
phenomenon	phenomena, phenomenons
spectrum	spectra, spectrums
stratum	strata
symposium	symposia, symposiums
vortex	vortexes, vortices

Seven To Watch

1. Myself used as a substitute for the word "me". Myself is a reflexive pronoun.

> Please give a copy of the agenda to Anne and me. (*not* to Anne and myself)
>
> I myself checked the agenda.
>
> Cheryl and I checked the agenda. (*not* Cheryl and myself checked the agenda)
>
> The agenda was checked by Barbara and me. (*not* by Barbara and myself)

2. Due to. "Due to", which means "attributable to", is used only to modify a noun or pronoun directly preceding it in the sentence or following a form of the verb "to be".

> Cutbacks due to decreased funding have left us without basic reference books.
>
> The accuracy of the prediction is due to a superior computer program.

3. A colon incorrectly placed between a verb and its complement or object.

> The four steps in the procedure are observation, analysis, documentation, and recommendation. (*not* The four steps in the procedure are: ...)
>
> The procedure requires four steps: observation, analysis, documentation, and recommendation.

4. Phenomenon, criterion. These are the singular forms; the plurals are phenomena and criteria.

> This phenomenon cannot be completely explained by the NMR results.
>
> This criterion is very important for generating reactive hydroxyl radicals.

5. The serial comma. Use a comma before the coordinating conjunction in a series of words, phrases, or clauses of equal rank containing three or more items.

> Mass, IR, and NMR spectra are reported.
>
> Most of the reactivity falls into three categories: addition of a nucleophile, thermal or photochemical activation, or a change in the coordination mode.

6. Check that each opening parenthesis has a matching closing parenthesis and vice versa.

7. Using "based on" and "on the basis of". Phrases starting with "based on" must modify a noun or pronoun that usually immediately precedes or follows the phrase. Use phrases starting with "on the basis of" to modify a verb.

> On the basis of the molecular orbital calculations, we propose a mechanism that can account for all the major features of alkali and alkaline earth catalyzed gasification reactions. (*not* Based on ...)

Bibliography

Bernstein, Theodore M. *The Careful Writer: A Modern Guide to English Usage;* Atheneum: New York, 1965.

Berry, T. E. *The Most Common Mistakes in English Usage;* McGraw-Hill: New York, 1971.

Copperud, Roy H. *American Usage and Style: The Consensus;* Van Nostrand-Reinhold: New York, 1980.

Flesch, Rudolf. *The ABC of Style: A Guide to Plain English;* Harper & Row: New York, 1966.

Fowler, H. W. *A Dictionary of Modern English Usage,* 2nd ed.; Oxford University: New York, 1983.

Hodges, John C.; Horner, Winifred B.; Webb, Suzanne S.; Miller, Robert K. *Harbrace College Handbook,* 12th ed.; Harcourt, Brace, Jovanovich: New York, 1994.

Perrin, P. G. *Writer's Guide and Index to English,* 6th ed.; Scott, Foresman: Glenview, IL, 1978.

Semmelmeyer, Madeline; Bolander, Donald O. *Instant English Handbook: An Authoritative Guide and Reference on Grammar, Correct Usage, and Punctuation;* Dell: New York, 1968.

The New York Public Library Writer's Guide to Style and Usage; HarperCollins: New York, 1994.

van Leunen, M.-C. *A Handbook for Scholars;* Alfred A. Knopf: New York, 1979.

CHAPTER 4

Editorial Style

This chapter presents recommended stylistic and editorial conventions, mainly but not solely for ACS publications. The style recommended by ACS is, for the most part, taken from established authoritative sources, such as *The Chicago Manual of Style*, *Words into Type*, and the *U.S. Government Printing Office Style Manual*.

Other points of style are discussed in Chapter 5, "Numbers, Mathematics, and Units of Measure"; Chapter 7, "Names and Numbers for Chemical Compounds"; and Chapter 8, "Conventions in Chemistry".

Hyphenation

Consult a dictionary to resolve hyphenation questions. *Merriam-Webster's Collegiate Dictionary* and *Webster's New World Dictionary of the American Language* are the desk dictionaries used by the ACS technical editing staff. ACS staff also use the unabridged *Webster's Third New International Dictionary*.

Prefixes

◆ Most prefixes are not hyphenated. Do not hyphenate the following prefixes when added to words that are not proper nouns.

after	bi	counter	down
ante	bio	cyber	electro
anti	by	de	extra
auto	co	di	hetero

homo	metalla	photo	supra
hyper	metallo	physico	techno
hypo	micro	poly	tele
in	mid	post	thermo
infra	mini	pre	trans
inter	mis	pro	tri
intra	mono	pseudo	ultra
intro	multi	re	un
iso	nano	retro	under
macro	neo	semi	uni
mega	non	stereo	up
meso	over	sub	video
meta	peri	super	visco

Examples

antibacterial	microorganism	pseudomorph
cooperation	multicolored	superacid
cyberspace	nonpolar	transactinide
extranuclear	photoredox	viscoelastic
interelectrode	polypeptide	
isospin	precooled	

Exceptions Hyphens are sometimes used (1) when letters are doubled, (2) when more than one prefix is present, or (3) when the unhyphenated form does not convey the intended meaning.

anti-infective	inter-ring	post-reorganization
anti-inflammatory	intra-ring	post-translational
bi-univalent	mid-infrared	pre-equilibrium
co-ion	non-native	sub-bandwidth
co-worker	non-nuclear	un-ionize

◆ Some prefixes may be hyphenated or not, depending on meaning.

recollect *or* re-collect
recover *or* re-cover
reform *or* re-form
retreat *or* re-treat

Rare exceptions

autoxidation	counter electrode	hetero group
homo nucleoside		

◆ Do not hyphenate multiplying prefixes.

hemi, mono, di, tri, tetra, penta, hexa, hepta, octa, ennea, nona, deca, deka, undeca, dodeca, etc.

semi, uni, sesqui, bi, ter, quadri, quater, quinque, sexi, septi, octi, novi, deci, etc.

bis, tris, tetrakis, pentakis, hexakis, heptakis, octakis, nonakis, decakis, etc.

Examples

divalent
heptacoordinate tetrahedron
hemihydrate
triethyl phosphate
1,4-bis(3-bromo-1-oxopropyl)piperazine
tris(ethylenediamine)cadmium dihydroxide
tetrakis(hydroxymethyl)methane
triatomic
hexachlorobenzene
2,2′-bipyridine
1,1′:3′,1″:3″,1‴-quaterphenyl

◆ Hyphenate a prefix to a two-word compound.

multi-million-dollar lawsuit
non-diffusion-controlled system
non-English-speaking colleagues
non-radiation-caused effects
non-tumor-bearing organ
pseudo-first-order reaction
pre-steady-state condition

◆ Hyphenate prefixes to chemical terms.

non-hydrogen bonding
non-phenyl atoms
non-alkane

◆ Hyphenate a prefix to a numeral.

pre-1900s

◆ Hyphenate prefixes to proper nouns and adjectives, and retain the capital letter.

anti-Markovnikov
non-Gaussian
non-Coulombic
non-Newtonian
oxy-Cope

Suffixes

◆ Most suffixes are not hyphenated. Do not hyphenate the following suffixes when added to words that are not proper nouns.

able
fold
ful
less
like
ment
ship
wide
wise

Examples

clockwise
fellowship
lifelike
multifold
rodlike
spoonful
statewide
worldwide

Exceptions

gel-like
shell-like
bell-like

◆ Hyphenate the suffixes "like" and "wide" when they are added to words of three or more syllables.

computer-like	resonance-like
radical-like	university-wide

◆ Hyphenate the suffix "like" in two-word compounds used as unit modifiers.

rare-earth-like	transition-metal-like	first-order-like

◆ Hyphenate the suffix "like" to chemical names.

adamantane-like	morphine-like	olefin-like

◆ Hyphenate a numeral and a suffix.

10-fold	25-fold

◆ Hyphenate suffixes to proper nouns, and retain the capital letter.

Kennedy-like	Claisen-type

Compound Words

Compound words are two or more terms used to express a single idea. Compound words in common usage are listed in most dictionaries. Many are hyphenated, but many are not.

back-reaction	half-life	son-in-law
cross-link	self-consistent	

◆ Hyphenate spelled-out fractions.

one-half	three-fourths	two-thirds
one-ninth		

◆ Hyphenate two-word verbs.

air-dry	freeze-dry	ring-expand
flame-seal	jump-start	vacuum-dry

◆ Do not hyphenate phrasal verbs. As unit modifiers or nouns, these words are often hyphenated or closed up; check a dictionary.

break down	mix up	stand by
build up	scale up	take off
grow up	set off	warm up
hand out	set up	wear out
line up	slow down	

◆ Do not hyphenate foreign phrases used as unit modifiers.

ab initio calculation	in situ evaluation
ad hoc committee	in vivo reactions

Exception Some foreign phrases are hyphenated in the original language: laissez-faire.

◆ People who have double surnames may choose to hyphenate them or use a space between them. When they are hyphenated, use a hyphen, not an en dash, between the two surnames in a person's name. Some combinations of two given names are also hyphenated.

Robert Baden-Powell	Joseph-Louis Gay-Lussac
David Ben-Gurion	Irene Joliot-Curie
Chen-Chou Fu	Jackie Joyner-Kersee
John Edward Lennard-Jones	Cecil Day-Lewis

Unit Modifiers

Unit modifiers are two words that together describe a noun; they are almost always hyphenated. Most unit modifiers consist of

- a noun and an adjective (e.g., time-dependent reaction, radiation-sensitive compound, water-soluble polymer, halogen-free oscillator)
- an adjective and a noun (e.g., high-frequency transition, small-volume method, first-order reaction, outer-sphere redox couple)
- an adjective and a participle (e.g., slow-growing tree, broad-based support, far-reaching influence)
- a noun and a participle (e.g., time-consuming project, earth-shaking news, peace-loving politicians, silver-coated electrode)
- an adverb and an adjective (e.g., above-average results, still-unproven technique)
- two nouns (e.g., ion-exchange resin, liquid-crystal polymers, transition-state modeling, charge-transfer reaction, gas-phase hydrolysis)

The following is a short list (by no means complete) of unit modifiers commonly seen in ACS publications. **These should be hyphenated when modifying a noun.**

air-dried	cost-effective	energy-transfer
air-equilibrated	diffusion-controlled	excited-state
back-bonding	double-bond	first-order
^{14}C-labeled	electron-diffraction	flame-ionization
charge-transfer	electron-transfer	fluorescence-quenching

free-energy
free-radical
gas-phase
gel-filtration
Gram-positive
halogen-free
high-energy
high-frequency
high-performance
high-pressure
high-resolution
high-temperature
inner-sphere
ion-exchange
ion-promoted
ion-selective
large-volume
laser-induced
least-squares
light-catalyzed
long-chain
long-lived
low-energy
low-frequency
low-pressure
low-resolution
low-temperature
moisture-sensitive
nearest-neighbor
oil-soluble
outer-sphere
radiation-caused
radiation-produced
radiation-sensitive
rate-limiting
reversed-phase
room-temperature
round-bottom
rubber-lined
second-harmonic
second-order
short-chain
short-lived
side-chain
size-dependent
small-volume
solid-phase
solid-state
species-specific
steady-state
structure-specific
temperature-dependent
thin-layer
three-phase
time-dependent
transition-metal
transition-state
two-dimensional
vapor-phase
water-soluble
weak-field

Exceptions

particle size distribution
water gas shift

◆ Hyphenate combinations of color terms used as unit modifiers.

silver-gray body
red-black precipitate
blue-green solution
bluish-purple solid

◆ Do not hyphenate unit modifiers if the first word is an adverb ending in "ly".

recently developed procedure
carefully planned experiment
accurately measured values
poorly written report

◆ Hyphenate unit modifiers containing the adverbs "well", "still", "ever", "ill", and "little".

ever-present danger
ever-rising costs
ill-fitting stopper
little-known hypothesis
still-new equipment
well-known scientist
well-trained assistants

◆ Do not hyphenate unit modifiers containing the adverbs "well", "still", "ever", "ill", and "little" if they are modified by another adverb.

very well studied hypothesis
most ill advised investment
very high density lipoprotein

◆ Hyphenate unit modifiers containing a comparative or superlative if the meaning could be different without the hyphen.

higher-temperature values	lowest-frequency wavelengths
nearest-neighbor interaction	best-known processes
best-loved adviser	least-squares analysis

◆ Hyphenate a number and a unit of time or measure used as a unit modifier.

12-min exposure	20-mL aliquot
10-mg sample	20th-century development

Exception 1 complex units of measurement such as M, N, *m*, mol dm^{-3}, and mol %

a 0.1 mol dm^{-3} solution   a 0.1 M NaOH solution

Exception 2 complex numbers such as 1.2×10^{-4} cm^{-1} peak

Exception 3 units of temperature: °C, K, °F, °B, and °R

a 37 °C water bath   25 K increments

Exception 4 units closed up to numbers: °, ′, ″, and %

25% decrease   45° angle   12′ angle

◆ Hyphenate converted units given in parentheses if they are also adjectival.

a 7-in. (17.8-cm) funnel   a 0.25-in. (6.4-mm) conductor

◆ When a unit modifier contains an en dash between numbers, use a hyphen between the last number and the unit of measure. Or, use the word “to”.

a 1–2-h sampling time	a 25–30-mL aliquot
a 1- to 2-h sampling time	a 25- to 30-mL aliquot

◆ When two or more unit modifiers with the same ending base (a word or unit of measure) modify one noun, use a hyphen after each element, and do not repeat the ending base.

100-, 200-, and 300-mL aliquots
25- to 50-mg samples
high-, medium-, and low-frequency measurements
first- and second-order reactions

◆ Do not hyphenate unit modifiers that are chemical names.

amino acid level	sodium hydroxide solution
barium sulfate precipitate	acetic anhydride concentration

◆ Hyphenate unit modifiers made up of a single letter or number and a noun or adjective.

α-helix	1-isomer	s-orbital
^{13}C-enriched	L-anomer	*t*-test
^{14}C-labeling	O-ring	U-band

D-configuration | 3-position | *x*-axis
γ-ray | π-electron | X-band

◆ Do not hyphenate unit modifiers if one of the words is a proper name.

Lewis acid catalysis
Fourier transform technique
Schiff base measurement

◆ Hyphenate unit modifiers that contain spelled-out numbers.

five-coordinate complex
one-electron transfer
seven-membered ring
three-dimensional model
three-neck flask
three-stage sampler
two-compartment model
two-phase system

◆ Hyphenate unit modifiers that contain a present or past participle.

air-equilibrated samples
English-speaking colleagues
fluorescence-quenching solution
hydrogen-bonding group
immobilized-phase method
ion-promoted reaction
laser-induced species
methyl-substituted intermediate
photon-induced conversion
problem-solving abilities
rate-limiting step
research-related discussion
steam-distilled sample

Caution Watch for cases where the participle forms a unit with the noun that follows: for example, "ligand binding site" should not be hyphenated.

◆ Hyphenate unit modifiers of three or more words.

head-to-head placement
high-molecular-weight compound
nine-membered-ring species
out-of-plane distance
root-mean-square analysis
signal-to-noise ratio
tried-and-true approach
voltage-to-frequency converter

◆ Hyphenate unit modifiers containing a number, a unit of measure, and a word.

3-year-old child
4-mm-thick layer
100-nm-diameter droplets

◆ Hyphenate unit modifiers containing three words when similar two-word modifiers are hyphenated.

acid-catalyzed reaction
metal-promoted reaction
general-acid-catalyzed reaction
transition-metal-promoted reaction

◆ Do not hyphenate unit modifiers containing three or more words, even if similar two-word modifiers are hyphenated, when doing so would break other rules. For example, do not hyphenate unit modifiers if one of the words is a proper name. Do not hyphenate unit modifiers that are two-word chemical names.

acid-catalyzed reactions, *but* Lewis acid catalyzed reactions
copper-to-iron ratio, *but* sodium chloride to iron ratio

◆ Hyphenate unit modifiers used as predicate adjectives. Usually, only unit modifiers that consist of nouns and adjectives or nouns and participles can be used as predicate adjectives.

In these cluster reactions, dehydrogenation is size-dependent.

All compounds were light-sensitive and were stored in the dark.

The reaction is first-order.

The complex is square-planar.

The antibody is species-specific.

The movie was thought-provoking.

◆ Hyphenate phrases also containing en dashes when they are used as unit modifiers.

alkyl–heavy-metal complexes
high-spin–low-spin transition
metal–metal-bonded complexes
Michaelis–Menten-like kinetics
retro-Diels–Alder reaction
transition-metal–chalcogen complexes

◆ Hyphenate phrases containing parenthetical expressions when they are used as unit modifiers.

element (silicon or tin)-centered radicals

◆ When the unit modifier in parentheses is a number and unit of measure, hyphenate them if they would ordinarily be hyphenated.

low-energy (2-J) bonds high-temperature (150 °C) reaction

Capitalization

In Text

Generally, in text keep all words lowercase, including chemical names and terms, except proper nouns and adjectives. However, there are many exceptions.

◆ Capitalize the words "figure", "table", "chart", and "scheme" only when they refer to a specific numbered item.

Figure 1 Chart 4 Table 2
Schemes 4–7

◆ Do not capitalize the "r" in "X-ray" at the beginning of a sentence.

◆ Capitalize parts of a book when they refer to a specific titled and numbered part.

Chapter 3	Appendix I	Section 4.2
but the preface	*and* the contents	

◆ Do not capitalize "page" with a number.

the photographs on page 3

◆ Capitalize only the name of the eponym, not the accompanying noun.

Avogadro's number	Lewis acid
Boltzmann constant	nuclear Overhauser effects
Einstein's theory	Raman spectroscopy
Graham's law	Schiff base
Hodgkin's disease	Stokes' law

Exception

Nobel Prize   Nobel Peace Prize

◆ Capitalize adjectives formed from proper names.

Boolean	Einsteinian	Lorentzian
Cartesian	Freudian	Mendelian
Copernican	Gaussian	Newtonian
Coulombic	Hamiltonian	
Darwinian	Laplacian	

◆ Capitalize the first word after a colon if the colon introduces more than one complete sentence, a quotation, or a formal statement.

Chemists find enzymes attractive as potentially useful synthetic tools for many reasons: Enzymes catalyze reactions with high regio- and stereoselectivity. They cause tremendous rate accelerations under mild reaction conditions. They reduce the need for protecting groups and give enantiomerically pure products.

An emulsion is a thermodynamically unstable system: it has a tendency to separate into two phases.

Two types of asymmetric reactions were conducted: synthesis of styrene oxide and reduction of olefinic ketones.

The editor wishes to make the following point: No papers will receive preferential treatment on the basis of artwork.

◆ Do not capitalize lowercase chemical descriptors hyphenated to chemical names when they are at the beginning of a sentence.

o-Dichlorobenzene was the solvent.

cis-4-Chloro-3-buten-2-one was obtained in 74% yield.

◆ When the first word of a sentence is a roman chemical descriptor that is not part of a chemical name, capitalize it.

> Syn hydroxylation of cycloalkenes was attempted.
>
> Trans hydroxyl groups are oxidized biochemically.
>
> Cis and trans isomers are used in pharmaceuticals and agrochemicals.
>
> Erythro diols were obtained in good yield.

◆ Do not capitalize chemical names or nonproprietary drug names unless they are at the beginning of a sentence or are in a title or heading. In such cases, capitalize the first letter of the English word, not the locant, stereoisomer descriptor, or positional prefix. (See Chapter 7, "Names and Numbers for Chemical Compounds".)

◆ Some reaction names are preceded by element symbols; they may be used as nouns or adjectives. When they are the first word of a sentence or appear in titles and headings, the first letter of the word is capitalized.

> N-Oxidation of the starting compounds yielded compounds **3–10**.
>
> N-Benzoylated amines undergo hydroxylation when incubated with yeast.
>
> Preparation of S-Methylated Derivatives
>
> O-Substituted Structural and Functional Analogs

◆ Always capitalize genus names as formal names, but never capitalize species names, even in titles. Do not capitalize the abbreviation for species, singular or plural (sp. and spp., respectively).

Bacillus subtilis	*Proteus vulgaris*
Cyanocitta cristata	*Pseudomonas aeruginosa*
Escherichia coli	*Staphylococcus aureus*
Pneumococcus aureus	*Streptococcus pneumoniae*
Salmonella sp.	*Polygonum* spp.

a bacterium of the genus *Salmonella*

◆ Do not capitalize genus names used as common nouns except at the beginning of a sentence or in a title or heading.

bacillus	klebsiella	streptococcus

hippopotamus, a member of the genus *Hippopotamus*
gorilla, a member of the genus *Gorilla*
pseudomonad, a member of the genus *Pseudomonas*

◆ Do not capitalize the adjectival or plural form of a genus name unless it is at the beginning of a sentence or in a title or heading.

pneumococcal	streptococcal	bacilli

◆ In text, do not capitalize polymer names that contain the names of the polymerizing species in parentheses following the prefix "poly". At the beginning of a sentence, capitalize only the "P" in "poly".

> Poly(vinyl chloride) is a less useful polymer than poly(ethylene glycol).

◆ Capitalize trademarks; use them as adjectives with the appropriate nouns.

Ficoll	Pyrex	Triton
Novocain (*but* novocaine)	Sephadex	Tween
Plexiglas (*but* plexiglass)	Styrofoam	Teflon

◆ Do not capitalize the word "model" with a number or code.

> γ counter (Beckman model 5500B)
> mass spectrometer (Perkin-Elmer model 240C)
> multichannel spectrometer (Otsuka model MCPD-1000)
> spectrometer (Varian model XL-200)
> Waters model 660 gradient controller

◆ Do not capitalize the common names of equipment.

electron-diffraction chamber	mass spectrometer
dynamic mechanical analyzer	mercury lamp
flame-ionization detector	spectrophotometer
gas chromatograph	temperature controller unit

◆ Use only an initial capital letter, not all capitals, for company names, which are not trademarks and are not protected by law.

◆ Capitalize the names of specific organizations or entities, including ACS sections, committees, and governing bodies, but not the general terms for them.

American Chemical Society	the society
the Milwaukee Section	a local section
ACS Board of Directors	the board
ACS Division of Fuel Chemistry	the division
ACS Committee on Nomenclature	an ACS committee
Clean Water Act	the act
Environmental Protection Agency	the agency
University of Michigan	the university

◆ Capitalize the names of specific titles when they appear with a person's name, but not the general terms for them.

the professor	the general	the mayor
Professor Carol Zachary	General James Shore	
Mayor Ralph Estes	Walter Baldwin, Professor of Chemistry	

> The well-known professor Carol Zachary will give a tutorial.
>
> James Shore, a general in the U.S. Army, will teach a graduate course.

Our speaker will be the retired general James Shore.

Isaac Bickford is an assistant professor.

Ralph Estes is the mayor of a small town in upstate New York.

◆ Capitalize the names of special events but not the general terms for them.

202nd ACS National Meeting
32nd ACS Western Regional Meeting
14th Biennial Conference on Chemical Education
Eastern Analytical Symposium
3rd World Conference of Chemical Engineers

the conference	the spring national meeting
the international symposium	the symposium
the national meeting	the world conference
the regional meeting	the workshop

◆ Capitalize sections of the country but not the corresponding adjectives.

the Northeast, *but* northeastern
the Midwest, *but* midwestern

◆ Do not capitalize the names of the four seasons: summer, fall, winter, spring, autumn.

◆ Capitalize Earth, Sun, and Moon only when used in an astronomical sense.

Venus and Mars are the closest planets to Earth.

The Earth rotates on its axis and revolves around the Sun.

The Moon is the only body that orbits the Earth.

But The sun is actually a globe of gas held together by its own gravity.

The earth's crust consists mostly of crystalline rock.

Water bodies on the earth's surface contain a variety of chromophoric substances.

Pollution occurs to some extent everywhere on earth.

The sun is the primary source of radiation that can cause chemical transformations.

The next full moon will be on Thursday.

In Titles and Headings

These guidelines apply to titles and headings at all levels; that is, they apply to subtitles and subheadings.

◆ In titles and headings that are typeset in capital and lowercase letters, capitalize the main words, which are nouns, pronouns, verbs, adjectives, adverbs, and subordinate conjunctions, regardless of the number of letters.

Do not capitalize coordinating conjunctions ("and", "but", "or", "nor", "yet", "so"), articles ("a", "an", "the"), or prepositions. Do capitalize the "to" in infinitives. Do capitalize the first and last words of a title or heading, regardless of part of speech, unless the word is mandated to be lowercase (e.g., pH, d Orbital).

Reactions of Catalyst Precursors with Hydrogen and Deuterium
Scope of the Investigations: The First Phase
Nickel-Catalyzed Addition of Grignard Reagents: Ring-Opening Reactions with Nucleophiles
Derivatives from a Chiral Borane–Amine Adduct
Properties of Organometallic Fragments in the Gas Phase
The Computer as a Tool To Improve Chemistry Teaching

Exception 1 In titles and headings, capitalize particles that are parts of phrasal verbs.

Break Down	Mix Up	Stand By
Build Up	Set Off	Take Off
Grow Up	Set Up	Warm Up
Hand Out	Slow Down	Wear Out
Line Up	Sort Out	

Exception 2 In titles and headings, capitalize particles that are parts of phrasal adjectives.

End-On Bonding   Side-On Bonding   In-Plane Atoms
but Out-of-Plane Vibrations (only the first preposition is capitalized)

◆ In titles and headings, capitalize "as" when it is used as a subordinating conjunction but not when it is used as a preposition.

Kinetics of Cyanocobalamin As Determined by Binding Capacity
Alumina as a Catalyst Support

◆ Do not capitalize the "r" in "X-ray" in titles and headings. Do capitalize the "r" in "γ ray" and the "p" in "α particle" and "β particle" in titles and headings.

◆ Do not capitalize lowercase chemical descriptors in titles or in headings, but do capitalize the first letter of the English word.

Reaction of *trans*-4-(Phenylsulfonyl)-3-buten-2-one

◆ When abbreviated units are acceptable in titles or headings, do not capitalize those that are ordinarily lowercase.

Determination of *N*-Nitrosodimethylamine at Concentrations <7 ng/L
Analysis of Milligram Amounts

◆ Always capitalize genus names, but never capitalize species names, in titles and headings.

Novel Metabolites of *Siphonaria pectinata*
Active-Site Nucleophile of *Bacillus circulans* Xylanase

◆ In titles and headings, capitalize all main words in a unit modifier.

High-Temperature System
Base-Catalyzed Cyclization
Cross-Linked Polymer
Deuterium-Labeling Experiment
Thyrotropin-Releasing Hormone
Non-Hydrogen-Bonding Molecules

◆ In titles and headings, capitalize each component of compound words if the component would be capitalized when standing alone.

Cross-Link
Quasi-Elastic
Half-Life

◆ Do not capitalize hyphenated suffixes.

Synthesis of Cubane-like Clusters

◆ In titles and headings, capitalize only the first letter ("P") of polymer names that contain the names of the polymerizing species in parentheses following the prefix "poly".

Reactions of Poly(methyl methacrylate)
New Uses for Poly(ethylene terephthalate)
Synthesis and Characterization of Poly(isobutylene-*b*-methyl vinyl ether)
Light-Scattering Studies of Poly(ethylene-*co*-butylene)
Polystyrene-*block*-poly(2-cinnamoylethyl methacrylate) Adsorption
IR Spectroscopic Analysis of Poly(1*H*,1*H*-fluoroalkyl α-fluoroacrylate)

◆ Capitalize only the first letter in a chemical name containing complex substituents in parentheses or brackets.

Structures of Tetrakis(methyl isocyanide)iron Complexes
Preparations of (Methyl isocyanide)iron Compounds

Surnames

Capitalization

Although a current trend is to lowercase the surnames of persons when these names are used as modifiers and have become very familiar, many are still capitalized. The following is a list (by no means complete) of names that should be capitalized.

Avogadro
Beckmann
Beilstein
Boltzmann
Bragg
Brønsted
Büchner
Bunsen
Claisen
Dewar benzene, flask
Dreiding
Erlenmeyer

Gram	Mahalanobis	Poisson
Kekulé	Mössbauer	Priestley
Kjeldahl	Petri	Scatchard
Markovnikov	Poiseuille	VandenHeuvel

Exceptions

de Broglie	van der Waals	van't Hoff

- Surnames that are used as units of measure are lowercase.

ampere	gauss	ohm
angstrom	gilbert	pascal
coulomb	gray	poise
curie	hartree	siemens
dalton	henry	sievert
darcy	hertz	stokes
debye	joule	tesla
einstein	kelvin	watt
erg	langmuir	weber
faraday	newton	

In the temperature–current curves, temperature is given in kelvins and current is shown in amperes.

NMR coupling constants are reported in hertz.

Hyphenation

- Hyphenation of double surnames is discussed on p 79.

- Hyphenate prefixes and suffixes to proper names as nouns and adjectives, and retain the capital letter.

anti-Markovnikov	non-Coulombic	oxy-Cope
hetero-Diels–Alder	non-Gaussian	retro-Diels–Alder
Kennedy-like	non-Newtonian	

Foreign Surnames

Some foreign surnames follow a format different from the American system. Most people are aware that the Chinese use their surnames first, followed by their given names. For example, Sun Yat-sen's surname is Sun. However, the problem of identifying surnames extends to many other cultures. This multiplicity of usage can create problems in bibliographic indexes and in reference citations. A reference citation in a bibliography should always list the surname first, followed by first name or initials. In a byline, the author names should be presented in standard American format (given names first and surnames last) to ensure consistency of citation practice. If a footnote would clarify the situation or eliminate any perceived confusion, use a footnote.

In most cultures, the surname is the family name, but it may not be the formal name, that is, the name or shortest string of names that are properly used following a title (Mr., Dr., Professor, etc.). Presented here are some cases in which different customs are used for the order of surnames, given names, and formal names. This list is by no means complete, but at least it will help you to be aware of these differences.

Arabic Often many names; the position of the surname is highly variable. The formal name often consists of two or three names including articles that can be joined. Examples: Ibn Saud, Abd al-Qadir.

Chinese Two or three names; the surname is first. Examples: Chiang Kai-shek is Dr. Chiang; Chou En-lai is Dr. Chou.

Hebrew Two or three names; the surname is the last one or two and is the formal name. Examples: David Ben-Gurion is Dr. Ben-Gurion; Moshe Bar-Even is Dr. Bar-Even.

Hungarian Two names; the surname is first, and it is the formal name. However, the second name is accepted as formal internationally.

Japanese Two names; the surname is the formal name. The surname is first in Japanese. However, when the names are translated into non-Asian languages, surnames appear last. Example: Taro Yamada is Dr. Yamada.

Korean Usually three names; the surname is first and is the formal name. In North Korean names, all three parts start with a capital letter. Examples: Kim Il Sung is Dr. Kim. In South Korean names, the two parts of the given name are hyphenated, and the second part is lowercase. Example: Kim Young-sum is Dr. Kim.

Spanish Frequently three or more names; the last two are surnames, sometimes connected by “y”. The second surname is often dropped or abbreviated. The formal name begins with the first surname and includes the second surname only in very formal usage. Example: Juan Perez Avelar is Dr. Perez or Dr. Perez Avelar, but never Dr. Avelar. The two surnames may also be hyphenated. Example: José Gregorio Angulo-Vivas is Dr. Angulo or Dr. Angulo-Vivas, but never Dr. Vivas.

Thai Two names; the surname is last, but the formal name is first.

Vietnamese Two or three names; the first is the surname and formal name.

Initials Some foreign names are abbreviated with two-letter initials that reflect transliteration from a non-Latin alphabet: Ch., Kh., Ph., Sh., Th., Ts., Ya., Ye., Yu., and Zh.

Some foreign names are abbreviated with hyphenated initials: C.-C. Yu.

Special Typefaces

Special typefaces help the reader quickly distinguish certain letters, words, or phrases from the rest of the text.

Italic Type

Chapter 5 describes the use of italic type in mathematical material, and Chapters 7 and 8 give guidelines for the use of italic type in chemical names and conventions in chemistry.

◆ Use italic type sparingly to emphasize a word or phrase. Do not use italics for long passages.

◆ Use italic type for a word being defined or for a newly introduced term the first time it appears in text.

> In an *outer-sphere transfer*, an electron moves from reductant to oxidant with no chemical alteration of the primary coordination spheres.

◆ Use italic type for the titles and abbreviations of periodicals, books, and newspapers. If "the" is the first word of the title, italicize and capitalize it.

> *Enough for One Lifetime* is the biography of Wallace Carothers.
>
> I read three articles on that new chiral compound in the *Journal of the American Chemical Society*.
>
> An article on a promising cholesterol biosynthesis inhibitor appeared in *The Journal of Organic Chemistry* this month.
>
> *The Washington Post* did a feature story on the president's daughter.

◆ Do not use italic type for common Latin terms and abbreviations.

ab initio	et al.	in vitro
ad hoc	etc.	in vivo
a priori	i.e.	status quo
ca.	in situ	vs
e.g.		

◆ Use italic type for genus and species names of all animals, plants, and microorganisms, but not when these names are used as singular or plural common nouns or when they are adjectival.

Staphylococcus aureus is the bacterium that causes staphylococcal infection.

Bacillus coagulans and *Bacillus dysenteriae* are two species of bacilli.

The red rhododendron, *Rhododendron arboreum*, needs bright sun.

◆ Do not use italic type for "pH"; "p" is always lowercase and "H" is always capitalized.

◆ Do not use italic type for M (molar) or N (normal). Do use italic type for *m* (molal).

Greek Letters

◆ Use Greek letters, not the spelled-out words, for chemical and physical terms.

γ radiation, *not* gamma radiation
β particle, *not* beta particle

Computer-Related Usage

◆ Capitalize the first letter of the names of computer languages.

AP
Basic
Cobol
Eiffel
Fortran
Java
Logo
Pascal
Perl
Python
Smalltalk

◆ Capitalize the first letter of the names of programs, and follow the manufacturer's or creator's usage within the name.

Alchemy
ChemDraw
ChemIntosh
ChemPlus
EasyPlot
EndNote Plus
FileMaker Pro
HyperChem
MacWrite
Microsoft Excel
Molecular Presentation Graphics
MULTAN78
SigmaPlot
SIMI4A
Symphony
TK Solver
Un-Plot-It
UniVersions
WordPerfect
Wordstar

◆ Use lowercase letters for the spelled-out forms of protocols, except as the first word of a sentence and in titles and headings.

network news-transfer protocol (NNTP)

List of Terms

On the following pages are the spelling, capitalization, and abbreviations of some common computer and Internet terms:

active matrix
anonymous FTP
applet
application
Archie
artificial intelligence (AI)
ASCII (American Standard Code for Information Interchange)
asynchronous
back up (verb)
backup (noun, adjective)
batch processing
baud
baud rate
BBS (bulletin board system)
bit
bitmap
Bitnet
bitstream
bps (bits per second)
browser
bulletin board
bulletin board system (BBS)
byte
C (programming language)
C++ (programming language)
CAD (computer-assisted design)
CAD/CAM (computer-assisted design and manufacturing)
CCD (charge-coupled device)
CD (compact disc)
CD key
CD-ROM (compact disc read-only memory)
central processing unit (CPU)
common gateway interface (CGI)
compact disc (CD)
CompuServe
computer graphics metafile (CGM)
CPU (central processing unit)
CRT (cathode ray tube)
cursor
CVC (color video controller)
cyberspace
daemon
data domain
data log
data parse, data parsing
data processing
data set
database
database management system (DBMS)
debug (verb)
DEC (Digital Equipment Corporation)
default
defragment
desktop
Dialog (search service)
dialog box
Digital Equipment Corporation (DEC)
digital signal
directory
disc (compact disc only)
disk
disk drive
disk space
diskette
DNS (domain name system or server)
DOS
double-click (as verb)
download
dpi (dots per inch)
DTD (document-type definition)
duplex
e-journal
e-mail
e-money
e-publish
e-zine (electronic magazine)
EBCDIC (Extended Binary-Coded Decimal Interchange Code)
electronic mail (e-mail)
end user (noun)
EPS (encapsulated PostScript)
Ethernet (*but* an ethernet)
fiber optics
file name
file transfer protocol (FTP)
filter
finger
firewall
floppy disk
flowchart
format, formatting, formatted

Fortran
FrameMaker
FreeNet (*but* a freenet)
freeware
front end
FTP (file transfer protocol)
gateway
GB (gigabyte or 1024 megabytes; always a space between number and GB)
GDDM (graphical data display manager)
GDI (graphics device interface)
general purpose interface bus (GPIB)
GIF (graphics interchange format)
4GL (fourth-generation language)
Gopher (*but* a gopher)
Gopherspace
graphic (noun)
graphical interface
graphics (adjective)
graphics conversion
graphics files
graphics terminal emulation
GUI (graphical user interface)
hard disk
hard disk drive
hardware
hardwired
Harvard Graphics
high-level-language compiler
home directory
home page (lowercase, but capitalized when part of a specific name, e.g., ACS Home Page)
hot key
hotline
HTML (hypertext markup language)
HTTP (hypertext transfer protocol)
HyperCard
hyperlink
hypermedia
hypertext
IBM-compatible
IBM PC
IBM PC/AT
IBM PC/XT
IBM PS/2
icon
iconization
iconize
information superhighway
input
input/output (I/O)
integrated circuit
interdomain conversion
Internet
intranet
IRC (Internet Relay Chat)
ISDN (Integrated Services Digital Network)
Java
joystick
JPEG (Joint Photographic Experts Group)
K (kilobyte, actually 1024 bytes; always closed up to number; as in 8K or 16K disk drive; kB is preferred)
kB (kilobyte; actually 1024 bytes)
KB (kilobyte; kB is preferred)
kbps (kilobits per second)
kBps (kilobytes per second)
keyboard
keypad
keystroke
kilobit
kilobyte
LAN (local area network)
laptop
LCD (liquid-crystal display)
Lexis
list-administration software
list-management software
list owner
list server
Listserv (the software)
local area network (LAN)
log in, logging in (verb)
log off, logging off (verb)
log on, logging on (verb)
log out, logging out (verb)
login name
logon name
Lotus 1-2-3
Macintosh

Macintoshes
macro, macros
mail-transfer agent (MTA)
mail-user agent (MUA)
mainframe
math coprocessor
MB (megabyte, actually 1024 kilobytes; always a space between number and MB)
meta-list
microchip
microcomputer
microcomputing
microprocessor
Microsoft Excel
Microsoft Windows
Microsoft Word
MIME (multipurpose Internet mail extension)
minicomputer
minifloppy disk
modem
monitor (noun)
Mosaic
motherboard
motif
mouse (plural: mouse devices)
MPEG (Motion Picture Experts Group)
MS-DOS (always hyphenated)
MTA (mail-transfer agent)
MUA (mail-user agent)
NCP (network control program)
NCSA (National Center for Supercomputing Applications)
Net (when referring to the Internet; lowercase when referring to any network)
Netscape
netware
network
newsgroup
NNTP (network news-transfer protocol)
node, nodes
Novell
NREN (National Research and Education Network)
OCR (optical character recognition)
off-site (always hyphenated)
offline (one word in computer context)
on-site (always hyphenated)
online (one word in computer context)
OS (operating system)
output
PageMaker
PAM (pulse amplitude modulation)
parallel port
parser
password
path
PC (personal computer, usually IBM-compatible)
pdb (Protein Data Bank) format
Perl (programming language)
personal computer (PC)
personal home page (PHP)
PIF (picture interchange format)
pixel
PL/1
plaintext
PNG (portable network graphics)
popup
port
PostScript
PPP (point-to-point protocol)
primary domain
print queue
programmer
programming
PROM (programmable read-only memory)
protocol
PS/2
pull-down (adjective)
QuarkXPress
queue
RAM (random access memory)
RDBMS (rotational database management system)
read-only memory (ROM)
read/write permission
real time (noun)
real-time (unit modifier)

remote job entry (RJE)
RFC (request for comments)
rich text
ROM (read-only memory)
RPG (report program generator)
run time (noun)
run-time (adjective)
scale up (verb)
scanner
screen dump
script
SCSI (small computer system interface)
serial communication
serial port
server
set up (verb)
setup (noun)
SGML (standard generalized markup language)
shared user
shareware
shortcut
shut down (verb)
shutdown (noun, adjective)
sign off (verb)
sign-off (noun, adjective)
Silicon Graphics
simplex
SLIP (serial-line Internet protocol)
SMB (server message block)
SMTP (simple mail transfer protocol)
software
source code
spelling checker
spreadsheet
stand-alone (always hyphenated)
start up (verb)
startup (noun)
STN Express
strikethrough
submenu
systems programs
T-1, T-3 (hyphen)
Tcl (pronounced "tickle"; programming language)
TCP/Connect II
TCP/IP (transmission control protocol/Internet protocol)
telecommute
Telnet
terminal emulation program
terminal server
throughput
TIFF (tagged image file format)
time-sharing (always hyphenated)
toolbar
toolbox
trackball
TSO (time-sharing option)
TTL (transistor–transistor logic)
TTY (teletype)
UGA (ultra graphics accelerator)
Unicode
Unify
UNIX
URC (uniform resource characteristic)
URL (uniform resource locator)
URN (uniform resource name)
Usenet
user id, user ids
utility program
VAX
Veronica
VGA (video graphics adapter)
video adapter
VisiCalc
VRML (virtual reality modeling language)
VT-100, VT-200, VT-220
WAIS (Wide Area Information Service)
WAN (wide-area network)
the Web
Web browser
Web page
Web server
Web site
webmaster
webzine
window (general term, not specific program)
Windows (Microsoft)
word-processing software
word processor
WordPerfect

Wordstar
wordwrap
workstation
World Wide Web (three words, no hyphens)
World Wide Web Consortium (W3O)
WORM (write once read many)
WWW (World Wide Web)
WYSIWYG (what you see is what you get)
X Windows
Xbase
Xenix
Xyvision
XyWrite

Breaking Computer Addresses at the End of a Line

Uniform Resource Locators

A typical uniform resource locator (URL), which is an address on the World Wide Web, takes the following forms:

http://www.domain.zone
http://www.domain.zone/name1/~name2/
http://domain.zone/name1/~name2/name3.html

The number of names varies. For example,

http://www.acs.org
http://pubs.acs.org
http://www.unitedmedia.com/comics/dilbert/

These examples are short, but URLs can be quite long, and in narrative text they often will need to be broken at the end of a line. Break URLs after a slash or a period, but not the last period if possible, and do not insert a hyphen or any other character.

E-Mail Addresses

A typical e-mail address usually takes one of these forms:

personname@company-name.zone
initial_surname@company-name.zone
surnameinitial@company-name.zone

for example, j_dodd@acs.org. All kinds of variations on the person's name and initials are possible, and besides the underscore, other types of punctuation are used. Long names are often truncated.

Break e-mail addresses in text after the @ or a period. Do not insert a hyphen or any other character.

Chapter 6, "References", presents the editorial style for electronic sources listed in reference lists and bibliographies.

Trademarks

A trademark is an adjective that describes a material or product (e.g., Teflon resin, Kleenex tissue). The term "brand name" is a synonym for trademark. In ACS publications, do not use the trademark symbols ™ and ® or the service mark symbol ℠. They are not necessary to ensure legal protection for the trademark.

◆ Capitalize trademarks; use them as adjectives with the appropriate nouns. Do not use them in titles.

Ficoll	Pyrex	Triton
Novocain (*but* novocaine)	Sephadex	Tween
Plexiglas (*but* plexiglass)	Styrofoam	Teflon

◆ In general, however, use generic names rather than trade names.

cross-linked dextran polymer beads, *not* Sephadex
diatomaceous earth, *not* Celite
4,4′-isopropylidenediphenol, *not* Bisphenol A
2-methoxyethanol, *not* Methyl Cellosolve
mineral oil, *not* Nujol
petroleum jelly, *not* Vaseline
photocopy, *not* Xerox
poly(ethylene glycol), *not* Carbowax
tensile testing machine, *not* Instron tester

◆ Use trademarks as adjectives only, never as nouns or verbs. Do not make a trademark plural.

Abbreviations and Acronyms

An abbreviation is a short form of a word; often the individual letters are pronounced; in an acronym, the letters always form a pronounceable word. ACS is an abbreviation; CASSI is an acronym.

A list of ACS-recommended abbreviations and acronyms is given in the supplement to this chapter. Check the list to find an abbreviation. If no abbreviation is listed for the term you are using, you may devise an abbreviation provided that (1) it is not identical to an abbreviation of a unit of measure, (2) it will not be confused with the symbol of an element or a group, (3) it does not hamper the reader's understanding, and (4) you do not use the same abbreviation for more than one spelled-out form.

◆ If a very long name or term is repeated many times throughout a paper, an abbreviation is warranted. Place the abbreviation in parentheses following the spelled-out form the first time it appears in the text. If it is used in the abstract, define it in the abstract and again in the text. After defining the abbreviation in the text, you may use it throughout the paper.

Exceptions The following list shows abbreviations that never need to be defined. Refer to the list at the end of the chapter for all other abbreviations.

a.m.	before noon (Latin ante meridiem)
anal.	analysis
at. wt	atomic weight
bp	boiling point
ca.	about (Latin circa)
cf.	compare (Latin confer)
CP	chemically pure
DNA	deoxyribonucleic acid
ed.	edition
Ed., Eds.	Editor, Editors
e.g.	for example (Latin exempli gratia)
eq(s)	equation(s) [with number(s)]
equiv	equivalent(s) [with number(s)]
equiv wt	equivalent weight
et al.	and others (Latin et alii)
etc.	and so forth (Latin et cetera)
fp	freezing point
GLC	gas–liquid chromatography
i.d.	inside diameter
i.e.	that is (Latin id est)
in.	inch, inches
IR	infrared
m	molal
M	molar
mmp	mixture melting point
mp	melting point
M_r	relative molecular mass (molecular weight)
N	normal
NMR	nuclear magnetic resonance
no., nos.	number, numbers
o.d.	outside diameter
p, pp	page, pages
p.m.	after noon (Latin post meridiem)
P.O.	Post Office (with Box and number)
ref(s)	reference(s) [with number(s)]
RNA	ribonucleic acid
sp., spp.	species, singular and plural
sp gr	specific gravity
sp ht	specific heat
sp vol	specific volume
U.K.	United Kingdom

U.S.	United States
USP	United States Pharmacopeia
UV	ultraviolet
vol	volume
vs	versus
v/v	volume per volume
wt	weight
w/v	weight per volume
w/w	weight per weight

◆ Avoid abbreviations in the title of a paper.

◆ For some, but not all, abbreviations, case is important; that is, if they are capitalized, they must never be made lowercase; if they are lowercase, they must never be capitalized. This guideline applies to abbreviations that would lose their meanings or change meanings if their forms are changed, such as units of measure (e.g., mg cannot be changed to Mg, min cannot be changed to Min), mathematical symbols (e.g., pH cannot be changed to PH or ph), and chemical symbols (e.g., *o* for ortho cannot be changed to *O*).

However, if the meaning would not be affected, some abbreviations can be capitalized at the beginning of a sentence and in titles and headings, especially if they are so common that they are more like words than abbreviations. For example, you could use "e-mail" in text and "E-mail" at the beginning of a sentence.

◆ Symbols for the chemical elements are not treated as abbreviations. They need not be defined, and they are typeset in roman type.

◆ Abbreviate units of measure and do not define them when they follow a number. Without a number, spell them out.

9 V/s or $9 \text{ V} \cdot \text{s}^{-1}$, *but* measured in volts per second

For exceptions, see Chapter 5, p 164.

◆ Abbreviations that are common to a specific field may be permitted without identification in books and journals in that field only, at the discretion of the editor.

◆ For genus and species names, spell out the full genus name in the title, in the abstract, and the first time it appears in text. Abbreviate it thereafter with the same species name, but spell it out again with each different species name. Form the abbreviation with the initial of the genus name. If the paper contains more than one genus name that starts with the same initial letter, devise abbreviations that distinguish them. Use italic type for all names and abbreviations.

First time	*Subsequently*
Bacillus subtilis	*B. subtilis*
Bacillus stearothermophilus	*B. stearothermophilus*
Escherichia coli	*E. coli*
Salmonella typhimurium	*S. typhimurium*
Staphylococcus aureus	*Staph. aureus*

◆ Use "e.g.", "i.e.", "vs", and "etc." only in figure captions, in tables, and in parentheses in text. Elsewhere, spell out "for example", "that is", "versus", and "and so forth".

◆ Do not confuse abbreviations and mathematical symbols. An abbreviation is usually two or more letters; a mathematical symbol should generally be only one letter, possibly with a subscript or superscript. An abbreviation may be used in narrative text but seldom appears in equations; a mathematical symbol is preferred in equations and may also be used in text. For example, in text with no equations, PE may be used for potential energy, but in mathematical text and equations, E_p is preferred. Abbreviations are typeset in roman type; most mathematical symbols are typeset in italic type.

◆ Do not abbreviate

- the words "day", "week", "month", and "year" (except in ACS journals in descriptions of experimental work)
- days of the week
- titles not used with a name
- states not used with a city

◆ In text, spell out all months with or without a specific day.

On August 3, 1996, we completed the second phase of the experiment.

The final results will be available in January 1997.

◆ Use the following abbreviations (with no periods) or spelled-out forms for months with a day or with a day and year in footnotes, tables, figure captions, bibliographies, and lists of literature cited.

Jan	April	July	Oct
Feb	May	Aug	Nov
March	June	Sept	Dec

◆ Use the abbreviations U.S. and U.K. as adjectives only; spell out United States and United Kingdom as the noun forms in text. Either United Kingdom or U.K. may be used in addresses.

U.S. science policy	chemical industry in the United States
U.K. educational system	educational system in the United Kingdom

◆ Form the plurals of multiletter, all-capital abbreviations and abbreviations ending in a capital letter by adding a lowercase "s" only, with no apostrophe.

HOMOs	PAHs	PCs
JPEGs	PCBs	pHs

◆ To avoid ambiguity or poor appearance, add an apostrophe and a lowercase "s" to form the plurals of lowercase abbreviations, single-capital-letter abbreviations, abbreviations ending in a subscript or superscript, and abbreviations ending in an italic letter.

cmc's	p*K*'s	pK_a's
T_g's	O's (or oxygens; Os is the symbol for osmium)	

◆ Use two-letter abbreviations for U.S. state and territory names and Canadian provinces and territories on all letters going through the U.S. Post Office and most express delivery services. Use them after the name of a city in text, footnotes, and references.

United States

AL	Alabama	MT	Montana
AK	Alaska	NE	Nebraska
AZ	Arizona	NV	Nevada
AR	Arkansas	NH	New Hampshire
CA	California	NJ	New Jersey
CO	Colorado	NM	New Mexico
CT	Connecticut	NY	New York
DE	Delaware	NC	North Carolina
DC	District of Columbia	ND	North Dakota
FL	Florida	OH	Ohio
GA	Georgia	OK	Oklahoma
HI	Hawaii	OR	Oregon
ID	Idaho	PA	Pennsylvania
IL	Illinois	RI	Rhode Island
IN	Indiana	SC	South Carolina
IA	Iowa	SD	South Dakota
KS	Kansas	TN	Tennessee
KY	Kentucky	TX	Texas
LA	Louisiana	UT	Utah
ME	Maine	VT	Vermont
MD	Maryland	VA	Virginia
MA	Massachusetts	WA	Washington
MI	Michigan	WV	West Virginia
MN	Minnesota	WI	Wisconsin
MS	Mississippi	WY	Wyoming
MO	Missouri		

U.S. Territories

GU	Guam	SM	Samoa
PR	Puerto Rico	VI	Virgin Islands

Canada

AB	Alberta	NS	Nova Scotia
BC	British Columbia	ON	Ontario
MB	Manitoba	PE	Prince Edward Island
NB	New Brunswick	PQ	Quebec
NF	Newfoundland	SK	Saskatchewan
NT	Northwest Territories	YT	Yukon Territory

◆ Spell out and capitalize "company" and "corporation" as part of company names when they appear in an author's affiliation. Abbreviate them elsewhere in text. After the first mention, drop Co. and Corp. and use only the company name.

Bibliography

AIP Style Manual, 4th ed.; American Institute of Physics: New York, 1990.

American Medical Association Manual of Style, 8th ed.; Williams & Wilkins: Baltimore, MD, 1989.

ASM Style Manual for Journals and Books; American Society for Microbiology: Washington, DC, 1991.

The Chicago Manual of Style, 14th ed.; University of Chicago Press: Chicago, IL, 1982.

The Microsoft Manual of Style for Technical Publications; Microsoft Press: Redmond, WA, 1995.

Publication Manual of the American Psychological Association, 4th ed.; American Psychological Association: Washington, DC, 1994.

Scientific Style and Format: The CBE Manual for Authors, Editors, and Publishers, 6th ed.; Cambridge University Press: New York, 1994.

U.S. Government Printing Office Style Manual; Government Printing Office: Washington, DC, 1984.

Webster's Standard American Style Manual; Merriam-Webster: Springfield, MA, 1985.

Wired Style: Principles of English Usage in the Digital Age; Hale, Constance, Ed.; HardWired: San Francisco, CA, 1996.

Words into Type, 3rd ed.; Prentice-Hall: Englewood Cliffs, NJ, 1974.

Ten To Watch

1. The prefix "non". Do not hyphenate the prefix "non" before a common adjective.

nonabrasive	noncoherent	nongaseous
nonorganic		

2. The suffix "like". Hyphenate the suffix "like" to words of three or more syllables, to two-word compounds used as unit modifiers, to chemical names, and to proper nouns and adjectives.

bacteria-like	cycloalkane-like	Claisen-like
ion-exchange-like		

3. Genus and species names. Always capitalize genus names, but never capitalize species names, even in titles. Always italicize genus and species names.

Bacillus subtilis	*Proteus vulgaris*
Cyanocitta cristata	*Pseudomonas aeruginosa*
Escherichia coli	*Staphylococcus aureus*
Pneumococcus aureus	*Streptococcus pneumoniae*

4. Names of persons used as units of measure. Surnames that are used as units of measure are lowercase.

ampere	debye	newton
angstrom	gauss	pascal
coulomb	hertz	poise
curie	joule	watt
dalton	kelvin	weber

5. Abbreviations that need never be defined.

DNA	RNA	GLC
IR	NMR	UV

6. Eponyms and nouns. Capitalize only the name of the eponym, not the accompanying noun.

Avogadro's number	Lewis acid
Boltzmann constant	nuclear Overhauser effects
Einstein's theory	Raman spectroscopy

Exceptions

Nobel Prize	Nobel Peace Prize

7. Forming the plural of all-capital abbreviations. Add a lowercase "s" only, with no apostrophe.

HOMOs	PCBs	PAHs

8. Do not use italic type for common Latin terms and abbreviations.

ab initio	et al.	in vivo
ad hoc	etc.	status quo
a priori	i.e.	vs
ca.	in situ	
e.g.	in vitro	

9. Use Greek letters, not the spelled-out words, for chemical and physical terms.

γ radiation	β particle	α helix

10. Hyphen versus en dash in surname combinations. Use a hyphen in the name of a person with two surnames (unless that person uses a space).

Irene Joliot-Curie	Jackie Joyner-Kersee

Use an en dash to link the names of two persons.

Bose–Einstein	Friedel–Crafts	Michaelis–Menten
Diels–Alder	Lineweaver–Burk	van't Hoff–Le Bel

Supplement: Abbreviations, Acronyms, and Symbols in All Categories

This list is not intended to be exclusive. Alternative choices, in many cases, are acceptable. Proscribed usages are specifically indicated.

A

α	fine structure constant
	rotation, specific rotation
	stereochemical descriptor
$[\alpha]^t_D$	specific rotation at temperature *t* and wavelength of sodium D line
$[\alpha]^t_\lambda$	specific rotation at temperature *t* and wavelength λ
a	antisymmetric
	are (unit of area, 100 m^2)
	atto (10^{-18})
	axial [*use* 2(a)-methyl in names]
a	*a* axis
	absorptivity
	axial chirality [as in (*aR*)-6,6′-dinitrodiphenic acid]
a_0	Bohr radius (0.52917 Å)
A	adenosine
	alanine
	ampere
	ring (italic in steroid names)
Å	angstrom
A	absorbance [as in $A = \log(1/T)$]
	anticlockwise (chirality symbol)
	Helmholtz energy
	mass number
AAS	atomic absorption spectroscopy
abs	absolute
ac	alternating current
ac	anticlinal
Ac	acetyl
	actinium
acac	acetylacetonato (ligand)
acam	acetamide (ligand)
AcCh	acetylcholine
AcChE	acetylcholinesterase
AcO	acetate
ACS	American Chemical Society
ACTH	adrenocorticotropin; adrenocorticotropic hormone
A.D.	anno Domini
Ade	adenine
Ado	adenosine

ADP	adenosine 5′-diphosphate
AEM	analytical electron microscopy
AES	atomic emission spectroscopy
	Auger electron spectroscopy
af	audio frequency
AFM	atomic force microscopy
AFS	atomic fluorescence spectroscopy
AGU	anhydroglucose unit
AI	artificial intelligence
AIChE	American Institute of Chemical Engineers
ala	alanyl in genetics
Ala	alanyl, alanine
alt	alternating, as in poly(A-*alt*-B)
a.m.	ante meridiem
AM	amplitude modulation
AMP	adenosine 5′-monophosphate
	adenosine 5′-phosphate
amu	atomic mass unit [amu, reference to oxygen, is deprecated; u (reference to mass of ^{12}C) should be used]
anal.	analysis (Anal. in combustion analysis presentations)
anhyd	anhydrous
ANN	artificial neural network
ANOVA	analysis of variance
Ans	ansyl
ANSI	American National Standards Institute
ansyl	8-anilino-1-naphthalenesulfonyl
antilog	antilogarithm
AO	atomic orbital
ap	antiperiplanar
AP	appearance potential
API	American Petroleum Institute
APIMS	atmospheric pressure ionization mass spectrometry
APS	appearance potential spectroscopy
aq	aqueous
A_r	relative atomic mass (atomic weight)
Ar	aryl
AR	analytical reagent (e.g., AR grade)
Ara	arabinose
ara-A	adenosine, with arabinose rather than ribose (arabinoadenosine, also ara-A, araA)
ara-C	cytidine, with arabinose rather than ribose (arabinocytidine, also ara-C, araC)
arb unit	arbitrary unit (clinical)
Arg	arginyl, arginine
ARPES	angle-resolved photoelectron spectroscopy
ARPS	angle-resolved photoelectron spectroscopy
ARS	Agricultural Research Service
as	asymmetrical

AS	absorption spectroscopy
Asa	β-carboxyaspartic acid
ASCII	American Standard Code for Information Interchange
ASIS	aromatic solvent-induced shift
Asn	asparaginyl, asparagine
Asp	aspartyl, aspartic acid
ASTM	American Society for Testing and Materials
Asx	"Asn or Asp"
asym	asymmetrical
ATCC	American Type Culture Collection
atm	atmosphere
atom %	atom percent
ATP	adenosine 5′-triphosphate
ATPase	adenosinetriphosphatase
ATR	attenuated total reflection
at. wt	atomic weight
au	atomic unit
AU	absorbance unit
	astronomical unit (length)
AUFS	absorbance units at full scale
av	average

B

β	stereochemical descriptor
b	barn (neutron capture area, 10^{-24} cm^2)
	bohr (unit of length)
	broad or broadened (spectra)
b	*b* axis
	block, as in poly(A-*b*-B)
B	"aspartic acid or asparagine"
	bel
	buckingham (10^{-26} esu cm^2)
	ring (italic in steroid names)
B	boat (conformation)
bar	unit of pressure; unit and abbreviation are the same
bbl	barrel
B.C.	before Christ
bcc	body-centered cubic (crystal structure)
bccub	body-centered cubic (crystal structure)
BCD	binary coded decimal
B.C.E.	before the common era
Bd	baud
BDH	British Drug House
BEHP	bis(2-ethylhexyl) phthalate
BET	Brunauer–Emmett–Teller (adsorption isotherm)
BeV	billion electronvolts
bGH	bovine growth hormone
Bi	biot

b.i.d.	twice a day
binap	2,2′-bis(diphenylphosphino)-1,1′-binaphthyl (ligand)
binol	1,1′-bi-2-naphthol (ligand)
biol	biological(ly)
bipy	2,2′-bipyridine, 2,2′-bipyridyl (bpy preferred)
	4,4′-bipyridine, 4,4′-bipyridyl (bpy preferred)
bis-Tris	[bis(2-hydroxyethyl)amino]tris(hydroxymethyl)methane (also bistris, Bis-Tris, bis-tris)
bit	binary digit
BL	bioluminescence
BM	Bohr magneton (*use* μ_B)
Bn	benzyl (also Bzl)
BN	bond number
BO	Born–Oppenheimer
BOD	biological oxygen demand
bp	base pair
	boiling point
bps	bits per second
Bps	bytes per second
bpy	2,2′-bipyridine, 2,2′-bipyridyl
	4,4′-bipyridine, 4,4′-bipyridyl
BPY	bipyramidal (coordination compounds)
Bq	becquerel
br	broad or broadened (spectra)
BSA	bovine serum albumin
Btu	British thermal unit
bu	bushel
Bu	butyl
BWR	Benedict–Webb–Rubin (equation)
Bz	benzoyl
Bzac	benzoylacetone
Bzl	benzyl (also Bn)

C

χ	magnetic susceptibility
c	candle
	centered (crystal structure)
	centi (10^{-2})
	cyclo [as in c-C_6H_{11}, c-Hx (cyclohexyl)]
c	*c* axis
	concentration, for rotation, e.g., $[\alpha]^{20}{}_{489}$ +25° (*c* 0.13, $CHCl_3$)
	cyclo [as in *c*-S_6 (*cyclo*-hexasulfur)]
	specific cytochrome (i.e., cytochrome *c*)
C	Celsius (*use* °C as unit abbreviation)
	coulomb
	cysteine
	cytidine
	ring (italic in steroid names)

C	chair (conformation)
	clockwise (chirality symbol)
ca.	circa, about [used before an approximate date or figure (ca. 1960)]
CAD	computer-assisted design
cal	calorie
$\mathrm{cal_{IT}}$	International Table calorie
calcd	calculated
CAM	computer-assisted manufacturing
cAMP	adenosine cyclic 3′,5′-phosphate
	adenosine 3′,5′-cyclic phosphate
CAN	ceric ammonium nitrate
CARS	coherent anti-Stokes Raman spectroscopy
CAT	computed axial tomography
	computer-averaged transients
cB	conjugate base, counterbase
CB	conjugate base, counterbase
Cbz	carbobenzoxy, carbobenzyloxy, (benzyloxy)carbonyl, benzyl-oxycarbonyl
cc	cubic centimeter (*do not use*; *use* $\mathrm{cm^3}$ or mL)
CCD	charge-coupled device
CCGC	capillary column gas chromatography
ccp	cubic close-packed (crystal structure)
cd	candela
	current density
CD	circular dichroism
	compact disc
CDC	Centers for Disease Control and Prevention
CDH	ceramide dihexoside [$\mathrm{Cer(Hex)_2}$]
cDNA	complementary DNA
CDP	cytidine 5′-diphosphate
CD-ROM	compact disc read-only memory
CE	Cotton effect
C.E.	common era
CE–MS	capillary electrophoresis–mass spectrometry
CERCLA	Comprehensive Environmental Response, Compensation, and Liability Act
cf.	compare
CFC	chlorofluorocarbon
cfm	cubic feet per minute
CFR	Code of Federal Regulations
CFSE	crystal field stabilization energy (also cfse)
cfu	colony-forming units (bacterial inocula)
cgs	centimeter–gram–second (as in cgs system)
cgsu	centimeter–gram–second unit(s)
ChE	cholinesterase
CHF	coupled Hartree–Fock
Ci	curie

CI	chemical ionization
	configuration interaction
CIDEP	chemically induced dynamic electron polarization
CIDNP	chemically induced dynamic nuclear polarization
CIMS	chemical ionization mass spectrometry
CL	cathodoluminescence
	chemiluminescence
c/m^2	candles per square meter
CM	carboxymethyl (as in CM-cellulose)
CMA	Chemical Manufacturers Association
cmc	critical micelle concentration
CMH	ceramide monohexoside [Cer(Hex)]
CMO	canonical molecular orbital
CMP	cytidine 5′-monophosphate, cytidine 5′-phosphate
cmr	carbon magnetic resonance (*do not use*; *use* ^{13}C NMR)
CMR	carbon magnetic resonance (*do not use*; *use* ^{13}C NMR)
CN	coordination number
CNDO	complete neglect of differential overlap
CNRS	Centre National de la Recherche Scientifique
CNS	central nervous system
co	copoly (as in A-*co*-B)
CoA	coenzyme A
cod	1,5-cyclooctadiene (ligand)
COD	chemical oxygen demand
coeff	coefficient
colog	cologarithm
compd	compound
con	conrotatory (may be italic)
concd	concentrated
concn	concentration
const	constant
cor	corrected
cos	cosine
cosh	hyperbolic cosine
COSY	correlation spectroscopy
cot	cotangent
	1,3,5,7-cyclooctatetraene (ligand)
coth	hyperbolic cotangent
counts/s	counts per second
C_p	heat capacity at constant pressure
cp	candlepower
cP	centipoise
Cp	cyclopentadienyl
CP	central processor
	chemically pure
	cross-polarization
cpd	contact potential difference

CPE	controlled-potential electrolysis
CPK	Corey–Pauling–Koltun (molecular models)
	creatine phosphokinase
CPL	circular polarization of luminescence
cpm	counts per minute
CP/MAS	cross-polarization/magic-angle spinning (also permitted: CP-MAS, CP–MAS, CPMAS, CP MAS)
cps	counts per second (*use* counts/s)
	cycles per second (*use* Hz or s^{-1})
CPSC	Consumer Product Safety Commission
CPU	central processing unit
CRAMPS	combined rotation and multiple-pulse spectroscopy
CRIMS	chemical reaction interface mass spectrometry
crit	critical
cRNA	complementary RNA
CRT	cathode ray tube
CRU	constitutional repeating unit
cryst	crystalline
csc	cosecant
csch	hyperbolic cosecant
CSIRO	Commonwealth Scientific and Industrial Research Organisation (Australia)
CT	charge transfer
CTEM	conventional transmission electron microscopy
CTH	ceramide trihexoside [$Cer(Hex)_3$]
CTP	cytidine 5′-triphosphate
CU-8	cubic, coordination number 8
cub	cubic (crystal structure)
C_v	heat capacity at constant volume
CV	coefficient of variation
	cyclic voltammetry
CVD	chemical vapor deposition
CW	constant width
	continuous wave (as in CW ESR)
cwt	hundredweight
Cy	cyclohexyl
cyclam	1,4,8,11-tetraazacyclotetradecane
Cyd	cytidine
Cys	cysteinyl, cysteine
cyt	cytochrome
Cyt	cytosine
cytRNA	cytoplasmic RNA
CZE	capillary zone electrophoresis

D

δ	NMR chemical shift in parts per million downfield from a standard
∂	partial differential

d	day (spelled-out form is preferred)
	deci (10^{-1})
	deoxy
	deuteron
	differential (mathematical)
	diffuse
	doublet (spectra)
d.	diameter, with i. and o. (inside and outside)
d	density
	dextrorotatory
	distance
	spacing (X-ray)
D	absolute configuration
D	aspartic acid
	debye
	deuterium
	ring (italic in steroid names)
D	diffusion coefficient ($cm^2\ s^{-1}$)
	symmetry group [e.g., D_3; also used in names, such as (+)-D_3-trishomocubane]
2-D	two-dimensional (also 2D)
3-D	three-dimensional (also 3D)
da	deca or deka (10)
Da	dalton
daf	dry ash free
dAMP	2′-deoxyadenosine 5′-monophosphate or phosphate (the A can be replaced with C, G, U, etc.)
dansyl	5-(dimethylamino)-1-naphthalenesulfonyl
dB	decibel
DBMS	database management system
dc	direct current
DD-8	dodecahedral, coordination number 8
DD NMR	dipolar decoupling NMR
DDT	1,1,1-trichloro-2,2-bis(*p*-chlorophenyl)ethane
de	diastereomeric excess
DEAE	(diethylamino)ethyl (as in DEAE-cellulose)
dec	decomposition
decomp	decompose
DEFT	driven equilibrium Fourier transform
deg	degree (*use* °B, degrees Baume; °C, °F, *but* K)
DEG	diethylene glycol
DEHP	bis(2-ethylhexyl) phthalate (BEHP is preferred)
DES	diethylstilbestrol
det	determinant
df	degrees of freedom
DF	degrees of freedom
diam	diameter
dil	dilute
dis	disrotatory (may be italic)

distd	distilled
DLVO	Derjaguin–Landau–Verwey–Overbeek
DMA	dynamic mechanical analyzer
DMBA	9,10-dimethylbenz[*a*]anthracene
DME	1,2-dimethoxyethane
	dropping mercury electrode
DMEM	Dulbecco's modified Eagle's medium
DMF	dimethylformamide
DMN	diaminomaleonitrile
dmr	deuterium magnetic resonance (*do not use; use* ^{2}H NMR)
DMR	deuterium magnetic resonance (*do not use; use* ^{2}H NMR)
DMSO	dimethyl sulfoxide (also Me_2SO)
DMTA	dynamic mechanical thermal analyzer
DNA	deoxyribonucleic acid
DNase	deoxyribonuclease
DNMR	dynamic nuclear magnetic resonance
DNP	deoxynucleoprotein
	dynamic nuclear polarization (NMR)
DNPH	(2,4-dinitrophenyl)hydrazine
Dns	dansyl
Dopa	3-(3,4-dihydroxyphenyl)alanine (also DOPA)
DP	degree of polymerization (also dp)
dpm	disintegrations per minute
DPN	diphosphopyridine nucleotide (NAD is preferred)
DPNH	reduced DPN (NADH is preferred)
DPPH	2,2-diphenyl-1-picrylhydrazyl
dps	disintegrations per second
Dq	crystal field splittings
DQF	double quantum filtered
DRIFT	diffuse reflectance Fourier transform
Ds	crystal field splittings
DSC	differential scanning calorimetry
dT	thymidine
Dt	crystal field splittings
DTA	differential thermal analysis
DTC	depolarization thermocurrent
	differential thermal calorimetry
dTDP	thymidine 5′-diphosphate
DTE	dithioerythritol
dThd	thymidine
dTMP	thymidine 5′-monophosphate, thymidine 5′-phosphate
DTT	dithiothreitol
dTTP	thymidine 5′-triphosphate
dyn	dyne

E

ε	molar absorptivity
	dielectric constant

ε^*	complex permittivity
η	hapto
	viscosity
e	base of natural logarithm
	electron
	equatorial [in names, e.g., 2(e)-methyl]
e_{aq}^-	hydrated electron
$e^-(aq)$	hydrated electron
e_s^-	solvated electron
$e^-(s)$	solvated electron
e	electronic charge
E	exa (10^{18})
	glutamic acid
E	electromotive force
	energy
	entgegen (configuration)
	envelope (conformation)
	potential energy
	specific extinction coefficient ($E_{280nm}^{1\%,1cm}$)
	Young's modulus
$E°$	standard electrode potential
	standard electromotive force
$E_{1/2}$	half-wave potential
E1	first-order elimination
E2	second-order elimination
E_a, E_A	Arrhenius or activation energy
ea_0	electronic charge in electrostatic units × Bohr radius or atomic units for dipole moment
EC	Enzyme Commission
	exclusion chromatography
ECD	electron-capture detector, detection
ECE	electrochemical, chemical, electrochemical (mechanisms)
ECG	electrocardiogram
ecl	electrochemical luminescence
ECL	electrochemical luminescence
ECP	effective core potential
ed.	edition, edited
Ed.	editor
ED	effective dose
ED_{50}	dose that is effective in 50% of test subjects (also ED50)
edda	ethylenediaminediacetato (ligand)
Eds.	editors
EDS	energy-dispersive system (or spectrometry)
edta	ethylenediaminetetraacetato (ligand)
EDTA	ethylenediaminetetraacetic acid, ethylenediaminetetraacetate
EDXS	energy-dispersive X-ray spectrometry
ee	enantiomeric excess
EEG	electroencephalogram
EELS	electron energy loss spectroscopy

EFG	electric field gradient
e.g.	for example
EGA	evolved gas analysis
EGD	evolved gas detection
EGR	exhaust gas recirculation
E_h	hartree (unit); Hartree energy
EH	extended Hückel
EI	electron impact
	electron ionization
EIA	enzyme immunoassay
E_k	kinetic energy
EKC	electrokinetic chromatography
EKG	electrocardiogram
EL	electroluminescence
ELISA	enzyme-linked immunosorbent (immunoadsorbent) assay
e/m	ratio of electron charge to mass
EM	electron microscopy
e-mail	electronic mail
EMC	equilibrium moisture content
emf	electromotive force
EMIS	electromagnetic isotope separation
emu	electromagnetic unit
en	ethylenediamine (ligand)
ENDOR	electron–nuclear double resonance
ent	reversal of stereo centers
E_p	potential energy
EPA	ether–isopentane–ethanol (solvent system)
	U.S. Environmental Protection Agency
epi	inversion of normal configuration (italic with a number, as in 15-*epi*-prostaglandin A)
EPMA	electron probe microanalysis
EPR	electron paramagnetic resonance
EPXMA	electron probe X-ray microanalysis
eq	equation
equiv	equivalent
equiv wt	equivalent weight
erf	error function
erfc	error function complement
erfc^{-1}	inverse error function complement
ESCA	electron spectroscopy for chemical analysis
esd	estimated standard deviation
ESE	electron spin echo
ESEEM	electron spin echo envelope modulation
ESI	electrospray ionization
ESIMS	electrospray ionization mass spectrometry
ESP	elimination of solvation procedure
ESR	electron spin resonance
esu	electrostatic unit
Et	ethyl

et al.	and others
etc.	and so forth
eu	entropy unit
EU	enzyme unit
eV	electronvolt
EXAFS	extended X-ray absorption fine structure
exch	exchangeable (spectra)
exp	exponential
expt	experiment
exptl	experimental

F

f	and page following (as in p 457 f)
	femto (10^{-15})
	fermi (unit of length, also fm)
	fine (spectral)
f	focal length
	frequency (in statistics)
	function [as in $f(x)$]
	furanose form
F	Fahrenheit (*use* °F as unit abbreviation)
	farad
	formal (*use* judiciously; M is preferred)
	phenylalanine
F	Faraday constant
	free energy
	variance ratio (in statistics)
FAAS	flame atomic absorption spectroscopy
FABMS	fast atom bombardment mass spectrometry
fac	facial
FAD	flavin adenine dinucleotide
FAES	flame atomic emission spectroscopy
FAFS	flame atomic fluorescence spectroscopy
FAS	flame absorption spectroscopy
fcc	face-centered cubic (crystal structure)
FCC	fluid catalytic cracking
Fd	ferredoxin
FDA	U.S. Food and Drug Administration
FEM	field emission microscopy or spectroscopy
FES	field emission spectroscopy
	flame emission spectrometry
ff	and pages following (as in p 457 ff)
FFEM	freeze-fracture electron microscopy
FFF	field flow fractionation
FFS	flame fluorescence spectroscopy
FFT	fast Fourier transform
FHT	Fisher–Hirschfelder–Taylor (space-filling models)
FI	field ionization

FIA	flow-injection analysis
	fluorescence immunoassay
fid	free induction decay (in Fourier transform work)
FID	flame ionization detector, detection
	free induction decay (in Fourier transform work)
FIK	field ionization kinetics
FIR	far-infrared
FLC	ferroelectric liquid crystal
fm	femtometer
	fermi (unit of length, also f)
FM	frequency modulation
FMN	flavin mononucleotide
FMO	frontier molecular orbital
FOPPA	first-order polarization propagator approach
fp	freezing point
FPC	fixed partial charge
FPT	finite perturbation theory
Fr	franklin
Fr	Froude number
Fru	fructose
FSGO	floating spherical Gaussian orbital
FSH	follicle-stimulating hormone
ft	foot
FT	Fourier transform
ft-c	foot-candle
FTICR	Fourier transform ion cyclotron resonance
FTIR	Fourier transform infrared (also FT/IR, FT-IR, and FT IR)
FTIRS	Fourier transform infrared spectroscopy
ft-lb	foot-pound
ft-lbf	foot-pound-force
FTP	file transfer protocol
FTS	Fourier transform spectroscopy
fw	formula weight
fwhh	full width at half-height
fwhm	full width at half-maximum

G

γ	microgram (*use* μg)
	photon
	surface tension
Γ	surface concentration
g	gas [as in $H_2O(g)$]
	gram
g	acceleration due to gravity (closed up to number preceding)
	splitting factor (ESR and NMR spectroscopy)
G	gauss
	generally labeled

G	giga (10^9)
	glycine
	guanosine
G	free energy (Gibbs)
	gravitational constant
Ga	Galileo number
gal	gallon
Gal	galactose
	galileo (unit of acceleration)
g-atom	gram-atom (*use* mol)
GB	gigabyte (1024 megabytes)
GC	gas chromatography
GDC	gas displacement chromatography
GDMS	glow discharge mass spectrometry
GDP	guanosine 5′-diphosphate
gem	geminal
GFAAS	graphite furnace atomic absorption spectroscopy
GFC	gas frontal chromatography
gfw	gram formula weight
GH	growth hormone (somatotropin)
GHz	gigahertz
Gi	gilbert
GIAO	gauge-invariant atomic orbital
Glc	glucose
GLC	gas–liquid chromatography
Gln	glutaminyl, glutamine
GLPC	gas–liquid partition chromatography
Glu	glutamyl, glutamic acid
Glx	"Gln or Glu"
gly	glycine (ligand)
Gly	glycyl, glycine
GMP	guanosine 5′-monophosphate, guanosine 5′-phosphate
GPC	gel permeation chromatography
gr	grain (unit of weight)
GSC	gas–solid chromatography
GSH	reduced glutathione
GSL	glycosphingolipid
GSSG	oxidized glutathione
GTP	guanosine 5′-triphosphate
Gua	guanine
Guo	guanosine
Gy	gray (international unit of absorbed dose)

H

h	hecto (10^2)
	helion
	hour

h	crystallographic index (*hkl*)
	Planck's constant
$\hbar$	Planck's constant divided by 2π
H	henry
	histidine
H	enthalpy
	half-chair (conformation)
	Hamiltonian
$\mathcal{H}$	Hamiltonian
^{1}H NMR	proton nuclear magnetic resonance
H_0	magnetic field (ESR and NMR spectroscopy)
ha	hectare
Hb	hemoglobin
Hbg	biguanide
HCG	human chorionic gonadotropin
hcp	hexagonal close-packed (crystal structure)
HCP	hexachlorophene
HCS	hazard communication standard
HDPE	high-density polyethylene
Hedta	ethylenediaminetetraacetate(3–) (-ato as ligand in full name)
H_2edta	ethylenediaminetetraacetate(2–) (-ato as ligand in full name)
H_3edta	ethylenediaminetetraacetate(1–) (-ato as ligand in full name)
H_4edta	ethylenediaminetetraacetic acid
HEEDTA	*N*-(2-hydroxyethyl)ethylenediaminetriacetate
Hepes	*N*-(2-hydroxyethyl)piperazine-*N*′-ethanesulfonic acid (also HEPES, hepes)
Hepps	*N*-(2-hydroxyethyl)piperazine-*N*′-propanesulfonic acid (also HEPPS, hepps)
hex	hexagonal (crystal structure)
HF	Hartree–Fock
hfs	hyperfine splitting
hfsc	hyperfine splitting constant
hGH	human growth hormone
HIPS	high-impact polystyrene
His	histidyl, histidine
HIV	human immunodeficiency virus
hkl	crystallographic index
HMDS	hexamethyldisilane
	hexamethyldisiloxane
HMO	Hückel molecular orbital
HMPA	hexamethylphosphoramide
HMPT	hexamethylphosphoric triamide
hnRNA	heterogeneous nuclear RNA
$h\nu$	indicates light; *h* is Planck's constant and ν is the photon frequency
HOHAHA	homonuclear Hartmann–Hahn
HOMO	highest occupied molecular orbital
H_2ox	oxalic acid
hp	horsepower
HPCE	high-performance capillary electrophoresis

HPLC	high-performance liquid chromatography
	high-pressure liquid chromatography
HREELS	high-resolution electron energy loss spectroscopy
HREM	high-resolution electron microscopy
HRMS	high-resolution mass spectrometry
HSP	heat shock protein
HTML	hypertext markup language
HTTP	hypertext transfer protocol
Hyl	hydroxylysyl, hydroxylysine
Hyp	hydroxyprolyl, hydroxyproline
	hypoxanthine
Hz	hertz

I

i	iso (as in *i*-Pr; *never use* *i*-propyl)
I	inosine
	isoleucine
I	electric current (also *i*)
	ionic strength
	moment of inertia
	spin quantum number (ESR and NMR spectroscopy)
ibid.	in the same place (in the reference cited; *use is discouraged*)
ic	intracerebrally
IC	integrated circuit
	ion chromatography
ICP	inductively coupled plasma
ICR	ion cyclotron resonance
ics	internal chemical shift
ICSH	interstitial-cell-stimulating hormone
ICT	International Critical Tables
i.d.	inside diameter
i_d	diffusion current
ID	infective dose
ID_{50}	dose that is infective in 50% of test subjects (also ID50)
IDAS	isotope dilution α spectrometry
IDMS	isotope dilution mass spectrometry
IDP	inosine 5′-diphosphate
i.e.	that is
IE	ionization energy
IEC	ion-exchange chromatography
IEEE	Institute of Electrical and Electronics Engineers
IEF	isoelectric focusing
IEP	isoelectric point
IETS	inelastic electron-tunneling spectroscopy
IFQ	interfacial fluorescence quenching
IKES	ion kinetic energy spectroscopy
Ile	isoleucyl, isoleucine
ILS	increased life span

im	intramuscularly
IMMA	ion microprobe mass analysis
IMP	inosine 5′-monophosphate, inosine 5′-phosphate
in.	inch
INDO	intermediate neglect of differential overlap
INDOR	internal nuclear double resonance
	internucleus (nucleus–nucleus) double resonance
INH	inhibitor
	isonicotinic acid hydrazide
Ino	inosine
INO	iterative natural orbital
insol	insoluble
I/O	input–output
ip	intraperitoneally
IP	ionization potential
ips	iron pipe size
ipso	position of substitution
IR	infrared
IRDO	intermediate retention of differential overlap
IRMA	immunoradiometric assay
IRMS	isotopic ratio mass spectrometry
IRP	internal reflection photolysis
IRRAS	infrared reflection–absorption spectroscopy
IRS	internal reflection spectroscopy
isc	intersystem crossing
ISCA	ionization spectroscopy for chemical analysis
ISE	ion-selective electrode
iso	inversion of normal chirality (not as in isopropyl, but in uses such as 8-*iso*-prostaglandin E_1; generally italic with a number)
ISO	International Organization for Standardization
ISS	ion-scattering spectroscopy
ITP	inosine 5′-triphosphate
	isotachophoresis
IU	international unit
IUPAC	International Union of Pure and Applied Chemistry
iv	intravenously

J

J	joule
J	coupling constant (NMR and ESR spectroscopy)
JT	Jahn–Teller

K

k	kilo (10^3)
k	Boltzmann constant (also k_B)
	crystallographic index (*hkl*)
	rate constant

K	1000 (as in 60K protein)
	kayser (*use* cm^{-1})
	kelvin (*do not use* °K)
	kilobyte (kB is preferred)
	lysine
K	equilibrium constant
Kα	spectral line
kat	katal (unit of enzyme catalytic activity)
Kβ	spectral line
k_B	Boltzmann constant
kb	kilobar (*use* kbar)
	kilobase
	kilobit
kB	kilobel
	kilobyte; 1024 bytes (kB is preferred; in computer terminology, K is often used, always closed up to the number, as in 8K disk drive)
kbar	kilobar
kbp	kilobase pair
kbps	kilobits per second
kBps	kilobytes per second
kD	kilodebye
kDa	kilodalton
KE	kinetic energy
kg	kilogram
kgf	kilogram-force
kHz	kilohertz
K_m	Michaelis constant
K_{oc}	carbon-referenced sediment partition coefficient
	organic chemicals partition coefficient
K_{ow}	octanol–water partition coefficient
K_{SP}	solubility product constant
K_w	autoionization constant

L

λ	absolute activity
	microliter (*use* μL)
	wavelength
λ_{ex}	excitation wavelength
λ_{max}	wavelength of maximum absorption
l	liquid [as in NH_3(l)]
l	crystallographic index (*hkl*)
	levorotatory
L	absolute configuration
L	leucine
	ligand
	liter
L_I	spectral line

L_{II}	spectral line
L_{III}	spectral line
Lac	lactose
LAMMA	laser microprobe mass spectrometry
LAN	local area network
lat	latitude
lb	pound
lbf	pound-force
LC	liquid chromatography
LCAO	linear combination of atomic orbitals
LCD	liquid-crystal display
LCICD	liquid-crystal-induced circular dichroism
LCVAO	linear combination of virtual atomic orbitals
LD	lethal dose
LD_{50}	dose that is lethal to 50% of test subjects
LDH	lactic dehydrogenase
LDMS	laser desorption mass spectrometry
LE	locally excited
LED	light-emitting diode
LEED	low-energy electron diffraction
LEEDS	low-energy electron diffraction spectroscopy
LEISS	low-energy ion-scattering spectroscopy
LEMF	local effective mole fraction
Leu	leucyl, leucine
LFER	linear free-energy relationship
LH	luteinizing hormone
LIF	laser-induced fluorescence
lim	limit
LIMS	laboratory information management system
LIS	lanthanide-induced shift
lit.	literature
LJ, L-J	Lennard-Jones
LLC	liquid–liquid chromatography
lm	lumen
LMCT	ligand-to-metal charge transfer
ln	natural logarithm
LNDO	local neglect of differential overlap
log	logarithm to the base 10
Log	principal logarithm
long.	longitude
Lp	Lorentz–polarization (effect)
Lp	Lorentz factor × polarization factor
LSC	liquid–solid chromatography
LSD	lysergic acid diethylamide
LSR	lanthanide shift reagent
LUMO	lowest unoccupied molecular orbital
lut	lutidine (ligand)
Lut	lutidine

lx	lux
Lys	lysyl, lysine

M

μ	chemical potential
	dipole moment
	electrophoretic mobility
	micro (10^{-6})
	micron (*do not use*; *use* μm or micrometer)
$\mu^{\pm}$	muon
μ_B	Bohr magneton
μ_N	nuclear magneton
μ_W	Weiss magneton
m	medium (spectra)
	meter
	mile (in mpg and mph; otherwise mi)
	milli (10^{-3})
	multiplet (spectra)
m	isotopic mass
	magnetic quantum number (ESR and NMR spectroscopy)
	meta
	molal (mol kg^{-1})
M	mega (10^6)
	mesomeric
	metal (*never* Me)
	methionine
	molar (mol dm^{-3}, mol L^{-1})
M	minus (left-handed helix)
[M]	molecular rotation
mAb	monoclonal antibody (also Mab, MAb)
Mal	maltose
Man	mannose
MAO	monoamine oxidase
MAS	magic-angle spinning
MASS	magic-angle sample spinning
max	maximum
Mb	myoglobin
MB	megabyte (1024 kilobytes)
MBE	molecular beam epitaxy
MCD	magnetic circular dichroism
mCi	millicurie
MCT	mercury cadmium telluride
MD	molecular dynamics
m_e	electron rest mass
m/*e*	mass-to-charge ratio (*m*/*z* is preferred)
Me	methyl (*never* metal)
MED	mean effective dose
MEKC	micellar electrokinetic capillary chromatography

MEM	minimum Eagle's essential medium
mequiv	milliequivalent
mer	polymer notation (as in 16-mer)
mer	meridional
Mes	mesylate, 2-morpholinoethanesulfonic acid, 2-morpholinoethane-sulfonate (also MES)
Met	methionyl, methionine
MetHb	methemoglobin
MetMb	metmyoglobin
MeV	million electronvolts
mho	reciprocal ohm (Ω^{-1} is preferred)
MHz	megahertz
mi	mile
min	minimum
	minute
MINDO	modified intermediate neglect of differential overlap
MIR	mid-infrared
MIRS	multiple internal reflection spectroscopy
ML	monolayer
MLCT	metal-to-ligand charge transfer
MLR	multiple linear regression
mmHg	millimeters of mercury (measure of pressure)
mmp	mixture melting point
mmu	millimass unit
m_n	neutron rest mass
M_n	number-average molecular weight
MO	molecular orbital
mol	mole
mol wt	molecular weight (M_r is preferred)
MOM	methoxymethyl
mon	monoclinic (crystal structure)
m_p	proton rest mass
mp	melting point
mpg	miles per gallon
mph	miles per hour
MPI	multiphoton ionization
MPV	Meerwein–Ponndorf–Verley
MQ ENDOR	multiple-quantum electron nuclear double resonance
M_r	relative molecular mass (molecular weight)
MR	molecular refraction
MRI	magnetic resonance imaging
mRNA	messenger RNA
MS	mass spectrometry
	mass spectrum
	microwave spectroscopy
MSDS	manufacturer's safety data sheet
	material safety data sheet
MSG	monosodium glutamate

MSH	melanocyte-stimulating hormone, melanotropin
Mt	megaton
MTD	mean therapeutic dose
mtDNA	mitochondrial DNA
mtRNA	mitochondrial RNA
mu	mass unit
MVA	mevalonic acid
MVS	multiple-variable storage
M_w	weight-average molecular weight
MW	molecular weight (M_r is preferred)
MWD	molecular weight distribution
Mx	maxwell
M_z	*z*-average molecular weight
m/z	mass-to-charge ratio

N

ν	frequency
$\tilde{\nu}$	wavenumber
$\nu_{1/2}$	full width at half-maximum height (NMR spectra)
ν_e	neutrino
ν_{max}	frequency of maximum absorption
n	nano (10^{-9})
	neutron
n	normal (as in *n*-butyl, *n*-Bu)
	refractive index ($n^{20}{}_{D}$, at 20 °C, Na D line)
	total number of individuals
N	asparagine
	newton
	normal (concentration)
	unspecified nucleoside
N_A	Avogadro's number
NAA	neutron activation analysis
[Na]ATPase	sodium ion activated ATPase (also Na-ATPase, NaATPase)
NAD	nicotinamide adenine dinucleotide
NADH	reduced nicotinamide adenine dinucleotide
NADP	nicotinamide adenine dinucleotide phosphate
NADPH	reduced nicotinamide adenine dinucleotide phosphate
[Na,K]ATPase	sodium and potassium ion activated ATPase (also Na,K-ATPase)
NAS	National Academy of Sciences
N.B.	nota bene (note well)
NBS	National Bureau of Standards
	N-bromosuccinimide
NCI	National Cancer Institute
NCSA	National Center for Supercomputing Applications
NDA	New Drug Application
NDDO	neglect of diatomic differential overlap
nDNA	nuclear DNA
NEMO	nonempirical molecular orbital

neut equiv	neutralization equivalent
NHE	normal hydrogen electrode
NIEHS	National Institute of Environmental Health Sciences
NIH	National Institutes of Health
NIOSH	National Institute for Occupational Safety and Health
NIR	near-infrared
NIST	National Institute of Standards and Technology
Nle	norleucyl, norleucine
NLM	National Library of Medicine
NLO	nonlinear optical (optics)
nm	nanometer
NM	nuclear magneton (*use* μ_N)
NMN	nicotinamide mononucleotide
NMR	nuclear magnetic resonance (*not* nmr)
no.	number
NO	natural orbital (as in CNDO/2-NO)
NOAA	National Oceanic and Atmospheric Administration
NOCOR	neglect of core orbitals
NOE	nuclear Overhauser effect
NOESY	nuclear Overhauser enhancement spectroscopy
NO_x	nitrogen oxides
Np	neper
NPR	net protein retention
NQR	nuclear quadrupole resonance
NRC	National Research Council
nRNA	nuclear RNA
NRTL	nonrandom two-liquid
NSOM	near-field scanning optical microscopy
NTP	National Toxicology Program
	normal temperature and pressure
	unspecified nucleoside 5′-triphosphate
Nuc	nucleoside (unspecified)
Nva	norvalyl, norvaline

O

ω	angular frequency
Ω	ohm
o	ortho
O	orotidine
OAc	acetate
obsd	observed
OC-6	octahedral, coordination number 6
OCR	optical character recognition
o.d.	outside diameter
OD	optical density
ODMR	optically detected magnetic resonance
ODU	optical density unit
Oe	oersted

OES	optical emission spectroscopy
OFDR	off-frequency decoupling resonance
OMVPE	organometallic vapor-phase epitaxy
Ord	orotidine
ORD	optical rotary dispersion
o-rh	orthorhombic (crystal structure)
Orn	ornithyl, ornithine
Oro	orotic acid
ORTEP	Oak Ridge thermal ellipsoid plot
OSHA	Occupational Safety and Health Administration
osm	osmolar
osM	osmolar
Osm	osmolar
OTTLE	optically transparent thin-layer electrode
O/W	oil in water (emulsion)
o/w	oil in water (emulsion)
ox	oxalato (ligand)
	oxidized or oxidation (in subscripts and superscripts)
oxidn	oxidation
oz	ounce

P

%	percent
‰	per thousand (parts per thousand)
π	pros (near) in NMR measurements (as in N^{π} of histidine)
	type of orbital, electron
$\pi^{\pm}$	pion
π^{0}	pion
ψ	pseudouridine
ψrd	pseudouridine
p	negative logarithm (as in pH)
	page
	pico (10^{-12})
	proton
p	angular momentum (ESR and NMR spectroscopy)
	para
	probability (in statistics)
	pyranose form
P	peta (10^{15})
	poise
	proline
P	plus (right-handed helix)
	probability (in statistics)
P450	specific cytochrome designation (i.e., cytochrome P450)
P-450	specific cytochrome designation (i.e., cytochrome P-450)
P_{450}	specific cytochrome designation (i.e., cytochrome P_{450})
^{31}P NMR	phosphorus-31 nuclear magnetic resonance
Pa	pascal

PAC	perturbed angular correlation
PAD	perturbed angular distribution
PAGE	polyacrylamide gel electrophoresis
$\mathrm{p}a_{\mathrm{H}}$	negative logarithm of hydrogen ion activity
PAH	polycyclic aromatic hydrocarbon
PAN	polyacrylonitrile
PBS	phosphate-buffered saline
pc	parsec (unit of length)
PC	paper chromatography
	personal computer
	planar chromatography
PCB	polychlorobiphenyl, polychlorinated biphenyl
PCDD	polychlorodibenzo-*p*-dioxin
	polychlorinated dibenzo-*p*-dioxin
PCDF	polychlorodibenzofuran
PCILO	perturbed configuration interaction with localized orbitals
PCP	pentachlorophenol
PCR	polymerase chain reaction
PCTFE	poly(chlorotrifluoroethylene)
PDL	pumped dye laser
PDMS	plasma desorption mass spectrometry
PE	polyethylene
	potential energy
PEG	poly(ethylene glycol)
PEL	permissible exposure limit
PEO	poly(ethylene oxide)
PES	photoelectron spectroscopy
PET	positron emission tomography
PETP	poly(ethylene terephthalate)
PFU	plaque-forming unit
PG	prostaglandin
pH	negative logarithm of hydrogen ion concentration
Ph	phenyl (for C_6H_5 only)
Phe	phenylalanyl, phenylalanine
phen	1,10-phenanthroline, *o*-phenanthroline
phr	parts per hundred parts of resin (or rubber)
$\mathrm{P_i}$	inorganic phosphate
PIB	polyisobutylene
PIXE	proton-induced X-ray emission
p*K*	negative logarithm of equilibrium constant
$\mathrm{p}K_{\mathrm{a}}$	p*K* for association
PL	photoluminescence
PLOT	porous-layer open-tubular
p.m.	post meridiem
PMMA	poly(methyl methacrylate)
PMO	perturbational molecular orbital
PMR	phosphorus magnetic resonance (*do not use*; change to ^{31}P NMR)
	polymerization of monomeric reactants
	proton magnetic resonance (*do not use*; change to ^{1}H NMR)

PNA	polynuclear aromatic hydrocarbon
PNDO	partial neglect of differential overlap
po	per os (orally)
POM	poly(oxymethylene), polyformaldehyde
POPOP	1,4-bis(5-phenyl-2-oxazolyl)benzene
pp	pages
PP	polypropene
ppb	parts per billion
ppbv	parts per billion by volume
PP_i	inorganic pyrophosphate, phosphoric acid
ppm	parts per million
ppmv	parts per million by volume
PPO	2,5-diphenyloxazole
PPP	Pariser–Parr–Pople
PPS	photophoretic spectroscopy
ppt	parts per trillion
	precipitate
pptv	parts per trillion by volume
Pr	propyl
PRDDO	partial retention of diatomic differential overlap
prepn	preparation
PRF	Petroleum Research Fund
PRFT	partially relaxed Fourier transform
Pro	prolyl, proline
pro-R	stereochemical descriptor (also pro-*R*)
pro-S	stereochemical descriptor (also pro-*S*)
PRT	platinum resistance thermometer
Ps	positronium
PS	polystyrene
psi	pounds per square inch
psia	pounds per square inch absolute
psig	pounds per square inch gauge
pt	pint
	point
PTC	phase-transfer catalysis
PTFE	poly(tetrafluoroethylene)
PTH	parathyroid hormone
	phenylthiohydantoin
PTV	programmed-temperature vaporizer
PU	polyurethane
PVA	poly(vinyl alcohol)
PVAC	poly(vinyl acetate)
PVAL	poly(vinyl alcohol)
PVC	poly(vinyl chloride)
PVDC	poly(vinylidene dichloride)
PVDF	poly(vinylidene difluoride)
PVE	poly(vinyl ether)
PVF	poly(vinyl fluoride)

PXRD	powder X-ray diffraction
py	pyridine (ligand)
Py	pyridine
PY	pyramidal (coordination compounds)
Py–GC–MS	pyrolysis–gas chromatography–mass spectrometry
pyr	pyrazine (ligand)
pyrr	pyrrolidine (ligand)
pz	pyrazole (ligand)

Q

q	quartet (spectra)
Q	glutamine
q	heat, electric charge
Q	heat, electric charge
QCPE	Quantum Chemistry Program Exchange
QELS	quasi-elastic light scattering
QSAR	quantitative structure–activity relationship
qt	quart

R

ρ	density
r	correlation coefficient
R	arginine
	Rankine (temperature scale, *use* °R as unit abbreviation)
	roentgen
R	gas constant
	rectus (configurational)
	regression coefficient
	resistance
rac	racemic
rad	radian
	unit of radiation
RAM	random access memory
RBS	Rutherford backscattering spectrometry
rd	rad
RDE	rotating disk electrode
r_e	electron radius
re	stereochemical descriptor (as in the *re* face)
recryst	recrystallized
red	reduced or reduction (in subscripts and superscripts)
redn	reduction
redox	reduction–oxidation
ref	reference
rel	relative
rel	relative (stereochemical descriptor)
REL	recommended exposure limit
rem	roentgen equivalent man
REM	rapid eye movement

rep	roentgen equivalent physical
rf	radio frequency
R_f	retention factor (ratio of distance traveled by the center of a zone to the distance simultaneously traveled by the mobile phase)
RFC	request for comments
RH	relative humidity
RI	refractive index
RIA	radioimmunoassay
Rib	ribose
RIMS	resonance ionization mass spectrometry
RIS	resonance ionization spectrometry
rms	root mean square
RNA	ribonucleic acid
RNase	ribonuclease
ROA	Raman optical activity
ROM	read-only memory
RPLC	reversed-phase liquid chromatography
rpm	revolutions per minute
RQ	respiratory quotient
RRDE	rotating ring-disk electrode
RRKM	Rice–Ramsperger–Kassel–Marcus
rRNA	ribosomal RNA
RRS	resonance Raman spectroscopy
RRT	relative retention time
RS	Raman spectroscopy
RSD	relative standard deviation
	risk-specific dose
Ry	rydberg

S

σ	standard deviation
	surface charge density
	surface tension
	tensile strength
	type of orbital, electron
Σ	summation
s	second
	single bond [as in s-cis (italic in compound names)]
	singlet (spectra)
	solid [as in NaCl(s)]
	strong (spectra)
s	secondary (as in *s*-Bu; *but sec*-butyl)
	sedimentation coefficient
	standard deviation (analytical)
	symmetrical
$s^0_{20,\mathrm{w}}$	sedimentation coefficient measured at 20 °C in water and extrapolated to 0 °C
s^2	sample variance

S	serine
	siemens
S	entropy
	sinister (configurational)
	skew (conformation)
SANS	small-angle neutron scattering
SAPR-8	square antiprismatic, coordination number 8
sar	sarcosine (*N*-methylglycine) (ligand)
Sar	sarcosyl, sarcosine (*N*-methylglycine)
SAR	structure–activity relationship
SARISA	surface analysis by resonance ionization of sputtered atoms
SAXS	small-angle X-ray scattering (or spectroscopy)
sc	subcutaneously
sc	synclinal
sccm	standard cubic centimeters per minute
SCE	saturated calomel electrode
SCF	self-consistent field
scfh	standard cubic feet per hour
SCF–HF	self-consistent field, Hartree–Fock
SCOT	support-coated open-tubular
SD	standard deviation
SDS	sodium dodecyl sulfate
SE	standard error
S_E2	second-order electrophilic substitution
sec	secant
sec	secondary (as in *sec*-butyl, *sec*-Bu)
SEC	size exclusion chromatography
sech	hyperbolic secant
SECM	scanning electrochemical microscopy
SECS	simulation and evaluation of chemical synthesis
SEM	scanning electron microscopy
	standard error of the mean
Ser	seryl, serine
SERS	surface-enhanced Raman spectroscopy (or scattering)
SEW	surface electromagnetic wave
S_{ex}	exciplex substitution
SFC	supercritical-fluid chromatography
SGML	standard generalized markup language
sh	sharp (spectra)
	shoulder (spectra)
Sh	Sherwood number
SHC	shape and Hamiltonian consistent
SHE	standard hydrogen electrode
si	stereochemical descriptor (as in the *si* face)
SI	International System of Units (Système International)
	secondary ion (as in SIMS)
SIM	selected-ion monitoring
SIMS	secondary-ion mass spectrometry

sin	sine
sinh	hyperbolic sine
SLR	spin–lattice relaxation
SMOSS	surface Mössbauer
SMSI	strong metal support interaction
sn	stereospecific numbering
SN	separation number
S/N	signal-to-noise ratio
S_N1	first-order nucleophilic substitution
S_N2	second-order nucleophilic substitution
S_Ni	internal nucleophilic substitution
SNO	semiempirical natural orbital
sol	solid
soln	solution
sp	specific
sp.	species (singular)
sp	synperiplanar
SP-4	square planar, coordination number 4
SPECT	single-photon-emission computed tomography
sp gr	specific gravity
sp ht	specific heat
spp.	species (plural)
SPR	stroboscopic pulse radiolysis
sp vol	specific volume
SPY-5	square pyramidal, coordination number 5
sq	square
SQF	single quantum filtered
SQUID	superconducting quantum interference device
sr	steradian
$S_{RN}1$	first-order nucleophilic substitution triggered by electron transfer
SRS	stimulated Raman scattering
SSC	standard saline citrate (NaCl–citrate)
St	stokes
std	standard
STEM	scanning transmission electron microscopy
STM	scanning tunneling microscopy
STO	Slater-type orbital
STO-3G	Slater-type orbital, three Gaussian
STP	standard temperature and pressure
Suc	sucrose
Sv	sievert
	svedberg
SVL	single vibrational level
swg	standard wire gauge
sym	symmetrical

T

τ	tele (far) in NMR measurements (as in N^{τ} of histidine)
θ	angle

[θ]	ORD measurement, deg cm^2/dmol
Θ	temperature (e.g., in Curie–Weiss expressions)
t	metric ton
	triplet (spectra)
	triton
t	Student distribution (the Student *t* test in statistics)
	temperature (in degrees Celsius)
	tertiary (as in *t*-Bu; but *tert*-butyl)
$t_{1/2}$	half-life
T	ribosylthymine
	tautomeric
	tera (10^{12})
	tesla
	threonine
	tritium
T	temperature (in kelvins)
	twist (conformation)
T-4	tetrahedral, coordination number 4
tan	tangent
tan δ	mechanical loss factor
tanh	hyperbolic tangent
TBP	tri-*n*-butyl phosphate
TBPY-5	trigonal bipyramidal, coordination number 5
T/C	treated vs cured
TCA	tricarboxylic acid cycle (citric acid cycle, Krebs cycle)
	trichloroacetic acid
TCD	thermal conductivity detector
TCP/IP	transmission control protocol/Internet protocol
TDS	total dissolved solids
TEA	tetraethylammonium
	transversely excited atmospheric
TEAE	triethylaminoethyl (as in TEAE-cellulose)
TEM	transmission electron microscopy
temp	temperature
tert	tertiary (as in *tert*-butyl; but *t*-Bu)
tetr	tetragonal (crystal structure)
TFA	trifluoroacetyl
T_g	glass-transition temperature
TGA	thermogravimetric analysis
Tham	tris(hydroxymethyl)aminomethane (also Tris)
THC	tetrahydrocannabinol
Thd	ribosylthymine
theor	theoretical
THF	tetrahydrofuran
Thr	threonyl, threonine
Thy	thymine
TIMS	thermal ionization mass spectrometry
TIP	temperature-independent paramagnetism
TL	triboluminescence

TLC	thin-layer chromatography
TMA	thermomechanical analysis
TMS	tetramethylsilane
	trimethylsilyl
TMV	tobacco mosaic virus
TnL	tunnel luminescence
TOC	total organic carbon
TOD	total oxygen demand
TOFMS	time-of-flight mass spectrometry
tol	tolyl (also Tol)
TOM	transmitted optical microscopy
Torr	torr
tosyl	4-toluenesulfonyl (also Ts)
TPD	temperature-programmed desorption
TPDE	temperature-programmed decomposition
TPN	triphosphopyridine nucleotide (*use* NADP)
TPNH	reduced TPN (*use* NADPH)
TPR	temperature-programmed reduction
TPR-6	trigonal prismatic, coordination number 6
TQMS	triple-quadrupole mass spectrometry
t_R	retention time
tr	trace
Tr	trace
tric	triclinic (crystal structure)
triflate	trifluoromethanesulfonate
trig	trigonal (crystal structure)
TRIR	time-resolved infrared
Tris	tris(hydroxymethyl)aminomethane (also Tham)
tRNA	transfer RNA
Trp	tryptophyl, tryptophan
Ts	tosyl (4-toluenesulfonyl)
TSC	thermal stimulated current
TSH	thyroid-stimulating hormone
tu	thiourea (ligand)
TVA	thermal volatilization analysis
Tyr	tyrosyl, tyrosine

U

u	unified atomic mass unit
U	uniformly labeled
	uridine
U	internal energy
UCST	upper critical solution temperature
UDP	uridine 5′-diphosphate
uhf	ultrahigh frequency
UHF	ultrahigh frequency
	unrestricted Hartree–Fock
UHV	ultrahigh vacuum

ULSI	ultra-large-scale integration
UMP	uridine 5′-monophosphate, uridine 5′-phosphate
uncor	uncorrected
uns	unsymmetrical
UPS	ultraviolet photoelectron spectroscopy
ur	urea (ligand)
Ura	uracil
Urd	uridine
URL	uniform resource locator
USDA	U.S. Department of Agriculture
USP	United States Pharmacopeial Convention
USP	*The United States Pharmacopeia*
UTP	uridine 5′-triphosphate
UV	ultraviolet
UV PES	ultraviolet photoelectron spectroscopy
UV–vis	ultraviolet–visible

V

v	vendeko (10^{-30})
v	scan rate
	velocity
V	valine
	vendeca (10^{30})
	volt
Val	valyl, valine
VASS	variable-angle sample spinning
VB	valence bond
VCD	vibrational circular dichroism
VDT	video display terminal
VEELS	vibrational energy loss electron spectroscopy
VESCF	variable electronegativity self-consistent field
vhf	very high frequency
VHF	very high frequency
vic	vicinal
vis	visible
viz.	namely
VLE	vapor–liquid equilibrium
VLSI	very large scale integration
VOA	vibrational optical activity
VOC	volatile organic compound
vol	volume
vol %	volume percent
vp	vapor pressure
VPC	vapor-phase chromatography
VPO	vapor pressure osmometry
VRML	virtual reality modeling language
vs	versus (v in legal expressions)
	very strong (spectra)

VSIP	valence-state ionization potential
VUV	vacuum ultraviolet
v/v	volume per volume
VVk	Van Vleck
vw	very weak (spectra)

W

w	weak (spectra)
w	weighting factor
	work
W	tryptophan
	watt
W	work
WAN	wide-area network
WAXS	wide-angle X-ray scattering
Wb	weber
WCOT	wall-coated open-tubular
WDS	wavelength-dispersive spectroscopy
WHO	World Health Organization
WHSV	weight-hourly space velocity
WLF	Williams–Landel–Ferry (molecular models)
wt	weight
wt %	weight percent
w/v	weight per volume
w/w	weight per weight
WWW	World Wide Web
WYSIWYG	what you see is what you get

X

x	xenno (10^{-27})
x	*x* axis
X	xanthosine (*use* N for unknown nucleoside)
	xenna (10^{27})
Xan	xanthine
XANES	X-ray absorption near-edge spectroscopy
	X-ray absorption near-edge structure
Xao	xanthosine
XEDS	X-ray energy-dispersive spectrometry
XES	X-ray emission spectroscopy
XMP	xanthosine 5′-monophosphate, xanthosine 5′-phosphate
XPS	X-ray photoelectron spectroscopy
XRD	X-ray diffraction
XRDF	X-ray radial distance function
XRF	X-ray fluorescence
Xyl	xylose

Y

y	yocto (10^{-24})

y	*y* axis
Y	yotta (10^{24})
	tyrosine

Z

z	zepto (10^{-21})
z	charge number of an ion
	z axis
Z	zetta (10^{21})
	benzyloxycarbonyl
	"glutamic acid or glutamine"
Z	atomic number
	zusammen (configurational)
zfs	zero-field splitting
zfsc	zero-field-splitting constant

CHAPTER 5

Numbers, Mathematics, and Units of Measure

Numbers

Both numerals and words can be used to express numbers. The usage and style conventions for numerals and words are different for technical and nontechnical material.

Numeral and Word Usage

◆ Use numerals with units of time or measure, and use a space between the numeral and the unit, except %, $, and ° (angular degrees), ′ (angular minutes), and ″ (angular seconds).

6 min	25 mL	125 V/s
0.30 g	50%	$250
273 K	47°8′23″	180°, *but* 180 °C
90 °F	50 μg of compound/dL of water	

Exception Spell out numbers with units of measure used in a nontechnical sense.

If you take five minutes to read this article, you'll be surprised.

◆ With items other than units of time or measure, use words for cardinal numbers less than 10; use numerals for 10 and above. Spell out ordinals "first" through "ninth"; use numerals for 10th or greater.

three flasks	30 flasks	third flask
12th flask	seven trees	10 trees

eighth example	33rd example	first century
21st century	sixfold	20-fold

Exception 1 Use all numerals in a series or range containing numbers 10 or greater, even in nontechnical text.

5, 8, and 12 experiments   2nd and 20th samples   5–15 repetitions

Exception 2 Use all numerals for numbers modifying nouns in parallel construction in the same sentence if one of the numbers is 10 or greater.

Activity was reduced in 2 pairs, not significantly changed in 11 pairs, and increased in 6 pairs.

We present new results pertaining to 12 phenanthrolines and 3 porphyrins.

Exception 3 For very large numbers used in a nontechnical sense, use a combination of numerals and words.

1 billion tons	180 million people	2 million pounds (*not* lb)
4.5 billion years	$15 million (*not* 15 million dollars)	

◆ When a sentence starts with a specific quantity, spell out the number as well as the unit of measure.

Twelve species were evaluated in this study.

Twenty slides of each blood sample were prepared.

Fifteen milliliters of supernate was added to the reaction vessel.

Twenty-five milliliters of acetone was added, and the mixture was centrifuged.

However, if possible, recast the sentence.

Acetone (25 mL) was added, and the mixture was centrifuged.

A 25-mL portion of acetone was added, and the mixture was centrifuged.

◆ Even when a sentence starts with a spelled-out quantity, use numerals when appropriate in the rest of the sentence.

Twenty-five milliliters of acetone and 5 mL of HCl were added.

Three micrograms of sample was dissolved in 20 mL of acid.

Fifty samples were collected, but only 22 were tested.

◆ Use numerals for expressions used in a mathematical sense.

The incidence of disease increased by a factor of 4.

The yield of product was decreased by 6 orders of magnitude.

The efficiency of the reaction was increased 2-fold.

After 2 half-lives, the daughter product could be measured.

People who do not eat vegetables have 3 times the risk for colon cancer.

◆ When the suffix "fold" is used in a nonmathematical sense, spell out the accompanying number if it is less than 10.

The purpose of this discussion is twofold.

◆ When the word "times" is used in a nonmathematical sense, spell out the accompanying number if it is less than 10.

The beaker was rinsed four times.

◆ Use numerals in ratios.

a ratio of 1:10	a 1:1 (v/v) mixture
a ratio of 1/10	a 1/1 (v/v) mixture

◆ In dates, use numerals without ordinal endings.

January 3, Jan 3; *not* January 3rd, Jan 3rd
September 5, Sept 5; *not* September 5th, Sept 5th

◆ Use numerals for decades, and form their plurals by adding an "s". Do not use apostrophes in any position.

the 1960s, *not* the 1960's, *not* the '60s
values in the 90s, *not* the 90's
She is in her 20s. (*not* her 20's)

◆ Use numerals with a.m. and p.m.

12:15 a.m.	4:00 p.m.

◆ Spell out and hyphenate fractions whose terms are both less than 10; use a piece fraction if one of the terms is 10 or greater.

one-quarter of the experiments	two-thirds of the results
$^{1}/_{20}$ of the subjects	$^{1}/_{12}$ of the volume

◆ Use numerals to label figures, tables, schemes, structures, charts, equations, and references. Number sequentially; do not skip numbers or number out of sequence. Use arabic numerals for references, but for the other items, the use of arabic and roman numbers varies in publications; consult a recent issue to determine what system is preferred.

◆ In journal articles and book chapters, instead of repeating chemical names over and over, use numerals in boldface (not italic) type to identify chemical species. Use these numbers only in text, not in article or chapter titles, and number consecutively.

This paper describes the syntheses, structures, and stereodynamic behavior of the novel hexacoordinate silicon complexes **1–4**.

The cyclization of 1,3,5-hexatriene (**6**) to 1,3-cyclohexadiene (**7**) is predicted to proceed more rapidly in an electrostatic field.

Complexes **8–12**, in the presence of monoamine oxidase, produce active catalysts for propylene polymerization.

Primary amines **2–5**, **7**, and **9** gave the same Cotton effect signs, depending on the configuration.

Monomer **III** reacts with the initiator (**I**, Ar = 2,6-diisopropylphenyl) via a ring-opening metathesis polymerization mechanism.

- Numerals may be used to name members of a series.

Sample 1 contained a high level of contamination, but samples 2 and 3 were relatively pure.

Methods 1 and 2 were used for water-soluble compounds, and methods 3 and 4 were used for oil-soluble compounds.

- When numerals are used as names and not enumerators, form their plurals by adding an apostrophe and "s" to avoid confusion with mathematical expressions and to make it clear that the "s" is not part of the name.

Many 6's were registered.

Intel 486's are not as fast as Pentium processors.

Boeing 747's are among the largest airplanes.

- Arabic numerals in parentheses may be used to enumerate a list of phrases or sentences in text. Always use an opening and a closing parenthesis, not one alone.

Some advantages of these materials are (1) their electrical properties after pyrolysis, (2) their ability to be modified chemically before pyrolysis, and (3) their abundance and low cost.

The major conclusions are the following: (1) We have further validated the utility of molecular mechanical methods in simulating the kinetics of these reactions. (2) A comparison of the calculated structures with available X-ray structures revealed satisfactory agreement. (3) The combined use of different theoretical approaches permitted characterization of the properties of a new isomer.

- Arabic numerals followed by periods or enclosed in parentheses may be used to enumerate a displayed list of sentences or to number paragraphs. Here are two acceptable ways to format a list.

These results suggest the following:

1. Ketones are more acidic than esters.
2. Cyclic carboxylic acids are more acidic than their acyclic analogues.
3. Alkylation of the active methylene carbon reduces the acidity.

These results suggest the following:
(1) Ketones are more acidic than esters.
(2) Cyclic carboxylic acids are more acidic than their acyclic analogues.
(3) Alkylation of the active methylene carbon reduces the acidity.

Style for Numerals

◆ For very large numbers with units of measure, use scientific notation or choose an appropriate multiplying prefix for the unit to avoid numerals larger than four digits.

1.2×10^6 s　　3.0×10^4 kg　　5.8×10^{-5} M *or* 58 μM
42.3 L, *not* 42,300 mL *or* 42 300 mL

Exception 1 In tables, use the same unit and multiplying prefix for all entries in a column, even if some entries thereby require four or more digits.

Exception 2 Use the preferred unit of a discipline, even when the numerals require four or more digits:

g/L	for mass density of fluids
kg/m^3	for mass density of solids
GPa	for modulus of elasticity
kPa	for fluid pressure
MPa	for stress

◆ In four-digit numbers, use no commas or spaces.

Exception Spaces or commas are inserted in four-digit numbers when alignment is needed in a column containing numbers of five or more digits.

◆ When a long number cannot be written in scientific notation, the digits must be grouped. For grouping of digits in long numbers (five digits or greater), check the publication in which the manuscript will appear. Two styles are possible.

- **Style 1:** In some publications, for numbers with five or more digits, the digits are grouped with commas placed between groups of three counting to the left of the decimal point.

  4837　　10,000　　930,582
  6,398,210　　85,798.62578

- **Style 2:** In some publications (including ACS journals), for numbers with five digits or greater, the digits are grouped with a thin space between groups of three, counting both to the left and to the right of the decimal point.

9319.4 74 183.0629 0.508 27
501 736.293 810 4

Exceptions

- U.S. monetary values are always written with commas: $425,000.
- U.S. patent numbers are always written with commas: U.S. Patent 5,376,421. The patent numbers of other countries should be presented as on the original patent document.
- Page numbers in reference citations are always printed solid: p 11597.

◆ Use the period as the decimal point, never a comma.

◆ Use numbers before and after a decimal point.

0.25, *not* .25 78.0 *or* 78, *not* 78.

◆ Use a decimal and a zero following a numeral only when such usage truly represents the precision of the measurement: 27.0 °C and 27 °C are not interchangeable.

◆ Use decimals rather than fractions with units of time or measure, except when doing so would imply an unwarranted accuracy.

3.5 h, *not* 3½ 5.25 g, *not* 5¼ g

◆ Standard deviation, standard error, or degree of accuracy can be given in two ways:

- with only the deviation in the least significant digit(s) placed in parentheses following the main numeral and closed up to it
- with all digits preceded by a ± and following the main numeral. Spaces are left on each side of the ±.

2.0089(1) means 2.0089 ± 0.0001

1.4793(23) means 1.4793 ± 0.0023

The shorter version is better in tables. Always specify which measure (e.g., standard deviation or standard error) of uncertainty is being used.

◆ When two numbered items are cited in narrative, use "and".

Figures 1 and 2 refs 23 and 24 compounds **I** and **II**

◆ Use a comma between two reference callouts in parentheses or as superscripts.

Lewis (*12, 13*) found Lewis^{12,13} found

When the reference numbers are on the line, the comma is followed by a space; when the numbers are superscripts, the comma is not followed by a space.

◆ Use an en dash in ranges or series of three or more numbered items, whether on the line or in a superscript.

43–49 325–372 1981–1983
Tables 1–4 temperatures of 100–125 °C
refs 3–5 aliquots of 50–100 mL
eqs 6–9 samples 5–10
past results (*27–31*) past results[27–31]
pages 237–239 pp 165–172

Exception 1 Do not use an en dash in expressions with the words "from ... to" or "between ... and".

from 20 to 80, *not* from 20–80
between 50 and 100 mL, *not* between 50–100 mL

Exception 2 When either one or both numbers are negative or include a symbol that modifies the number, use the word "to" or "through", not the en dash.

–20 to +120 K –145 to –30 °C ~50 to 60
10 to >600 mL <5 to 15 mg

◆ For ranges in scientific notation, retain all parts of all numbers or avoid ambiguity by use of parentheses or other enclosing marks.

9.2×10^{-3} to 12.6×10^{-3} *or* $(9.2–12.6) \times 10^{-3}$, *not* 9.2 to 12.6×10^{-3}

Use forms like $9.2 \times 10^{-3}–12.6 \times 10^{-3}$ with caution to avoid mistaking the dash for a minus sign.

◆ For very large numbers in ranges, retain all parts of all numbers.

26 million to 35 million

◆ Do not use e or E to mean "multiplied by the power of 10".

3.7×10^{5}, *not* 3.7e5

Mathematics

Mathematical Concepts

Variable A variable is a quantity that changes in value, substance, or amount, such as V for volume, m for mass, and t for time.

Constant A constant is a quantity that has a fixed value, such as h for the Planck constant and F for the Faraday constant.

Function The function $f(x) = y$ represents a rule that assigns a unique value of y to every x. The *argument* of the function is x.

Operator An operator is a symbol, such as a function (d, derivative; ln, logarithm; and $\mathcal{H}$, the Hamiltonian operator) or an arithmetic sign (+, −, =, and ×), denoting an operation to be performed.

Physical Quantity A physical quantity is a product of a numerical value (a pure number) and a unit. Physical quantities may be scalars or vectors, variables or constants.

Scalar A scalar is an ordinary number without direction, such as length, temperature, or mass. Any quantity not a vector quantity is a scalar quantity.

Vector A vector is a quantity with both magnitude and direction, such as force or velocity. For the vector $\mathbf{V} = [a, b]$, a and b are the *components* of **V**.

Tensor A tensor represents a generalized vector with more than two components.

Matrix A matrix is represented by a rectangular array of *elements*; an *array* consists of rows and columns. The elements of matrix **U** are u_{11}, u_{12}, etc.

$$\mathbf{U} = \begin{bmatrix} u_{11} & \cdots & u_{1n} \\ \vdots & \ddots & \vdots \\ u_{n1} & \cdots & u_{nn} \end{bmatrix}$$

Determinant The determinant of a matrix is a function that assigns a number to a matrix. For example, the determinant of the $n \times n$ matrix **B** is represented by

$$\det \mathbf{B} = \begin{vmatrix} b_{11} & \cdots & b_{1n} \\ \vdots & & \vdots \\ b_{n1} & \cdots & b_{nn} \end{vmatrix}$$

Index An index is a subscript or superscript character in an element of a matrix, vector, or tensor; indices usually represent numbers. For example, i and j are indices in b_{ij}.

Do not confuse abbreviations and mathematical symbols. An abbreviation is usually two or more letters; a mathematical symbol is generally only one letter, possibly with a subscript or superscript. An abbreviation is used in narrative text but seldom appears in equations; a mathematical symbol is preferred in equations and may also be used in text. For example, in text with no equations, PE for potential energy is acceptable, but in mathematical text and equations, E_p is preferred.

Usage and Style for Symbols

◆ Define all symbols for mathematical constants, variables, and unknown quantities the first time you use them in the text. If you use them in the abstract, define them there and then again at their first appearance in text. Do not define standard mathematical constants such as π, i, and e.

◆ Form the plurals of mathematical symbols by adding an apostrophe and "s" if you cannot use a word such as "values" or "levels".

> at *r* values greater than *is better than* at *r*'s greater than

◆ Do not use an equal sign as an abbreviation for the word "is" or the word "equals" in narrative text.

> $PV = nRT$, where P is pressure, *not* where P = pressure
> when the temperature is 50 °C, *not* when the temperature = 50 °C

◆ Do not use a plus sign as an abbreviation for the word "and" in narrative text.

> a mixture of A and B, *not* a mixture of A + B

◆ Do not use an asterisk to indicate multiplication except in computer language expressions.

Italic Type

Use italic type for

- variables: T for temperature, x for mole fraction, r for rate
- axes: the y axis
- planes: plane P
- components of vectors and tensors: $a_1 + b_1$
- elements of determinants and matrices: g_n
- constants: k_B, the Boltzmann constant; g, the acceleration due to gravity
- functions that describe variables: $f(x)$

◆ Even when you use mathematical constants, variables, and unknown quantities in adjective combinations, retain the italic type.

> In this equation, V_i is the frequency of the *i*th mode.

In eq 4, n is the number of extractions and M is the mass remaining after the nth extraction.

- Use italic for two-letter variables defining transport properties.

Al	Alfvén number	*Ma*	Mach number
Bi	Biot number	*Nu*	Nusselt number
Co	Cowling number	*Pe*	Péclet number
Da	Damkohler number	*Pr*	Prandtl number
Eu	Euler number	*Ra*	Rayleigh number
Fo	Fourier number	*Re*	Reynolds number
Fr	Froude number	*Sc*	Schmidt number
Ga	Galileo number	*Sh*	Sherwood number
Gr	Grashof number	*Sr*	Strouhal number
Ha	Hartmann number	*St*	Stanton number
Kn	Knudsen number	*We*	Weber number
Le	Lewis number	*Wi*	Weissenberg number

Roman Type

Use roman type for

- numerals
- punctuation and enclosing marks such as square brackets, parentheses, and braces
- most operators
- units of measure and time: mg, milligram; K, kelvin; Pa, pascal; mmHg, millimeters of mercury
- nonmathematical quantities or symbols: R, radical in chemical nomenclature; $\mathrm{S_1}$, molecular state; s, atomic orbital
- multiple-letter abbreviations for variables: IP, ionization potential; cmc, critical micelle concentration
- mathematical constants:

  e, the base of the natural logarithm, 2.71828…
  i, the imaginary number, $(-1)^{1/2}$
  π, 3.14159…

- transposes of matrices: $\mathbf{A}^{\mathrm{T}}$ (T is the transpose of matrix **A**)
- points and lines: point A, line $\overline{\mathrm{AB}}$
- determinants: det **A** is the determinant of matrix **A**
- trigonometric and other functions:

Ad	adjoint	cl	closure
Ai	Airy function	Coker	cokernel
arg	argument	cos	cosine
Bd	bound	cosh	hyperbolic cosine

cot	cotangent	log	logarithm (base 10)
coth	hyperbolic cotangent	Log	principal logarithm
csc	cosecant	lub	least upper bound
csch	hyperbolic cosecant	max	maximum
det	determinant	min	minimum
dim	dimension	mod	modulus
div	divergence	P	property
erf	error function	Re	real
erfc	complement of error function	sec	secant
exp	exponential	sech	hyperbolic secant
GL	general linear	sign, sgn	sign
glb	greater lower bound	sin	sine
grad	gradient	sinh	hyperbolic sine
hom	homology	SL	special linear
Im	imaginary	sp	spin
inf	inferior	Sp	symplectic
int	interior	sup	superior
ker	kernel	Sz(g)	Suzuki group
lim	limit	tan	tangent
lim inf	limit inferior	tanh	hyperbolic tangent
lim sup	limit superior	tr	trace
ln	natural logarithm (base e)	wr	wreath

Boldface Type

Use boldface type for

- vectors
- tensors
- matrices
- multidimensional physical quantities: **H**, magnetic field strength

Greek Letters

Greek letters (lightface or boldface) can be used for variables, constants, and vectors and anywhere a Latin letter can be used.

The Greek Alphabet

Name	*Uppercase*	*Lowercase*
Alpha	A	α
Beta	B	β
Gamma	Γ	γ
Delta	Δ	δ
Epsilon	E	ε
Zeta	Z	ζ
Eta	H	η

Name	*Uppercase*	*Lowercase*
Theta	Θ	θ, ϑ
Iota	I	ι
Kappa	K	κ
Lambda	Λ	λ
Mu	M	μ
Nu	N	ν
Xi	Ξ	ξ
Omicron	O	ο
Pi	Π	π
Rho	P	ρ
Sigma	Σ	σ
Tau	T	τ
Upsilon	Y	υ
Phi	Φ	ϕ, φ
Chi	X	χ
Psi	Ψ	ψ
Omega	Ω	ω

Spacing

◆ Leave a space before and after functions set in roman type, unless the argument is enclosed in parentheses, brackets, or braces.

$\log 2$	$-\log x$	$4 \sin \theta$
$\tan^2 y$	$\exp(-x)$	$\cosh(\beta e_0 \phi)$
$4 \tan(2y)$	$\mathrm{erfc}(y)$	

◆ Leave a space before and after mathematical operators that function as verbs or conjunctions; that is, they have numbers on both sides or a symbol for a variable on one side and a numeral on the other.

$20 \pm 2\%$	3.24 ± 0.01	4×5 cm
8×10^{-4}	$k \geq 420\ \mathrm{s}^{-1}$	$p < 0.01$
$T_g = 176\ °\mathrm{C}$	$n = 25$	1 in. = 2.54 cm

Exception 1 Leave no space around mathematical operators in subscripts and superscripts.

ΔH^{n-1}	$E_{\lambda>353}$	$\mathrm{M}^{(x+y)+}$

Exception 2 Leave no space around a slash (a/b), a ratio colon (1:10), or a center dot ($\mathbf{P}_{\mathrm{M}}{\cdot}\mathbf{V}$).

◆ Leave no space between simple variables being multiplied: xy. Do not use a center dot or the times sign (×) with single-letter scalar variables.

◆ In multiplication involving the two-letter symbols for transport properties, use a space, enclose them in parentheses, or use the times sign. When superscripts or subscripts are present, the symbols can be closed up.

$Re\ Nu$ $(Re)(Nu)$ $Re \times Nu$
$Re_x Nu_y$

◆ Use a space for simple multiplication of functions of the type $f(x)$ (one-dimensional) or $g(y, z)$ (multidimensional). Close up multipliers to such functions where applicable. You may also use additional enclosing marks instead of spaces.

$W = 2f(x)\ g(y, z)$ $W = 2[f(x)][g(y, z)]$

◆ When mathematical symbols are used as adjectives, that is, with one number that is not part of a mathematical operation, do not leave a space between the symbol and the number.

−12 °C 25 g (±1%) at 400× magnification
a conversion of >50% a probability of <0.01
The level can vary from −15 to +25 m.

Enclosing Marks

◆ Use enclosing marks (parentheses, brackets, and braces, also called fences) in accordance with the rules of mathematics. Enclose parentheses within square brackets, and square brackets within braces: {[( )]}.

◆ Use enclosing marks around arguments when necessary for clarity.

$\sin(x + 1)$ $\sin[2\pi(x - y)/n]$ $\log[-V(r)/kT]$

◆ Do not use square brackets, parentheses, or braces around the symbol for a quantity to make it represent any other quantity.

Incorrect where V is volume and (V) is volume at equilibrium
Correct where V is volume and V_e is volume at equilibrium

Subscripts and Superscripts

◆ Use italic type for subscripts and superscripts that are themselves symbols for physical quantities or numbers. Use roman type for subscripts and superscripts that are abbreviations and not symbols.

C_p for heat capacity at constant pressure
C_B for heat capacity of substance B

C_g where g is gas
E_i for energy of the *i*th level, where *i* is a number
g_n where n is normal
μ_r where r is relative
E_k where k is kinetic
ξ_e where e is electric

◆ In most cases, staggered subscripts and superscripts are preferred. Exponents should follow subscripts.

$x_1{}^2$ $\quad C_x{}^{1/2}$ $\quad T_{2m}{}^{-1}$
$\Delta H_1{}^{\ddagger}$ $\quad E_{ads}{}^{\circ}$

◆ Use a slash (/) in all subscript and superscript fractions, with no space on either side.

$t_{1/2}$ $\quad x^{1/2}$ $\quad M^{2/3}$
$f_{a/b}$

◆ Leave no space around operators in subscripts and superscripts.

$M^{(2-n)+}$ $\quad E_{T+\theta}$

◆ Leave no space around other expressions in subscripts and superscripts, unless confusion or misreading would result.

$Q_{\text{n-Bu(750°C)}}$ $\quad \beta_{\text{zero level}}$ $\quad E^{\text{365nm}}$

◆ The terms e^a and exp *a* have the same meaning and can be interchanged. When an exponent to the base e is very long or complicated, replace the e with exp and place the exponent on line and in enclosing marks. Leave no space between exp and the opening enclosing mark.

$\exp(\int y\ dt)$, *not* $e^{\int y dt}$
$\exp\{½kT[Y(a+b)-Z]\}$, *not* $e^{½kT[Y(a+b)-Z]}$

◆ In running text, do not use the radical sign ($\sqrt{\ }$)with long terms. Use enclosing marks around the term and a superscript 1/2, 1/3, 1/4 (etc.) for square, cube, fourth root (etc.), respectively.

$(x-y^2)^{1/3}$ $\quad [\sinh^2 u + (\cosh u - 1)^2]^{1/2}$

Abbreviations and Symbols

◆ Certain abbreviations are used only in the context of mathematical equations. Define all of these the first time they are used.

lhs	left-hand side (of an equation)
rhs	right-hand side (of an equation)
o.d.e.	ordinary differential equation
rms	root mean square

rmsd	root-mean-square deviation
s.t.	subject to
w.r.t.	with respect to

◆ Some standard usages and symbols for mathematical operations and constants need never be defined. They include the following:

e	natural base (approximately 2.7183)
$\exp x$, e^x	exponential of x
i	imaginary number
$\ln x$	natural logarithm of x
$\log x$	logarithm to the base 10 of x
$\log_a x$	logarithm to the base a of x
$\approx$	approximately equal to
$\sim$	approximately, asymptotically equal to
$\propto$	proportional to
$\rightarrow$	approaches (tends to)
$\equiv$	identically equal to
∞	infinity
Σ	summation
Π	product
$\cup$	union
$\int$	integral
$\oint$	line integral around a closed path
∇	del (or nabla) operator, gradient
∇^2	Laplacian operator
$<$	less than
$\leq$	less than or equal to
$\ll$	much less than
$>$	greater than
$\geq$	greater than or equal to
$\gg$	much greater than
$\neq$	not equal to
$\parallel$	parallel to
$\perp$	perpendicular to
$\|a\|$	absolute magnitude of a
$a^{1/2}$, $\sqrt{a}$	square root of a
$a^{1/n}$, $\sqrt[n]{a}$	nth root of a
$\bar{a}$, $<a>$	mean value of a
Δx	finite increment of x
∂x	partial differential, infinitesimal increment of x
dx	total differential of x
$f(x)$	function of x
$\int y\,dx$	integral of y with respect to x
$\int_a^b y\,dx$	integral of y from $x = a$ to $x = b$
$\mathbf{A}$	vector of magnitude A
$\mathbf{A}\cdot\mathbf{B}$	scalar product of **A** and **B**
$\mathbf{A}\times\mathbf{B}$, $\mathbf{AB}$	vector product of **A** and **B**
$\overline{AB}$	length of line from A to B

Equations

Mathematical equations can be presented within running text or displayed on lines by themselves. Follow the guidelines for style and usage just described under "Usage and Style for Symbols" (starting on p 151).

- Leave a space
  - before and after mathematical signs used as operators ($=$, $\neq$, $\equiv$, $\sim$, $\approx$, $\cong$, $>$, $+$, $-$, $\times$, $\div$, $\cup$, $\supset$, $\subset$, $\in$, etc., but not slash, ratio colon, or center dot), except when they appear in superscripts or subscripts
  - before trigonometric and other functions set in roman type
  - after trigonometric and other functions set in roman type when their arguments are not in enclosing marks
  - before and after derivatives: $\iint f(x)\ \mathrm{d}x\ f(y)\ \mathrm{d}y$
  - between built-up fractions as components of products:

    $$\frac{a}{b}\ \frac{c}{d}$$

    or clarify with enclosing marks and no space: $(a/b)(c/d)$
  - between functions as components of products: $W = 2f(x)\ g(y, z)$
- Leave no space
  - between single-item variables being multiplied
  - in any part of a superscript or subscript, unless confusion or misreading would result
  - between any character and its own superscript, prime, or subscript
  - on either side of a colon used for a ratio
  - on either side of a centered dot
  - on either side of a slash
  - after mathematical operators used as adjectives: -10
  - after functions when the argument is in parentheses: $\tanh(\lambda/2)$
  - between an opening parenthesis, bracket, or brace and the next character: $(2x)y$

- between a closing parenthesis, bracket, or brace and the previous character: $2(xy)$
- between back-to-back parentheses, brackets, and braces, e.g.,](
- between nested parentheses, brackets, and braces, e.g., [(
- in any part of limits to summations, products, and integrals
- in any part of lower limits to min, max, lim, and inf

◆ Use or do not use spaces around ellipses depending on the treatment of other items in the series.

no spaces: $a_n a_{n+1} a_{n+2} \cdots a_{n+36}$
spaces: $a_n + a_{n+1} + a_{n+2} + \ldots + a_{n+36}$
space before: $a, b, \ldots, x$

◆ Use enclosing marks in accordance with the rules of mathematics. If the slash (/) is used in division and if there is any doubt where the numerator ends or where the denominator starts, use enclosing marks for one or the other or both.

$(x + y)/(3x - y)$
$(a/b)/c$, *or* $a/(b/c)$, *but never* $a/b/c$
$\frac{x + y}{2} = z$ *would be better as* $(x + y)/2 = z$
$\frac{x + y}{z} + 2a$ *would be better as* $[(x + y)/z] + 2a$

◆ If an equation is very short and will not be referred to again, you may run it into the text.

> A fluid is said to be Newtonian when it obeys Newton's law of viscosity, given by $\tau = \eta\gamma$, where τ is the shear stress, η is the fluid dynamic constant, and γ is the shear rate.

◆ You may use mathematical expressions as part of a sentence when the subject, verb, and object are all part of the mathematical expression.

When $V = 12$, eq 15 is valid.
(V is the subject, = is the verb, and 12 is the object.)

◆ When an equation is too long to fit on one line, break it *after* an operator that is not within an enclosing mark (parentheses, brackets, or braces) or break it between sets of enclosing marks. Do not break equations after integral, product, and summation signs; after trigonometric and other functions set in roman type; or before derivatives.

◆ Number displayed equations by using any consistent system of sequencing.

> 1, 2, 3, …
> 1a, 1b, 2, …
> I, II, III, …
> A, B, C, …
> A-1, A-2, A-3, …
> B.1, B.2, B.3, …
> C1, C2, C3, …

◆ Use equation identifiers in the proper sequence according to appearance in text. Do not skip numbers or letters in the sequence.

◆ Place identifiers in parentheses, flush right on the same line as the equation.

$$V = 64\pi kT\gamma^2 \exp(-\kappa h) \qquad (3)$$

◆ Do not use any closing punctuation on the line with displayed equations.

◆ When introducing a displayed equation, do not automatically use a colon; in most cases a colon is incorrect because the equation finishes a phrase or sentence.

> An ideal gas law analogy is

$$\pi A = nRT$$

> If the principal radii are R_1 and R_2, then

$$\Delta = R_1 - R_2$$

> The area per adsorbed molecule can be calculated from

$$a = N_A\Gamma_S$$

> The attractive energy can be approximated by

$$V_A = Ar(12H)^{-1}$$

> The simplest method is to use a mapping potential of the form

$$\varepsilon_m = (1 - \lambda_m)\varepsilon_A + \lambda_m\varepsilon_B$$

> Marshall developed an equation for rapid coagulation:

$$n = 1 + S\pi Drt$$

◆ Following a displayed equation that is part of a sentence, punctuate the text as if it were a continuation of a sentence including the equation.

> As can be seen in the equation for the confidence region (J)

$$J = 1 + F(N - p)^2$$

> F is the value of the distribution, N is the number of measurements, and p is the number of parameters.

◆ To cite an equation in text, use the abbreviation "eq" if it is not the first word of the sentence. Spell out "equation" when it is the first word of a sentence or when it is not accompanied by a number. The plural of "eq" is "eqs".

> The number of independent points can be calculated from eq 3.
>
> The number of independent points can be calculated from eqs 3 and 4.
>
> Equation 1 is not accurate for distances greater than 10 μm.
>
> Equations 1 and 2 are not accurate for distances greater than 10 μm.

Specialized Notation

Ratio and Mixture Notation

Use either a colon or a slash to represent a ratio, but not an en dash. Use either a slash or an en dash between components of a mixture, but not a colon.

> dissolved in 5:1 glycerin/water
>
> dissolved in 5:1 glycerin–water
>
> the metal/ligand (1:1) reaction mixture
>
> the metal–ligand (1:1) reaction mixture
>
> the metal–ligand (1/1) reaction mixture
>
> the methane/oxygen/argon (1/50/450) matrix
>
> the methane/oxygen/argon (1:50:450) matrix

Set Notation

The following symbols are used in set notation. Leave a space before and after all operators, but not before and after braces.

$A = \{a, b\}$	set *A*; *A* is italic; braces are used
$A \cup B$	union of sets *A* and *B*
$A \cap B$	intersection of sets *A* and *B*
$A \in B$	*A* is a member (element) of *B*
$A \notin B$	*A* is not a member (element) of *B*
$A \subset B$	*A* is contained in *B*
$A \not\subset B$	*A* is not contained in *B*
$A \supset B$	*A* contains *B*
$A \not\supset B$	*A* does not contain *B*
$\forall A$	for all (every) *A*
$\exists$	there exists
$\ni$	such that
$\therefore$	therefore

Geometric Notation

Leave no spaces around geometric notation. Use italic type for planes and axes and roman type for points and lines.

$X \perp Y$	*X* is perpendicular to *Y*
$X \| Y$	*X* is parallel to *Y*
$\angle \mathrm{AB}$	the angle between A and B
$\overline{\mathrm{AB}}$	length of line from A to B

Statistics

Certain statistical symbols are standard.

Σ	summation
σ, SD	standard deviation
CV	coefficient of variation
df, DF	degrees of freedom
f	frequency
F	variance ratio
n, N	total number of individuals or random variables
p, P	probability
r	correlation coefficient
R	regression coefficient
RSD	relative standard deviation
s^2	sample variance
SE	standard error
SEM	standard error of the mean
t	Student distribution (the Student *t* test)
$\bar{x}$	arithmetic mean

A common statistical measurement is the Student *t* test or Student's *t* test. Student was the pseudonym of W. Gossett, an eminent mathematician.

Units of Measure

Usage

◆ Where possible, use metric and SI units (discussed in the next section) in all technical documents. The following conventions apply to all units of measure:

- Abbreviate units of measure when they accompany numerals.
- Leave a space between a numeral and its unit of measure, except when they form a unit modifier, in which case use a hyphen between them in simple situations.

- Do not use a period after an abbreviated unit of measure (exception: in. for inch).
- Do not define units of measure.

500 mL	3 min	4 Å
9 V/s	$9\ \mathrm{V\ s^{-1}}$	$9\ \mathrm{V{\cdot}s^{-1}}$
200 mV	$4.14 \times 10^{-9}\ \mathrm{m^2/(V\ s)}$	2.6×10^4 J
3-min interval	2-μm droplet	500-mL flask

Exception 1 Do not leave a space between a number and the percent, angular degree, angular minute, or angular second symbols.

50%	90°	75′
18″		

Exception 2 Never hyphenate numbers and units of concentration or temperature, even as parts of unit modifiers, but leave a space between the number and the unit.

0.1 M NaCl	3 N HCl	a 0.1 mol $\mathrm{dm^{-3}}$ solution
20 °C difference	5 K isotherm	

◆ Use °C with a space after a number, but no space between the degree symbol and the capital C: 15 °C.

◆ Do not add an "s" to make the plural of any abbreviated units of measure. The abbreviations are used as both singular and plural.

50 mg, *not* 50 mgs
3 mol, *not* 3 mols

◆ Write abbreviated compound units with a center dot or a space between the units to indicate multiplication and a slash (/) or negative exponent for division. Enclose compound units following a slash in parentheses.

watt per meter-kelvin is $\mathrm{W{\cdot}m^{-1}{\cdot}K^{-1}}$ *or* W/(m·K) *or* $\mathrm{W\ m^{-1}\ K^{-1}}$ *or* $\mathrm{W\ (m\ K)^{-1}}$ *or* W/(m K)

cubic decimeter per mole-second is $\mathrm{dm^3{\cdot}mol^{-1}{\cdot}s^{-1}}$ *or* $\mathrm{dm^3/(mol{\cdot}s)}$ *or* $\mathrm{dm^3\ mol^{-1}\ s^{-1}}$ *or* $\mathrm{dm^3\ (mol\ s)^{-1}}$ *or* $\mathrm{dm^3/(mol\ s)}$

joules per mole-kelvin is $\mathrm{J{\cdot}mol^{-1}{\cdot}K^{-1}}$ *or* J/(mol·K) *or* $\mathrm{J\ mol^{-1}\ K^{-1}}$ *or* $\mathrm{J\ (mol\ K)^{-1}}$ *or* J/(mol K)

◆ Spell out units of measure that do not follow a numeral. Do not capitalize them unless they are at the beginning of a sentence or in a title.

several milligrams (*not* several mg)	a few milliliters (*not* a few mL)
degrees Celsius	reciprocal seconds
milligrams per kilogram	volts per square meter

Exception 1 Abbreviate units of measure in parentheses after the definitions of variables directly following an equation.

$$L = D/P_{\mathrm{O}}$$

where L is the distance between particles (cm), D is the particle density (g/cm^3), and P_{O} is the partial pressure of oxygen (kPa).

Exception 2 Spell out the words "day", "week", "month", and "year", even with a numeral (except in ACS journals in descriptions of experimental work).

3 days	10 weeks	13 months
2 years		

Exception 3 Certain units of measure have no abbreviations: bar, darcy, einstein, erg, faraday, and langmuir. The symbol for the unit torr is Torr. The unit rad is abbreviated rd; the unit radian is abbreviated rad.

Exception 4 In column headings of tables and in axis labels of figures, abbreviate units of measure, even without numerals.

◆ Add an "s" to form the plural of spelled-out units: milligrams, poises, kelvins, amperes, watts, newtons, and so on.

Exceptions bar, hertz, lux, stokes, siemens, and torr remain unchanged; darcy becomes darcies; henry becomes henries.

◆ Do not capitalize surnames that are used as units of measure.

ampere	franklin	newton
angstrom	gauss	ohm
coulomb	gilbert	pascal
curie	gray	poise
dalton	hartree	siemens
darcy	henry	sievert
debye	hertz	stokes
einstein	joule	tesla
erg	kelvin	watt
faraday	langmuir	weber

Celsius and Fahrenheit are always capitalized. They are not themselves units; they are the names of temperature scales.

◆ Do not use a slash in spelled-out units of measure. Use the word "per".

Results are reported in meters per second.

The fluid density is given in kilograms per cubic meter.

◆ Do not mix abbreviations and spelled-out units within units of measure.

newtons per meter, *not* N per meter
100 F/m, *not* 100 farad/m

Exception in more complex situations

50 mL of water and 20 mg of NaOH per gram of compound

◆ Use a slash (/), not the word "per", before the abbreviation for a unit in complex expressions.

50 μg of peptide/mL 25 mg of drug/kg of body weight

◆ When the first part of a unit of measure is a word that is not itself a unit of measure, use a slash (/) before the final abbreviated unit.

10 counts/s 12 domains/cm^3
2×10^3 ions/min 125 conversions/mm^2

◆ When the last part of a unit of measure is a word that is not itself a unit of measure, use either a slash (/) or the word "per" before the word that is not a unit.

0.8 keV/channel 0.8 keV per channel
7 μ_B/boron 7 μ_B per boron

◆ Leave no space between the multiplicative prefix and the unit, whether abbreviated or spelled out.

kilojoule or kJ milligram or mg microampere or μA

◆ Use only one multiplicative prefix per unit.

nm, *not* mμm

◆ In ranges and series, retain only the final unit of measure.

10–12 mg 5, 10, and 20 kV 60–90°
between 25 and 50 mL from 10 to 15 min

◆ Do not use the degree symbol with kelvin: 115 K.

◆ In titles and headings, do not capitalize abbreviated units of measure that are ordinarily lowercase.

Analysis of 2-mg Samples
A 50-kDa Protein To Modulate Guanine Nucleotide Binding

The International System of Units (SI)

Before the 1960s, four systems of units were commonly used in the scientific literature: the English system (centuries old, using yard and pound), the metric system (dating from the 18th century, using meter and kilogram as standard units), the CGS system (based on the metric, using centimeter, gram, and second as base units), and the MKSA or Giorgi system (using meter, kilogram, second, and ampere as base units).

The International System of Units (SI, Système International d'Unités) is the most recent effort to develop a coherent system of units. It is coherent because there is only one unit for each base physical quantity, and units for all other quantities are derived from these base units by simple equations. It has been adopted as a universal system to simplify communication of numerical data and to restrict proliferation of systems. SI units are used by the National Institute of Standards and Technology (NIST).

The SI is constructed from seven base units for independent quantities (meter, kilogram, second, ampere, kelvin, mole, and candela) plus two supplementary units for plane and solid angles (radian and steradian). Most physicochemical measurements can be expressed in terms of these units.

Certain units not part of the SI are so widely used that it is impractical to abandon them (e.g., liter, minute, and hour) or are so well established that the International Committee on Weights and Measures has authorized their continued use (e.g., bar, curie, and angstrom). In addition, quantities that are expressed in terms of the fundamental constants of nature, such as elementary charge, proton mass, Bohr magneton, speed of light, and Planck constant, are also acceptable. However, broad terms such as "atomic units" are not acceptable, although atomic mass unit, u, is acceptable and relevant to chemistry.

Follow all usage conventions given for units of measure. Use the abbreviations for SI units with capital and lowercase letters exactly as they appear in Tables 1–6.

Table 1. SI Units

Name	*Symbol*	*Physical Quantity*
Base units		
ampere	A	electric current
candela	cd	luminous intensity
kelvin	K	thermodynamic temperature
kilogram	kg	mass
meter	m	length
mole	mol	amount of substance
second	s	time
Supplementary units		
radian	rad	plane angle
steradian	sr	solid angle

Table 2. Multiplying Prefixes

Factor	*Prefix*	*Symbol*	*Factor*	*Prefix*	*Symbol*
10^{-30}	vendeko	v	10^{1}	deca	da
10^{-27}	xenno	x	10^{2}	hecto	h
10^{-24}	yocto	y	10^{3}	kilo	k
10^{-21}	zepto	z	10^{6}	mega	M
10^{-18}	atto	a	10^{9}	giga	G
10^{-15}	femto	f	10^{12}	tera	T
10^{-12}	pico	p	10^{15}	peta	P
10^{-9}	nano	n	10^{18}	exa	E
10^{-6}	micro	μ	10^{21}	zetta	Z
10^{-3}	milli	m	10^{24}	yotta	Y
10^{-2}	centi	c	10^{27}	xenna	X
10^{-1}	deci	d	10^{30}	vendeca	V

NOTE: Any of these prefixes may be combined with any of the symbols permitted within the SI. Thus, kPa and GPa will both be common combinations in measurements of pressure, as will mL and cm for measurements of volume and length. As a general rule, however, the prefix chosen should be 10 raised to that multiple of 3 that will bring the numerical value of the quantity to a positive value less than 1000.

Table 3. SI-Derived Units

Name	*Symbol*	*Quantity*	*In Terms of Other Units*	*In Terms of SI Base Units*
becquerel	Bq	activity (of a radionuclide)		s^{-1}
coulomb	C	quantity of electricity, electric charge		A•s, s•A
farad	F	capacitance	C/V	$m^{-2}\cdot kg^{-1}\cdot s^{4}\cdot A^{2}$
gray	Gy	absorbed dose, kerma, specific energy imparted	J/kg	$m^{2}\cdot s^{-2}$
henry	H	inductance	Wb/A	$m^{2}\cdot kg\cdot s^{-2}\cdot A^{-2}$
hertz	Hz	frequency		s^{-1}
joule	J	energy, work, quantity of heat	N•m	$m^{2}\cdot kg\cdot s^{-2}$
lumen	lm	luminous flux	lm	cd•sr
lux	lx	illuminance	lm/m^{2}	$m^{-2}\cdot cd\cdot sr$
newton	N	force		$m\cdot kg\cdot s^{-2}$
ohm	Ω	electric resistance	V/A	$m^{2}\cdot kg\cdot s^{-3}\cdot A^{-2}$
pascal	Pa	pressure, stress	N/m^{2}	$m^{-1}\cdot kg\cdot s^{-2}$
siemens	S	conductance	A/V	$m^{-2}\cdot kg^{-1}\cdot s^{3}\cdot A^{2}$
sievert	Sv	dose equivalent	J/kg	$m^{2}\cdot s^{-2}$
tesla	T	magnetic flux density	Wb/m^{2}	$kg\cdot s^{-2}\cdot A^{-1}$
volt	V	electric potential, potential difference, electromotive force	W/A	$m^{2}\cdot kg\cdot s^{-3}\cdot A^{-1}$
watt	W	power, radiant flux	J/s	$m^{2}\cdot kg\cdot s^{-3}$
weber	Wb	magnetic flux	V•s	$m^{2}\cdot kg\cdot s^{-2}\cdot A^{-1}$

Table 4. SI-Derived Compound Units

Name	*Symbol*	*Quantity*	*In Terms of Other Units*
ampere per meter	A/m	magnetic field strength	
ampere per square meter	A/m^2	current density	
candela per square meter	cd/m^2	luminance	
coulomb per cubic meter	C/m^3	electric charge density	$m^{-3}{\cdot}s{\cdot}A$
coulomb per kilogram	C/kg	exposure (X- and γ-rays)	
coulomb per square meter	C/m^2	electric flux density	$m^{-2}{\cdot}s{\cdot}A$
cubic meter	m^3	volume	
cubic meter per kilogram	m^3/kg	specific volume	
farad per meter	F/m	permittivity	$m^{-3}{\cdot}kg^{-1}{\cdot}s^4{\cdot}A^2$
henry per meter	H/m	permeability	$m{\cdot}kg{\cdot}s^{-2}{\cdot}A^{-2}$
joule per cubic meter	J/m^3	energy density	$m^{-1}{\cdot}kg{\cdot}s^{-2}$
joule per kelvin	J/K	heat capacity, entropy	$m^2{\cdot}kg{\cdot}s^{-2}{\cdot}K^{-1}$
joule per kilogram	J/kg	specific energy	$m^2{\cdot}s^{-2}$
joule per kilogram kelvin	J/(kg K)	specific heat capacity, specific entropy	$m^2{\cdot}s^{-2}{\cdot}K^{-1}$
joule per mole	J/mol	molar energy	$m^2{\cdot}kg{\cdot}s^{-2}{\cdot}mol^{-1}$
joule per mole kelvin	J/(mol K)	molar entropy, molar heat capacity	$m^2{\cdot}kg{\cdot}s^{-2}{\cdot}K^{-1}{\cdot}mol^{-1}$
kilogram per cubic meter	kg/m^3	density, mass density	
meter per second	m/s	speed, velocity	
meter per second squared	m/s^2	acceleration	
mole per cubic meter[a]	mol/m^3	concentration (amount of substance per volume)	
newton-meter	N·m	moment of force	$m^2{\cdot}kg{\cdot}s^{-2}$
newton per meter	N/m	surface tension	$kg{\cdot}s^{-2}$
pascal second	Pa·s	dynamic viscosity	$m^{-1}{\cdot}kg{\cdot}s^{-1}$
radian per second	rad/s	angular velocity	
radian per second squared	rad/s^2	angular acceleration	
reciprocal meter	m^{-1}	wavenumber	
reciprocal second	s^{-1}	frequency	
square meter	m^2	area	
square meter per second	m^2/s	kinematic viscosity	
volt per meter	V/m	electric field strength	$m{\cdot}kg{\cdot}s^{-3}{\cdot}A^{-1}$
watt per meter kelvin	W/(m K)	thermal conductivity	$m{\cdot}kg{\cdot}s^{-3}{\cdot}K^{-1}$
watt per square meter	W/m^2	heat flux density, irradiance	$kg{\cdot}s^{-3}$
watt per square meter steradian	$W/(m^2\ sr)$	radiance	
watt per steradian	W/sr	radiant intensity	

[a]Liter (L) is a special name for cubic decimeter. The symbol M is not an SI unit, but expressions such as 0.1 M, meaning a solution with concentration of 0.1 mol/L, are acceptable.

Table 5. Other Units

Name	*Symbol*	*Quantity*	*Value in SI Units*
angstrom	Å	distance	1 Å = 10^{-10} m = 0.1 nm
bar	bar	pressure	1 bar = 10^5 Pa = 100 kPa = 0.1 MPa
barn	b	area, cross section	1 b = 10^{-28} m^2 = 100 fm^2
bohr	b, a_0	length	1 b ≈ 5.291 77 × 10^{-11} m
curie[a]	Ci	activity	1 Ci = 3.7 × 10^{10} Bq
dalton	Da	atomic mass	1 Da = 1.660 540 × 10^{-27} kg
darcy[b]	darcy	permeability	
day	day	time	1 day = 24 h = 86 400 s
debye[c]	D	electric dipole moment	
degree	°	plane angle	1° = (π/180) rad
degree Celsius	°C	temperature	—[d]
dyne[e]	dyn	force	
einstein	einstein	light energy	—[f]
electronvolt	eV	—[g]	1 eV = 1.602 19 × 10^{-19} J
erg[h]	erg	energy or work	
faraday	faraday	electric charge	1 faraday = 96 485.31 C
fermi	f	length	1 f = 10^{-15} m
franklin	Fr	electric charge	1 Fr = 3.335 64 × 10^{-10} C
galileo	Gal	acceleration	1 Gal = 10^{-2} m s^{-2}
gauss	G	magnetic induction	1 G = 10^{-4} Wb/m^2
gilbert[i]	Gi	magnetomotive force	
hartree	hartree, E_h	energy	1 hartree = 4.359 75 × 10^{-18} J
hectare	ha	area	1 ha = 1 hm^2 = 10^4 m^2
hour	h	time	1 h = 60 min = 3600 s
liter	L	volume	1 L = 1 dm^3 = 10^{-3} m^3
metric ton	t	mass	1 t = 10^3 kg
minute	min	time	1 min = 60 s
minute	′	plane angle	1′ = (1/60)° = (π/10 800) rad
parsec	pc	length	1 pc ≈ 3.085 68 × 10^{16} m
poise[j]	P	dynamic viscosity	
rad	rad, rd[k]	absorbed dose	1 rad = 0.01 Gy = 1 cGy = 100 erg·g^{-1}
roentgen	R	exposure	1 R = 2.58 × 10^{-4} C·kg^{-1}
roentgen equivalent man[l]	rem	weighted absorbed dose	1 rem = 0.01 Sv
second	″	plane angle	1″ + (1/60)′ = (π/648 000) rad
stokes[m]	St	kinematic viscosity	
svedberg	Sv	time	1 Sv = 10^{-13} s
unified atomic mass unit	u	—[n]	1 u = 1.660 540 × 10^{-27} kg

[a]1 Ci = 2.2 × 10^{12} disintegrations per minute.

[b]1 darcy is the permeation achieved by the passage of 1 mL of fluid of 1-cP viscosity flowing in 1 s

under a pressure of 1 atm (101 kPa) through a porous medium that has a cross-sectional area of 1 cm^2 and a length of 1 cm.

[c]1 D = 10^{-18} Fr cm.

[d]Temperature intervals in kelvins and degrees Celsius are identical; however, temperature in kelvins equals temperature in degrees Celsius plus 273.15.

[e]1 dyn is equal to the force that imparts an acceleration of 1 cm/s^2 to a 1-g mass.

[f]1 einstein equals Avogadro's number times the energy of one photon of light at the frequency in question.

[g]The electronvolt is the kinetic energy acquired by an electron in passing through a potential difference of 1 V in vacuum.

[h]1 erg is the work done by a 1-dyn force when the point at which the force is applied is displaced by 1 cm in the direction of the force.

[i]1 Gi is the magnetomotive force of a closed loop of one turn in which there is a current of $(1/4\pi) \times 10$ A.

[j]1 P is the dynamic viscosity of a fluid in which there is a tangential force of 1 dyn/cm^2 resisting the flow of two parallel fluid layers past each other when their differential velocity is 1 cm/s per centimeter of separation.

[k]When there is a possibility of confusion with the symbol for radian, rd may be used as the symbol for rad.

[l]1 rem has the same biological effect as 1 rad of X-rays.

[m]1 St is the kinematic viscosity of a fluid with a dynamic viscosity of 1 P and a density of 1 g/cm^3.

[n]The unified atomic mass unit is equal to $^1/_{12}$ of the mass of an atom of the nuclide ^{12}C.

Table 6. Non-SI Units That Are Discouraged

Discouraged Unit	*Value in SI Units*
calorie (thermochemical)	4.184 J
conventional millimeter of mercury	133.322 Pa
grad	2π/400 rad
kilogram-force	9.806 65 N
metric carat	0.2 g
metric horsepower	735.499 W
mho	1 S
micron	1 μm
standard atmosphere	101.325 kPa
technical atmosphere	98.066 5 kPa
torr	133.322 Pa

Bibliography

Metric Practice; ANSI/IEEE 268–1992; American National Standards Institute: New York, 1992.

SI Units and Recommendations for the Use of Their Multiples and of Certain Other Units; ISO 1000:1992; International Standards Organization: Geneva, Switzerland, 1992.

Swanson, E. *Mathematics into Type: Copy Editing and Proofreading of Mathematics for Editorial Assistants and Authors;* American Mathematical Society: Providence, RI, 1991.

Use of the International System of Units (SI) (The Modernized Metric System); E380–93; American Society for Testing and Materials: West Conshohocken, PA, 1993.

CHAPTER 6

References

This chapter presents style conventions for citing references within a manuscript and for listing complete reference citations. Many of the references in the examples were created to illustrate a style point under discussion; they may not be real references.

Citing References in Text

In ACS publications, you may cite references in text in three ways:

1. By superscript numbers in *Accounts of Chemical Research*, *Analytical Chemistry* (except review issues), *Chemical Reviews*, *Chemistry of Materials*, *Energy & Fuels*, *Inorganic Chemistry*, the *Journal of the American Chemical Society*, the *Journal of Chemical Information and Computer Sciences*, the *Journal of Medicinal Chemistry*, the *Journal of Natural Products*, *The Journal of Organic Chemistry*, the *Journal of Pharmaceutical Sciences*, the *Journal of Physical and Chemical Reference Data*, *The Journal of Physical Chemistry A*, *The Journal of Physical Chemistry B*, *Langmuir*, *Macromolecules*, *Organic Process Research & Development*, and *Organometallics*. The superscript numbers appear outside the punctuation if the citation applies to a whole sentence or clause.

   Oscillation in the reaction of benzaldehyde with oxygen was reported previously.[3]

2. By italic numbers on the line and in parentheses inside the punctuation in *Analytical Chemistry* (review issues), *Biochemistry*, *Bioconjugate Chemistry* (one of two acceptable styles), *Chemical Health &*

Safety, *Chemical Research in Toxicology*, *CHEMTECH, Environmental Science & Technology*, the *Journal of Chemical Education,* and *Today's Chemist at Work*.

The mineralization of TCE by a pure culture of a methane-oxidizing organism has been reported (*6*).

3. By author name and year of publication in parentheses inside the punctuation in *Bioconjugate Chemistry* (one of two acceptable styles), *Biotechnology Progress*, *Industrial & Engineering Chemistry Research*, the *Journal of Agricultural and Food Chemistry,* and the *Journal of Chemical and Engineering Data*.

The primary structure of this enzyme has also been determined (Finnegan et al., 1996).

In ACS books, all three of these systems are used, depending on the subject matter and series.

◆ In all three systems, the author's name may be made part of the sentence. In such cases, in the name–year system, place only the year in parentheses.

The syntheses described by Fraser[8] take advantage of carbohydrate topology.

Jensen (*3*) reported oscillation in the reaction of benzaldehyde with oxygen.

According to Harris (1997), drug release is controlled by varying the hydrolytic stability of the ester bond.

◆ With numerical reference citations, start with 1 and number consecutively throughout the paper, including references in text and those in tables, figures, and other nontext components. If a reference is repeated, do not give it a new number; use the original reference number.

◆ Whenever authors are named, if a reference has two authors, give both names joined by the word "and". If a reference has more than two authors, give only the first name listed followed by "et al." Do not use a comma before et al.; always use a period after al.

Allison and Perez[12] Johnson et al. (*12*)
(O'Brien and Alenno, 1996) (Bachrach et al., 1997)

◆ To cite more than one reference by the same principal author and various coauthors, use the principal author's name followed by "and co-workers" or "and colleagues".

Pauling and co-workers[10,11] Cram and colleagues (*27–29*)

◆ When citing more than one reference at one place by number, list the numbers in ascending order and separate them by commas (without spaces as superscripts; with spaces on line), or if they are part of a consecutive series, use a dash to indicate a range of three or more.

in the literature[2,5,8]
in the literature (*2, 5, 8*)

were reported[3–5,10]
were reported (*3–5, 10*)

◆ When citing more than one reference at one place by the name–year system, list them alphabetically according to the first author's name, followed by a comma and the year. Use a semicolon to separate individual references.

(Axelrod, 1997; Cobbs and Stolman, 1996; Gerson et al., 1997)

◆ When citing more than one reference by the same author at one place by the name–year system, do not repeat the name. List the name followed by the year of each of the references in ascending order; separate the years by commas. If an author has more than one reference in the same year, add lowercase letters to the years to differentiate them. Add letters to all of the years, for example, 1996a, 1996b, etc., not 1996, 1996a, etc.

(Trapani, 1994, 1996; Zillman, 1995)
(Knauth, 1996a, 1996b)
(Fordham, 1995; Fordham and Rizzo, 1995)

◆ Cite the reference in a logical place in the sentence.

recent investigations (cite)
other developments (cite)
was reported recently (cite)
as described previously (cite)
previous results (cite)
recently were demonstrated (cite)
a molecular mechanics study (cite)
Marshall and Levitt's approach (cite)
the procedure of Lucas et al. (cite)

Style for Reference Lists

Authors are responsible for the accuracy and completeness of all references. Authors should check all parts of each reference listing against the original document.

A reference must include certain minimum data:

- Periodical references must include the author names, abbreviated journal title, year of publication, volume number (if any), and initial page of cited article (the complete span is better).

- Book references must include the author or editor names, book title, publisher, city of publication, and year of publication.
- For material other than books and journals, sufficient information must be provided so that the source can be identified and located.

In lists, references always end with a period.

Periodicals

Recommended Formats

Author 1; Author 2; Author 3; etc. Title of Article. *Journal Abbreviation* **Year,** *Volume,* Inclusive Pagination.

Author 1; Author 2; Author 3; etc. *Journal Abbreviation* **Year,** *Volume,* Inclusive Pagination.

An exception is *Biochemistry.* Consult this journal's Instructions to Authors for correct format.

Author Name Field

Include all author names in a reference citation. With multiple authors, separate the names from one another by semicolons. Always end the author field with a period (exception: *Biochemistry*). List the names in inverted form: last name first, then first initial or name, middle initial or name, and qualifiers (Jr., II). Some publications list the first 10 authors followed by a semicolon and et al.

Cotton, F. A.
Basconi, J.; Lin, P. B.
Chandler, J. P., III; Levine, S. M.
Schafer, Frances.
Schafer, Frank W., Jr.
Schafer, F. W., Jr.
Fishman, W., II.
Farhataziz. (a single name is uncommon, but does occur; no period in *Biochemistry*)
Inderjit; Fontana, M. J. (the first author has a single name)

Article Title Field

Article titles are not essential in reference citations, but they are considered desirable to highlight the contents of a paper and facilitate location in reference libraries. Some ACS publications include the article title in journal references, and some do not; check the publication itself. Article titles are set in roman type without quotation marks and end with a period. In ACS journals, capitalization follows that of the original publication; in other publications, the main words are capitalized.

> Hill, M.; Fott, P. Kinetics of gasification of Czech brown coals. *Fuel* **1993,** *72,* 525–529.
>
> Klingenberg, B.; Vannice, M. A. Influence of Pretreatment on Lanthanum Nitrate, Carbonate, and Oxide Powders. *Chem. Mater.* **1996,** *8*, 2755–2768.

Journal Abbreviation Field

The journal name is an essential component of a periodical reference citation. Abbreviate the name according to the *Chemical Abstracts Service Source Index* (CASSI) and italicize it. One-word journal names are not abbreviated (e.g., *Biochemistry*, *Macromolecules*, *Nature*, *Science*). No punctuation is added to end this field; thus, a period will be there with an abbreviation but not with a spelled-out word.

CASSI and its quarterly supplements provide an extensive list of recommended journal abbreviations. The supplement to this chapter is a list of CASSI abbreviations for more than 1000 of the most commonly cited journals. ACS publication names, their abbreviations, and their volume numbers for 1997 are given in Table 1. Note that in some cases, the word "the" is part of the title.

Information Found in CASSI

Entries are arranged in CASSI alphabetically according to the abbreviated form of the title. Abbreviations are based on the standards of the International Organization for Standardization (ISO). Recommended abbreviations are indicated in boldface type (see p 180).

Using CASSI Abbreviations

◆ The boldface components of the publication title form the abbreviated title. Use a period after each abbreviation, and maintain the punctuation shown in CASSI.

> **Int**_ernational_ **J**_ournal of_ **Radiat**_ion_ **Oncol**_ogy,_ **Biol**_ogy,_ **Phys**_ics_
>
> *Int. J. Radiat. Oncol., Biol., Phys.*

Table 1. ACS Periodicals

Names As Registered in the U.S. Patent and Trademark Office	*1997 Vol.*	*CASSI Abbreviation*
Accounts of Chemical Research	30	Acc. Chem. Res.
Advance ACS Abstracts	5	Adv. ACS Abstr.
Analytical Chemistry	69	Anal. Chem.
Biochemistry	36	Biochemistry
Bioconjugate Chemistry	8	Bioconjugate Chem.
Biotechnology Progress	13	Biotechnol. Prog.
Chemical & Engineering News	75	Chem. Eng. News
Chemical Health & Safety	4	Chem. Health Saf.
Chemical Research in Toxicology	10	Chem. Res. Toxicol.
Chemical Reviews	97	Chem. Rev.
Chemistry of Materials	9	Chem. Mater.
CHEMTECH	27	CHEMTECH
Energy & Fuels	11	Energy Fuels
Environmental Science & Technology	31	Environ. Sci. Technol.
Industrial & Engineering Chemistry Research	36	Ind. Eng. Chem. Res.
Inorganic Chemistry	36	Inorg. Chem.
Journal of Agricultural and Food Chemistry	45	J. Agric. Food Chem.
Journal of the American Chemical Society	119	J. Am. Chem. Soc.
Journal of Chemical and Engineering Data	42	J. Chem. Eng. Data
Journal of Chemical Information and Computer Sciences	37	J. Chem. Inf. Comput. Sci.
Journal of Medicinal Chemistry	40	J. Med. Chem.
Journal of Natural Products	60	J. Nat. Prod.
The Journal of Organic Chemistry	62	J. Org. Chem.
Journal of Pharmaceutical Sciences	86	J. Pharm. Sci.
Journal of Physical and Chemical Reference Data	26	J. Phys. Chem. Ref. Data
The Journal of Physical Chemistry A	101	J. Phys. Chem. A
The Journal of Physical Chemistry B	101	J. Phys. Chem. B
Langmuir	13	Langmuir
Macromolecules	30	Macromolecules
Organic Process Research & Development	1	Org. Process Res. Dev.
Organometallics	16	Organometallics
Today's Chemist at Work	6	Today's Chemist at Work

◆ Maintain the word spacing shown in CASSI, except for U.S., U.S.A., D.C., and N.Y.

Ann. N.Y. Acad. Sci.
Chem. Eng. (N.Y.)
J. Res. Natl. Bur. Stand. (U.S.)
Proc. Natl. Acad. Sci. U.S.A.

◆ Use a terminal period only if the last word of the periodical title is abbreviated.

Int*ernational* **J***ournal of* **Sports Med***icine*
Int. J. Sports Med. (last word is abbreviated; period is used)

Int*ernational* **J***ournal of* **Cancer**
Int. J. Cancer (last word is not abbreviated; no period)

◆ If the periodical abbreviation in CASSI shows a hyphen with spaces on both sides, change the hyphen to an em dash closed up on each side.

Bull*etin* **Tech***nique* - **Valorisation** *et* **Util***isation des* **Combust***ibles,* **Inst***itut* **Nat***ional des* **Ind***ustries* **Extr***actives*
Bull. Tech.—Valorisation Util. Combust., Inst. Natl. Ind. Extr.

◆ If a boldface **n** precedes the volume number in CASSI, use the abbreviation "No." before the volume number in italics in the entry.

Br*itish* **Med***ical* **J***ournal* ... **n6372 1983**
Br. Med. J. **1983,** *No. 6372.*

Spec*ial* **Publ***ication* - **R***oyal* **Soc***iety of* **Chem***istry* ... **n37 1980**
Spec. Publ.—R. Soc. Chem. **1980,** *No. 37.*

Include all the information shown for volume in italics, especially for references to government publications and reports.

Brookhaven National **Lab**oratory, **[Rep**ort**] BNL** ... **BNL 51706 1983**
Brookhaven Natl. Lab., [Rep.] BNL **1983,** *BNL 51706.*

Los Alamos National **Lab**oratory, **[Rep**ort**] LA** (**U**nited **S**tates) ... **LA–9757–MS 1983**
Los Alamos Natl. Lab., [Rep.] LA (U.S.) **1983,** *LA-9757-MS.*

Exceptions to the Rules of CASSI Abbreviations

◆ Strict rules for CASSI abbreviations can be modified for periodicals whose titles include multiple parts, sections, and series.

CASSI abbreviation:

Acta Crystallogr., Sect. B: Struct. Crystallogr. Cryst. Chem. **1978,** *B34,* 110–115.

A Sample CASSI Entry

*J*ournal *of the* **Am***erican* **Chem***ical* **Soc***iety*. JACSAT. ISSN 0002–7863 (Absorbed Am. Chem. J.). In Eng; Eng sum. v1 1879+. *w* **118 1996**. *ACS Journals*.

AMERICAN CHEMICAL SOCIETY. JOURNAL. WASHINGTON, D. C.

Doc. Supplier: CAS.

AAP; AB 1905+; ABSR; ARaS; ATVA; AU–M 1893–1918,1920–1926,1928+; AkU 1879–1906,1919+; ArU; ArU–M 1923+; AzTeS; AzU 1889+; C; CL; CLSU; CLSU–M 1895–1897,1905,1908+; CLU–M; CLU–P; CMenSR 1916+; CPT; CSf; CSt; CSt–L; CU; CU–A; CU–I 1920+; CU–M; CU–Riv 1907+; CU–RivA; CU–RivP; CU–S; CU–SB; [etc.]

In this example,

- *J*ournal *of the* **Am***erican* **Chem***ical* **Soc***iety* is the complete publication title with its abbreviated form indicated by boldface type (*J. Am. Chem. Soc.*).
- **JACSAT** is the CODEN, a six-character, unique title abbreviation used to represent titles in manual or machine-based information systems. The CODEN source is the *International CODEN Directory,* administered by Chemical Abstracts Service. The sixth character of each CODEN is a computer-calculated check character that ensures the reliability of the CODEN in computer-based systems.
- **ISSN 0002–7863** is the International Standard Serial Number (ISSN), assigned by the Library of Congress.
- **Absorbed Am. Chem. J.** is a reference to former titles and to any variant forms of the selected title.
- **In Eng; Eng sum.** is the language of the publication, summaries, and tables of contents.
- **v1 1879+** is the history of the publication. Volume 1 began in 1879. The + following the year indicates that the publication is still in existence under that title.
- *w* means weekly. The frequency of publication could also be *a* for annually, *ba* for biennially (every two years), *bm* for bimonthly (every two months), *bw* for biweekly (every two weeks), *d* for

daily, *m* for monthly, *q* for quarterly, *sa* for semiannually (two times per year), *sm* for semimonthly (two times per month), or *sw* for semiweekly (two times per week).

- **118 1996** is the volume–year correlation (i.e., the first volume number of that year, which is the most recent covered by that edition of CASSI; volume 118 was the first volume number of 1996).
- ***ACS Journals*** is the publisher or source address or abbreviation.
- **AMERICAN CHEMICAL SOCIETY. JOURNAL. WASHINGTON, D. C.** is the AACR entry. This is the abbreviated entry as catalogued according to the *Anglo-American Cataloguing Rules* (2nd ed.). It is included here because of its predominance in library collection records.
- **Doc. Supplier: CAS** means that articles are available through the CAS Document Delivery Service.
- **AAP; AB 1905+; ABSR; etc.,** is the library holdings information. Libraries are identified by their *National Union Catalog* symbols, and holdings are shown by inclusive years.

Acceptable variation—The section title need not be named:

Acta Crystallogr., Sect. B **1978,** *34,* 110–115.

Acceptable variation—The section can be indicated by the volume number:

Acta Crystallogr. **1978,** *B34,* 110–115.

◆ For some periodicals whose CASSI abbreviation includes a place of publication, you need not add the place of publication unless its omission would create ambiguity. If CASSI lists only one journal with a given main title, there is no ambiguity in omitting the place of publication.

Use	*Not necessarily*
Science	Science (Washington, D.C.)
Nature	Nature (London)
Clin. Chem.	Clin. Chem. (Washington, D.C.)

Omission of the place of publication would create ambiguity for different journals having the same main title.

Transition Met. Chem. (N.Y.)
Transition Met. Chem. (London)

Sometimes the place of publication of a journal changes. These are given as "see" references in CASSI. In these cases, use the "see" reference place of publication, not the original.

For some journals of the Royal Society of Chemistry (formerly the Chemical Society), you must include parts or sections for the years 1966 and thereafter. Use the journal title at the time of publication of the article you are citing. Table 2 is an outline of the publication history of the Chemical Society journals.

Year of Publication Field

The year of publication is essential information in a periodical citation. The year is set in boldface type, followed by a comma.

Publication Volume Field

The volume number is important information and is recommended for all periodical citations; it is essential for publications having more than one volume per year (such as the *Journal of Chemical Physics*). The volume number is set in italic type and is separated from the pagination information by a comma.

◆ For periodicals in which each issue begins with page 1, include issue information (either the number or the date) in the publication volume field. Issue information is set in roman type, enclosed in parentheses, and spaced from the volume number, which it directly follows.

Issue Number:

> Rouhi, A. M. *Chem. Eng. News* **1996,** *74* (33), 37–42.

Date of Issue:

> Rouhi, A. M. *Chem. Eng. News* **1996,** *74* (Aug 12), 37–42.

◆ For publications that have supplements, the following form is recommended.

> Grote, T.; Modiano, M.; Gandara, D. *Eur. J. Cancer* **1995,** *31A* (Suppl. 5), S255.
>
> Sear, J. W.; Fisher, A.; Summerfield, R. J. *Eur. J. Anaesthesiol.* **1987,** *4* (Suppl. 1), 55–61.
>
> Taylor, V. L.; Kristiansen, M. *Tetrahedron* **1991,** *47* (Suppl. 1), 125–141.

◆ For journals that have no volume numbers, include issue numbers, especially when the pagination of each issue begins with page 1. Use the following form. Note that the issue number is not italicized.

> Wills, M. R.; Savory, J. *Lancet* **1983,** No. 2, 29.

Table 2. Journals of the Royal Society of Chemistry (Formerly the Chemical Society)

Journal Abbreviation	*History*
Chem. Commun.	1965–1968 (changed to J. Chem. Soc. D)
Chem. Commun. (Cambridge)	1996–present
Chem. Soc. Rev.	1972–present
Discuss. Faraday Soc.	1947–1971
Faraday Discuss. Chem. Soc.	1972–present
Faraday Symp. Chem. Soc.	1972–1984
J. Chem. Soc.	1849–1877 and 1926–1965; published 1878–1925 as J. Chem. Soc., Abstr. (even-numbered volumes; changed to Br. Chem. Abstr., A) and J. Chem. Soc., Trans. (odd-numbered volumes)
J. Chem. Soc. A	1966–1971 (changed to J. Chem. Soc., Dalton Trans. and J. Chem. Soc., Faraday Trans. 1 and 2)
J. Chem. Soc. B	1966–1971 (changed to J. Chem. Soc., Perkin Trans. 2)
J. Chem. Soc. C	1966–1971 (changed to J. Chem. Soc., Perkin Trans. 1)
J. Chem. Soc., Chem. Commun.	1972–1995
J. Chem. Soc. D	1969–1971 (changed to J. Chem. Soc., Chem. Commun.)
J. Chem. Soc., Dalton Trans.	1972–present
J. Chem. Soc., Faraday Trans. 1, J. Chem. Soc., Faraday Trans. 2	1972–1989 [I and II, not 1 and 2, are the correct forms of the spelled-out titles, although the official abbreviations sanctioned by the Royal Society (which uses CASSI) use 1 and 2 to accommodate computer alphabetization schemes.]
J. Chem. Soc., Faraday Trans.	1990–present (merged J. Chem. Soc., Faraday Trans. 1 and 2)
J. Chem. Soc., Perkin Trans. 1	1972–present
J. Chem. Soc., Perkin Trans. 2	1972–present
Q. Rev., Chem. Soc.	1947–1971 (changed to Chem. Soc. Rev.)
Trans. Faraday Soc.	1905–1971

Pagination Field

Pagination is an essential element of a reference citation. The complete page range is preferable, but initial page numbers are acceptable.

- In page spans, use all digits, closed up, with no commas or spaces.

2–15	12–19	44–49
103–107	108–117	234–236
345–359	1376–1382	2022–2134
11771–11779		

In ACS journals, condensed ranges (e.g., 12–9, 345–59, 2022–134) are acceptable. Do not, however, mix styles of presentation within a paper.

◆ You may also indicate pagination in reference citations by "f" or "ff", which means "and following" page or pages, respectively. The f or ff is set in roman type and is spaced from the preceding number:

> 60 f (indicates page 60 and the page following—pages 60 and 61)
>
> 60 ff (indicates page 60 and pages following)
>
> 58–60 ff (indicates pages 58 through 60 and pages following—essentially the same as 58 ff except that the three pages enumerated contain the most pertinent information and other relevant information is scattered on the rest of the pages)

◆ The pagination field may also include terms such as "and references therein" and similar expressions (especially in references to review articles). This phrase follows the page numbers and is not separated by a comma.

> Pagano, A. R.; Lajewski, W. M.; Jones, R. A. *J. Am. Chem. Soc.* **1995,** *117,* 11669–11672 and references therein.

◆ Some publications use article numbering, rather than page numbering, where each article starts on page 1. Use the article number in the pagination field.

> Brosset, C. *Ark. Kemi, Mineral. Geol.* **1945,** *20A,* No. 7.

Use of Punctuation To Indicate Repeating Fields of Information

Within a reference, use a semicolon or period to indicate repeating information.

1. Same author in multiple publications:

   Forster, T. *Naturwissenschaften* **1946,** *33,* 166; *Ann. Phys.* **1948,** *2,* 55; *Z. Naturforsch.* **1949,** *417,* 321.

2. Same author in multiple publications, but with letters to separate the references (the semicolon in the previous example is changed to a period):

   (a) Berger, B.; Heitefuss, R. *Fresenius' J. Anal. Chem.* **1989,** *334,* 360–362. (b) *Weed Res.* **1991,** *31,* 9–18.

3. Same author of multiple articles in the same journal:

   Alaiz, M.; Zamora, R.; Hidalgo, F. J. *J. Agric. Food Chem.* **1995,** *43,* 795–800; **1996,** *44,* 686–691.

When the year and volume are the same:

Clay, S. A.; Koskinen, W. C. *Weed Sci.* **1990,** *38,* 74–80, 262–266.

When the year is the same but the volumes are different:

Ferretti, A.; Lami, A. *Chem. Phys.* **1994,** *181,* 107; **1994,** *186,* 143.

References to Chemical Abstracts

Use a semicolon to separate the periodical citation from a reference to its abstract (*Chemical Abstracts*).

Ohta, G. Constituents of Rice Bran Oil. III. Structure of Oryzanol-C. *Chem. Pharm. Bull.* **1960,** *8,* 5–9; *Chem. Abstr.* **1961,** *55,* 5570.

Tamura, T.; Sakaedmani, N.; Matsumoto, T. Isolation of Dihydrositosterol Ferulate from Corngerm Oil. *Nippon Kagaku Zasshi* **1958,** *79,* 1011–1014; *Chem. Abstr.* **1960,** *54,* 24857m.

Chemical Abstracts routinely contains more than one abstract per page. In the previous example, a letter follows the abstract number to indicate the location of the referenced abstract on the cited page of *Chemical Abstracts.* The method of distinguishing which abstract was being cited has changed over the years. Three variations are worth noting.

1. The column (two columns per page) in which the abstract occurs followed by a superscript number:

   Chem. Abstr. **1946,** *40,* 4463^8. (This is the 8th abstract in column 4463.)

2. The column (two columns per page) in which the abstract occurs followed by a letter, either on the line or superscript (generally italic):

   Chem. Abstr. **1953,** *47,* 1167*f.* (This is abstract f in column 1167.)
   Chem. Abstr. **1947,** *41,* 571^d. (This is abstract d in column 571.)

3. The abstract number itself followed by an on-line letter (roman), often a computer check character.

   Chem. Abstr. **1989,** *110,* 8215j. (This is abstract number 8215.)

Special Situations

- You may treat Beilstein references as periodical references.

  Beilstein, 4th Ed. **1950,** *12,* 237.

◆ Cite journals published in a foreign language either by the actual non-English title or by a translated form.

Nippon Ishikai Zasshi or *J. Jpn. Med. Assoc.*
Nouv. J. Chim. or *New J. Chem.*

◆ When citing an article printed in the English translation of a foreign-language journal, include reference to the original article, if possible, and use a semicolon to separate the two citations.

Aksenov, A. B. *J. Gen. Chem. USSR (Engl. Transl.)* **1995,** *65,* 2050; *Zh. Obshch. Khim.* **1995,** *65,* 2100.

◆ Separate two or more companion publications with a semicolon.

Clear, J. M.; Kelly, J. M.; O'Connell, C. M.; Vos, J. G. *J. Chem. Res., Miniprint* **1981,** 3038; *J. Chem. Res., Synop.* **1981,** 260.

Nonscientific Magazines and Newspapers

Recommended Format

Author 1; Author 2; Author 3; etc. Title of Article. *Title of Periodical,* Complete Date, Pagination.

For nonscientific magazines and other periodicals that are not abstracted by Chemical Abstracts Service, give the authors' names in inverted form ending with a period, the article title in roman type with main words capitalized and ending with a period, the full magazine title in italic type followed by a comma, the complete date of the issue ending with a comma, and the pagination.

Byrne, John A. The Best & Worst Boards. *Business Week,* Nov 25, 1996, pp 82–98.
Stengel, Richard; Pooley, Eric. Masters of the Message. *Time,* Nov 18, 1996, p 77.
Suplee, Curt. Infinitesimal Carbon Structures May Hold Gigantic Potential. *The Washington Post,* Dec 2, 1996, p A3.

Books

Some ACS publications include the chapter title in book references; some do not; check the publication itself. Also, consult the Instructions to Authors in *Biochemistry* for exceptions to the format presented here and elsewhere in this chapter.

Recommended Formats for Books without Editors

Author 1; Author 2; Author 3; etc. Chapter Title. *Book Title,* Edition Number; Series Information (if any); Publisher: Place of Publication, Year; Volume Number, Pagination.

Author 1; Author 2; Author 3; etc. *Book Title;* Series Information (if any); Publisher: Place of Publication, Year; Volume Number, Pagination.

When a book has authors and no editors, it means either that the entire book was written by one author or that two or more authors collaborated on the entire book.

Bersuker, I. B. *Electronic Structure and Properties of Transition Metal Compounds: Introduction to the Theory;* Wiley & Sons: New York, 1996; Chapter 6.

Dresselhaus, M. S.; Dresselhaus, G.; Eklund, P. C. *Science of Fullerenes and Carbon Nanotubes;* Academic: New York, 1996; pp 126–141.

Hansch, C.; Leo, A. *Exploring QSAR: Fundamentals and Applications in Chemistry and Biology;* American Chemical Society: Washington, DC, 1995.

Rogers, D. W. *Computational Chemistry Using the PC,* 2nd ed.; VCH: New York, 1996.

Shore, B. W. *The Theory of Coherent Atomic Excitation: Simple Atoms and Fields;* Wiley & Sons: New York, 1990; Vol. 1.

Urban, M. W. *Attenuated Total Reflectance Spectroscopy of Polymers: Theory and Practice;* American Chemical Society: Washington, DC, 1996; Chapter 5.

Walker, J. M.; Cox, M. *The Language of Biotechnology: A Dictionary of Terms,* 2nd ed.; American Chemical Society: Washington, DC, 1996; p 56.

Recommended Formats for Books with Editors

Author 1; Author 2; Author 3; etc. Chapter Title. In *Book Title,* Edition Number; Editor 1, Editor 2, etc., Eds.; Series Information (if any); Publisher: Place of Publication, Year; Volume Number, Pagination.

Author 1; Author 2; Author 3; etc. In *Book Title,* Edition Number; Editor 1, Editor 2, etc., Eds.; Series Information (if any); Publisher: Place of Publication, Year; Volume Number, Pagination.

When a book has editors, it means that different authors wrote various parts of the book independently of each other. The word "In" before the book title indicates that the authors mentioned wrote only a part of the book, not the entire book.

Adams, M. R.; Garton, A. Far-Ultraviolet Degradation of Selected Polymers. In *Polymer Durability: Degradation, Stabilization, and Lifetime Prediction;* Clough, R. L., Billingham, N. C., Gillen, K. T., Eds.; Advances in Chemistry Series 249; American Chemical Society: Washington, DC, 1996; pp 139–158.

Almlof, J.; Gropen, O. Relativistic Effects in Chemistry. In *Reviews in Computational Chemistry;* Lipkowitz, K. B., Boyd, D. B., Eds.; VCH: New York, 1996; Vol. 8, pp 206–210.

Hillman, L. W. In *Dye Laser Principles with Applications;* Duarte, F. J., Hillman, L. W., Eds.; Academic: New York, 1990; Chapter 2.

Weaver, J. C.; Astumian, R. D. In *Electromagnetic Fields: Biological Interactions and Mechanisms;* Blank, M., Ed.; Advances in Chemistry Series 250; American Chemical Society: Washington, DC, 1996; pp 79–96.

If the book as a whole is being referenced, the author names may not appear.

Agricultural Materials as Renewable Resources: Nonfood and Industrial Applications; Fuller, G., McKeron, T. A., Bills, D. D., Eds.; ACS Symposium Series 647; American Chemical Society: Washington, DC, 1996.

Bioorganic Chemistry: Nucleic Acids; Hecht, S. M., Ed.; Oxford University Press: New York, 1996.

Author Name Field

◆ Separate the names of multiple authors by semicolons, and always end the author field with a period (except in *Biochemistry*). List names in inverted form: last name first, then first initial or name, middle initial or name, and qualifiers (Jr., II).

◆ If a book has no primary authors because each chapter was written by a different author, you may place the editor names in the author name field (especially for lists in alphabetical order). Separate editor names by commas, and in this case, the period after the abbreviation Ed. or Eds. terminates the field.

Bachrach, S. M., Ed. *The Internet: A Guide for Chemists;* American Chemical Society: Washington, DC, 1996.

◆ A book may have no named authors because it was compiled by a committee or organization. These books are discussed under the section "Works Written by an Organization or a Committee", p 198.

Chapter Title Field

Chapter titles are not essential, but they are considered desirable components in reference citations because they highlight the contents of a paper and facilitate its location in reference libraries. Chapter titles are set in roman type and end with a period.

Cieplak, P.; Kollman, P. A.; Radomski, J. P. Molecular Design of Fluorine-Containing Peptide Mimetics. In *Biomedical Frontiers of Fluorine Chemistry;*

Ojima, I., McCarthy, J. R., Welch, J. T., Eds.; ACS Symposium Series 639; American Chemical Society: Washington, DC, 1996; pp 143–156.

Book Title Field

Book titles are essential elements in book reference citations. In general, do not abbreviate book titles. They are set in italic type and are separated from the next field of the reference by a semicolon.

◆ The edition number (in ordinal form) and the abbreviation "ed." follow the book title, set off by a comma; they are set in roman type. The edition information is separated from the next field of the reference by a semicolon.

Chemical Carcinogens, 2nd ed.;

◆ When both authors and editors are given, use the word "In" (set in roman type) immediately before the title of the book to indicate that the cited authors wrote only part of the book.

Editor Name Field

For books with editors, list the names of the editors, after title and edition information, in inverted form as described under "Author Name Field", separated from one another by commas. The names are denoted as editors by including the abbreviation "Eds." or "Ed." after the final name. The editor field is set in roman type and ends with a semicolon (unless it is used in the author field location).

The Chemistry of the Atmosphere: Oxidants and Oxidation in the Earth's Atmosphere; Bandy, A. R., Ed.; Royal Society of Chemistry: Cambridge, England, 1995.

Heterogeneous Hydrocarbon Oxidation; Warren, B. K., Oyama, S. T., Eds.; ACS Symposium Series 638; American Chemical Society: Washington, DC, 1996.

In books that have no primary authors, the names of the editors may appear in either the author name field (especially for lists in alphabetical order) or the editor name field. When the editor names appear in the author name field, they are separated by commas and the field ends with a period.

Bandy, A. R., Ed. *The Chemistry of the Atmosphere: Oxidants and Oxidation in the Earth's Atmosphere;* Royal Society of Chemistry: Cambridge, England, 1995.

Warren, B. K., Oyama, S. T., Eds. *Heterogeneous Hydrocarbon Oxidation;* ACS Symposium Series 638; American Chemical Society: Washington, DC, 1996.

Publication Information Field

The name of the publisher, place of publication, and year of publication are essential elements in a book reference.

Name of Publisher

A sample listing of names and cities of frequently cited publishers is given on pp 192–194. Names and addresses of publishers are also listed in *Chemical Abstracts Service Source Index, 1907–1994 Cumulative,* pp 211–391.

◆ Generally, do not abbreviate publishers' names.

> American Chemical Society, *not* Am. Chem. Soc. *or* ACS
> American Ceramic Society, *not* Am. Ceram. Soc.

Exception You may use well-known acronyms or abbreviations created by the publishers themselves.

> AIChE *or* American Institute of Chemical Engineers
> ASTM *or* American Society for Testing and Materials
> IUPAC *or* International Union of Pure and Applied Chemistry

◆ In some publisher's names, words such as Co., Inc., Publisher, and Press are not essential.

> Academic Press: New York *or* Academic: New York

◆ Expanded names are also not essential.

> John Wiley & Sons *or* John Wiley *or* Wiley

◆ It is not necessary to repeat the publisher's name for a book compiled by the organization that published it.

> American Society of Brewing Chemists. *Methods of Analysis,* 8th ed.; St. Paul, MN, 1992.

Place of Publication

For the place of publication, give a city and state for U.S. cities or city and country for all others. The country or state is not needed if the city is considered a major city in the world and could not be confused easily with other cities of the same name (e.g., London, Paris, New York, and Rome). Use the two-letter postal abbreviations for states. Spell out names of countries unless they have standard abbreviations, such as U.K. for United Kingdom.

Birmingham, U.K.	Boca Raton, FL	Cambridge, MA
Cambridge, U.K.	Chichester, U.K.	Dordrecht, The Netherlands
Elmsford, NY	Englewood Cliffs, NJ	London

New York	Princeton, NJ	Springfield, IL
Springfield, MA	Washington, DC	

Year of Publication

In book references, the year is set in lightface (not bold) roman type, following the place of publication. Terminate the field with a period or with a semicolon if further information is given.

Jeffrey, G. A. *An Introduction to Hydrogen Bonding;* Oxford University Press: New York, 1997.
Cohen, S. S. *Crystallographic Methods;* Springer-Verlag: New York, 1996; pp 55–61.

Volume and Pagination Field

Volume Information

◆ The volume field contains specific information, such as volume number and chapter number. Use the following abbreviations and spelled-out forms with the capitalization, spelling, and punctuation shown:

Abstract
Chapter
No.
Paper
Part
Vol. (for specific volumes, Vol. 4; Vols. 1, 2; Vols. 1 and 2; Vols. 3–5)
vols. (for a number of volumes, 4 vols.)

Annual Review of Physical Chemistry; Strauss, H. L., Ed.; Annual Reviews: Palo Alto, CA, 1996; Vol. 47.
Treatise on Analytical Chemistry: Part I, Theory and Practice, 2nd ed.; Elving, P. J., Ed.; Wiley: New York, 1986; Vol. 14.

◆ If a volume or part number refers to the volume or part of an entire series of books, this information is placed where a series number would normally appear and not in the volume field for the specific book being cited.

Wiberg, K. In *Investigations of Rates and Mechanisms of Reactions;* Lewis, E. S., Ed.; Techniques of Chemistry, Vol. VI, Part I; Wiley: New York, 1974; p 764.

◆ If the book or set of books as a whole is the reference, do not include individual volume information.

McGraw-Hill Encyclopedia of Science and Technology, 7th ed.; McGraw-Hill: New York, 1992; 20 vols.

Publishers Often Cited

Check the title page, front and back, for the publisher's name and location. Some publishers have more than one place of publication. This list contains common names and locations.

Academic Press: New York
Addison-Wesley: Reading, MA
Allyn & Bacon: Needham Heights, MA
American Association for the Advancement of Science Press: Washington, DC
American Geophysical Union: Washington, DC
American Institute of Physics: Woodbury, NY
American Mathematical Society: Providence, RI
American Medical Association: Chicago, IL
American National Standards Institute: New York
American Psychological Association: Washington, DC
American Society for Microbiology: Washington, DC
American Society for Testing and Materials: Philadelphia, PA
American Society for Training and Development: Alexandria, VA
Edward Arnold: London
Atheneum Publishers: New York
Avon Books: New York
A. A. Balkema: Brookfield, VT
Ballantine Books: New York
Bantam Books: New York
Basic Books: New York
Berkley Publishing Group: New York
Blackwell Science: Cambridge, MA
R. R. Bowker: New Providence, NJ
Brookings Institution: Washington, DC
Butterworth-Heinemann: Woburn, MA
Butterworths: Markham, ON, Canada
Cambridge University Press: New York
Chapman & Hall: New York
Cold Spring Harbor Laboratory Press: Plainview, NY
Collins: New York
Columbia University Press: New York
Computational Mechanics: Billerica, MA
Cornell University Press: Ithaca, NY
CRC Press: Boca Raton, FL
Walter de Gruyter: Hawthorne, NY
Marcel Dekker: New York
Dell Publishing: New York
Doubleday: New York
Dover Publications: Mineola, NY

Elsevier Biomedical: Amsterdam
Elsevier/North Holland: New York
Elsevier Science: New York
Lawrence Erlbaum: Mahwah, NJ
Free Press: New York
W. H. Freeman: New York
Garland Publishing: New York
Geological Society of America: Boulder, CO
Gordon & Breach: Langhorne, PA
Harcourt Brace Jovanovich: Orlando, FL
Harper & Row: New York
HarperCollins: New York
Harvard University Press: Cambridge, MA
Ellis Harwood: Chichester, U.K.
Haworth Press: Binghamton, NY
Hemisphere Publishing Corp.: Bristol, PA
Houghton-Mifflin: Boston, MA
Humana Press: Totowa, NJ
IDG Books: Foster City, CA
Indiana University Press: Bloomington, IN
Interscience Publishers: New York
Iowa State University Press: Ames, IA
ISI Press: Philadelphia, PA
Johns Hopkins University Press: Baltimore, MD
Jones and Bartlett Publishers: Sudbury, MA
S. Karger: Basel, Switzerland
Kluwer Academic Publishers: Norwell, MA
Alfred A. Knopf: New York
Krieger Publishing: Malabar, FL
Lea & Febiger: Philadelphia, PA
Alan R. Liss: New York
Little, Brown: Boston, MA
Longman Group: Harlow, U.K.
Longman Scientific & Technical: Essex, U.K.
Macmillan Publishing: New York
McGraw-Hill: New York
Merck & Co.: Rahway, NJ
Merriam-Webster: Springfield, MA
Methuen: New York
MIT Press: Cambridge, MA
William Morrow & Co.: New York
Munksgaard International: Copenhagen
National Academy Press: Washington, DC
National Book Co.: Portland, OR
Naval Institute Press: Annapolis, MD
New York Academy of Sciences: New York

New York Botanical Garden: Bronx, NY
W. W. Norton: New York
Noyes Publications: Park Ridge, NJ
Osborne/McGraw-Hill: Berkeley, CA
Oxford University Press: New York
Pergamon Press: Elmsford, NY
Plenum Publishing: New York
Portland Press: London
Prentice Hall: New York
Princeton University Press: Princeton, NJ
Prometheus Books: Buffalo, NY
Random House: New York
Raven Press: New York
D. Reidel: Dordrecht, The Netherlands
Resources for the Future: Washington, DC
Routledge, Chapman & Hall: New York
Rutgers University Press: New Brunswick, NJ
W. B. Saunders: Philadelphia, PA
Charles Scribner's Sons: New York
Simon & Schuster: New York
Sinauer Associates: Sunderland, MA
Smithsonian Institution Press: Washington, DC
Springer Publishing: New York
Springer-Verlag: New York
St. Martin's Press: New York
Stanford University Press: Stanford, CA
State University of New York Press: Albany, NY
Charles C. Thomas: Springfield, IL
Timber Press: Portland, OR
University of California Press: Berkeley, CA
University of Chicago Press: Chicago, IL
University of Illinois Press: Champaign, IL
University of Michigan Press: Ann Arbor, MI
University of Oklahoma Press: Norman, OK
University of Texas Press: Austin, TX
University of Washington Press: Seattle, WA
University of Wisconsin Press: Madison, WI
U.S. Department of Energy, Office of Scientific and Technical Information: Oak Ridge, TN
Van Nostrand Reinhold: New York
VCH Publishers: New York
Viking: New York
Wiley & Sons: New York
Williams and Wilkins: Baltimore, MD
World Scientific: River Edge, NJ
Yale University Press: New Haven, CT

Pagination Information

◆ If you are citing a chapter, the complete page range is best, but initial page numbers are acceptable. Pagination may also be indicated by "f" or "ff" notation (meaning "and following" page or pages, respectively). The f or ff is set in roman type and is spaced from the preceding number. These points are illustrated under the "Pagination Field" heading for periodicals.

◆ Pagination information is set in roman type and ends with a period, except when miscellaneous information follows it, in which case it should end with a semicolon (see the next section). Use the abbreviations "p" and "pp" to indicate single and multiple pages, respectively.

p 57
p 93 f
pp 48–51
pp 30, 52, 76
pp 30, 52, 76 ff
pp 30, 52, and 76
pp 562–569
pp 562–9 (acceptable in journals)

1995; Vol. 2, p 35. 1996; pp 55–61.

◆ If the book as a whole is the reference, page numbers need not be given.

Miscellaneous Information

If you wish to include additional information about a book that is important for the reader to know, you may add it at the end of the reference with or without parentheses, append it to the title in parentheses before the semicolon, or place it between the title and the publisher.

AOCS. *Official Methods and Recommended Practices of the American Oil Chemists' Society;* Link, W. E., Ed.; Champaign, IL, 1958 (revised 1973).
Brown, H. C. *The Nonclassical Ion Problem;* Plenum: New York, 1977; Chapter 5 (with comments by P. v. R. Schleyer).
Otsu, T.; Kinoshita, M. *Experimental Methods of Polymer Synthesis* (in Japanese); Kagakudojin: Kyoto, Japan, 1972; p 72.
Tessier, J. Structure, Synthesis and Physical–Chemical Properties of Deltamethrin. *Deltamethrin Monograph;* Roussel-Uclaf: Paris, 1982; pp 37–66; translated by B. V. d. G. Walden.
Tessier, J. Structure, Synthesis and Physical–Chemical Properties of Deltamethrin. *Deltamethrin Monograph;* Walden, B. V. d. G., Translator; Roussel-Uclaf: Paris, 1982.
Volatile Compounds in Foods and Beverages; Maarse, H., Ed.; Dekker: New York, 1991; see also references therein.

Special Situations

- *Organic Syntheses* collective volumes should be treated as books.

  Organic Syntheses; Wiley & Sons: New York, Year; Collect. Vol. No., Pagination.

Year	*Collective Volume No.*
1941	I
1943	II
1955	III
1963	IV
1973	V
1988	VI
1990	VII
1993	VIII

- For references to the *Kirk-Othmer Encyclopedia*, include the article title followed by a period, similar to the citation of a chapter title.

  Chlorocarbons and Chlorohydrocarbons. *Kirk-Othmer Encyclopedia of Chemical Technology,* 4th ed.; Wiley & Sons: New York, 1993; Vol. 5, pp 1017–1072.

Series Publications

Publications such as book series that are periodical in nature, but are not journals, may be styled as either journals or books. CASSI lists every document abstracted and indexed by the Chemical Abstracts Service; hence, book titles are included and abbreviated. Key words to look for with these types of publications include "Advances", "Methods", "Progress", and "Series".

Use the regular citation format for a book reference, but include information pertaining to the series. That information follows the listing of editors and is set in roman type.

Recommended Format for Citation as a Book

Author 1; Author 2; Author 3; etc. In *Title;* Editor 1, Editor 2, Eds.; Series Title and Number; Publisher: Place of Publication, Year; Pagination.

Recommended Format for Citation as a Journal

Author 1; Author 2; Author 3; etc. *Abbreviation* **Year,** *Volume,* Pagination.

ACS Book Series

ACS book series are the ACS Monographs; ACS Symposium Series; Advances in Chemistry Series; Conference Proceedings Series; History of

Modern Chemical Sciences; and Profiles, Pathways, and Dreams. Some have the word "series" as part of the official title; some do not. Use the regular citation format for a book reference.

> *Aquatic Chemistry: Interfacial and Interspecies Processes;* Huang, C. P., O'Melia, C. R., Morgan, J. J., Eds.; Advances in Chemistry Series 244; American Chemical Society: Washington, DC, 1995.
>
> *Designing Safer Chemicals: Green Chemistry for Pollution Prevention;* DeVito, S. C., Garrett, R. L., Eds.; ACS Symposium Series 640; American Chemical Society: Washington, DC, 1996.
>
> Hudlický, M. *Oxidations in Organic Chemistry;* ACS Monograph 186; American Chemical Society: Washington, DC, 1990.

As for any book, you may cite specific chapters.

> Buttery, R. G.; Ling, L. C. Methods for Isolating Food and Plant Volatiles. In *Biotechnology for Improved Foods and Flavors;* Takeoka, G. R., Teranishi, R., Williams, P. J., Kobayashi, A., Eds.; ACS Symposium Series 637; American Chemical Society: Washington, DC, 1996; pp 240–248.

Non-ACS Serial Publications

◆ Either book or periodical reference format is acceptable. In book format, the series title is spelled out and is set in roman type.

> Kebarle, P. In *Techniques for the Study of Ion–Molecule Reactions;* Saunders, W., Farrar, J. M., Eds.; Techniques of Chemistry Series 20; Wiley & Sons: New York, 1988; p 125.
>
> *Radiationless Processes;* DiBartolo, D., Ed.; NATO Advanced Study Institute Series B62; Plenum: New York, 1979.
>
> Staerk, H.; Treichel, R.; Weller, A. In *Biophysical Effects of Steady Magnetic Fields;* Maret, G., Kiepenheuer, J., Boccara, N., Eds.; Springer Proceedings in Physics 11; Springer-Verlag: Berlin, 1986; p 85.

◆ In journal format, the series title is used as a journal title, abbreviated according to CASSI, and italicized, and the series number is used as a journal volume number.

> Kebarle, P. *Tech. Chem. (N.Y.)* **1988,** *20,* 125.
>
> *NATO Adv. Study Inst. Ser., Ser. B* **1979,** *62.*
>
> Staerk, H.; Treichel, R.; Weller, A. *Springer Proc. Phys.* **1986,** *11,* 85.

◆ If a volume or part number is given for a series of books instead of a series number, cite this information where a series number would normally appear because it refers to a volume or part of the entire series, not a volume or part of the specific book.

> Wiberg, K. In *Investigations of Rates and Mechanisms of Reactions;* Lewis, E. S., Ed.; Techniques of Chemistry, Vol. VI, Part I; Wiley & Sons: New York, 1974; p 764.

Works Written by an Organization or a Committee

An organization or a committee may be the author of a book or periodical article. Acronyms for very well known organizations may be used. It is not necessary to repeat the publisher's name for a work compiled by the organization that published it.

Book Format

American Society of Brewing Chemists. *Methods of Analysis,* 8th ed.; St. Paul, MN, 1992.

National Academy of Sciences, Committee on Science, Engineering, and Public Policy. *Report of the Research Briefing Panel on Biological Control in Managed Ecosystems;* National Academy Press: Washington, DC, 1987.

Regulatory Considerations: Genetically Engineered Plants; Boyce Thompson Institute for Plant Research: Ithaca, NY, 1988.

World Health Organization. *Guidelines for Drinking Water Quality,* 2nd ed.; Recommendations, 176; Geneva, Switzerland, 1993; Vol. 1.

Periodical Format

ASTM D 2892. *Annu. Book ASTM Stand.* **1978,** Part 24, Annex A1.

International Union of Pure and Applied Chemistry, Physical Chemistry Division, Commission on Molecular Structure and Spectroscopy. Presentation of Molecular Parameter Values for Infrared and Raman Intensity Measurements. *Pure Appl. Chem.* **1988,** *60,* 1385–1388.

IUPAC. Molecular Absorption Spectroscopy, Ultraviolet and Visible (UV/VIS). *Pure Appl. Chem.* **1988,** *60,* 1449–1460.

Meetings and Conferences

References to work presented at conferences and meetings must be treated on a case-by-case basis. At least three types of citations are possible:

1. Full citations of published abstracts and proceedings. The format resembles that of a book citation.

2. CASSI citations of published abstracts and proceedings. The format is that of a periodical citation.

3. References to oral presentations, posters, or demonstrations at technical meetings, possibly accompanied by handouts or brochures. These references contain no publication information.

Published Materials

Full Citations

Recommended Format for Full Citations

Author 1; Author 2; Author 3; etc. Title of Presentation. In *Title of the Collected Work,* Proceedings of the Name of the Meeting, Location of Meeting, Date of Meeting; Editor 1, Editor 2, etc., Eds.; Publisher: Place of Publication, Year; Abstract Number, Pagination.

The format resembles that of a book citation. The title field, however, includes additional information on the meeting title, location, and dates. The actual title of the book (collected work) is set in italic type and is separated from the meeting information by a comma. The information on meeting location is set in roman type, but it is not repeated if it is included in the book title. The entire field ends with a semicolon.

Garrone, E.; Ugliengo, P. In *Structure and Reactivity of Surfaces,* Proceedings of the European Conference, Trieste, Italy, Sept 13–20, 1988; Zecchina, A., Costa, G., Morterra, C., Eds.; Elsevier: Amsterdam, 1988.

Siegbahn, P. E. M. In *Current Aspects of Quantum Chemistry,* Proceedings of the International Congress, Barcelona, Spain, 1981; Carbo, R., Ed.; Elsevier: Amsterdam, 1981.

Uvdal, K.; Hasan, M. A.; Nilsson, J. O.; Salaneck, W. R.; Lundstrom, I.; MacDiarmid, A. G.; Ray, A.; Angelopoulos, M. In *Proceedings of Winter School on Conducting Polymers;* Kuzmany, H., Mehring, M., Roth, S., Eds.; Springer-Verlag: Berlin, 1987.

Abstracts are slightly different in that they usually do not have editors. The word "in" is not used before the book title.

Goolsby, D. A.; Battaglin, W. A.; Fallon, J. D. *Abstracts of the Technical Meeting,* U.S. Geological Survey Toxic Substances Hydrology Program, Sept 1993; U.S. Geological Survey: Reston, VA, 1994; p 83.

Prasad, A.; Jackson, P. *Abstracts of Papers, Part 2,* 212th National Meeting of the American Chemical Society, Orlando, FL, Aug 25–29, 1996; American Chemical Society: Washington, DC, 1996; PMSE 189.

When the phrase "Proceedings of" is part of the reference, include the publisher and place of publication. When a society sponsors a meeting, the society is assumed to be the publisher. If the place of the meeting and the place of publication are the same, additional publisher and place information is not required. However, many organizations such as the ACS sponsor meetings in various cities.

Harwood, J. S. Direct Detection of Volatile Metabolites Produced by Microorganisms. *Proceedings of the 36th ASMS Conference on Mass Spectrometry and Allied Topics,* San Francisco, CA, June 5–10, 1988.

CASSI Citations

Proceedings and abstracts of meetings and conferences are indexed in CASSI. The reference format follows that for periodicals.

Abstr. Pap.—Am. Chem. Soc. **1989**, *198*.
Proc.—IEEE Symp. Fusion Eng. **1989,** *13*.

CASSI gives the number of a meeting in ordinal form. Convert this number to an italic cardinal number and use it as a volume number in the citation, unless CASSI has already indicated another volume number.

Journal format can be used for references to preprint papers.

Jones, J.; Oferdahl, K. *Natl. Meet.—Am. Chem. Soc., Div. Environ. Chem.* **1989,** *29* (2), ENVR 22 (or Paper 22).

Material That Has No Publication Information

Recommended Format

Author 1; Author 2; Author 3; etc. Title of Presentation (if any). Presented at Conference Title, Place, Date; Paper Number.

List the data concerning the conference (name, place, and date) separated by commas and followed by a semicolon and the paper number (if any). The entire citation is set in roman type.

Castro, M. E.; Russell, D. H. Presented at the 32nd Annual Conference on Mass Spectrometry and Allied Topics, San Antonio, TX, 1984.

Ford, W. T. Presented at the 189th National Meeting of the American Chemical Society, Miami, FL, April 1985; Paper ORGN 79.

Killday, K. B.; Tempesta, M. S.; Bailey, M. E. Isolation and Characterization of Heme Pigments from Cured Meat. Presented at the 20th Midwest Regional Meeting of the American Chemical Society, Carbondale, IL, 1985.

Melchior, M. T. Presented at the 22nd Experimental NMR Conference, Asilomar, CA, 1981; Poster B29.

Nokihara, K.; Ando, E.; Forssmann, W. G. Presented at the 17th FEBS Meeting, Berlin, 1986.

Roos, I. A. G.; Stokes, K. H. Presented at the Third International Conference on the Chemistry of the Platinum Group Metals, Sheffield, U.K., July 1987; Paper O-36.

Rose, J. J. Presented at the Pittsburgh Conference, Atlantic City, NJ, March 1983; Paper 707.

Wilkins, C. L. Presented at the Pacific Conference on Chemistry and Spectroscopy, Pasadena, CA, Oct 1983.

Theses

Recommended Format

Author. Title of Thesis. Level of Thesis, Degree-Granting University, Location of University, Date of Completion.

References to theses should be as specific as practical, including, at a minimum, the degree-granting institution and date.

Cotruvo, J. Kinetic Model for Chlorophyll Degradation. Ph.D. Thesis, Massachusetts Institute of Technology, Cambridge, MA, June 1996.
Antigone, T. M.S. Thesis, Princeton University, 1997.

Author Name Field

Cite the name in inverted form: last name first, then first initial or name, middle initial or name, and qualifiers (Jr., II). End the field with a period.

Title Field

Thesis titles are not essential, but they are informative. They are set in roman type and end with a period.

Chandrakanth, J. S. Effects of Ozone on the Colloidal Stability of Particles Coated with Natural Organic Matter. Ph.D. Dissertation, University of Colorado, Boulder, CO, 1994.
Fischetti, W. Palladium-catalyzed syntheses of polyenes. M.S. Thesis, University of Delaware, Newark, DE, Dec 1982.

Thesis Level Field

Work done at a master's level is often called a thesis. Work toward the Ph.D. (Doctor of Philosophy) may be called a thesis or a dissertation, depending on the policy of the degree-granting institution. The following abbreviations are standard for U.S. degrees. Many variations exist for degrees from institutions of other countries.

A.B., B.A., B.S.
A.M., M.A., M.S., M.B.A.
Ph.D., M.D.

Bérard, Paula. M.S. Thesis, University of Massachusetts, Amherst, MA, 1966.
Vergoten, G. Doctoral Dissertation, University of Lille, Lille, France, 1977.

University Name and Location Field

The name of the degree-granting university is the minimum requirement for an acceptable citation. You may also include the city and state or city and country location. Use the two-letter postal abbreviations for states. Spell out names of countries unless they have standard abbreviations, such as U.K. for United Kingdom.

Brines, J. L. Ph.D. Thesis, Rockefeller University, New York, 1989.

Pollter, C. Isolierung und Identifizierung von Reaktionsprodukten aus gebundenen Aromavorstufen in *Majorana hortensis* Moench. Diploma Thesis, Technical University, Munich, Germany, 1986.

Waldo, G. S. Ph.D. Thesis, University of Michigan, Ann Arbor, MI, 1991.

Date of Completion Field

Indicate the date the thesis was completed by year only; month and year; or month, day, and year.

Fleissner, W. Ph.D. Dissertation, University of Tennessee, Chattanooga, TN, 1994.

Marshall, M. Ph.D. Thesis, University of California, San Francisco, CA, Dec 1995.

Patents

Recommended Format

Patent Owner 1; Patent Owner 2; etc. Title of Patent. Patent Number, Date.

The minimum data required for an acceptable citation are the name(s) of the patent owner(s), the patent number, and the date. Ensure that the patent stage (Patent, Patent Application, etc.) is indicated and that the pattern of the number (e.g., spaces, commas, dashes) follows that of the original patent document. If possible, include the title and the *Chemical Abstracts* reference (preceded by a semicolon) as well.

Bernson, S. W. Conversion of Methane. U.S. Patent 4,199,533, April 22, 1980.

Fritzberg, A. R. Metal Radionuclide Labeled Proteins for Diagnosis and Therapy. Eur. Pat. Appl. 188256, Jan 13, 1986.

Gordon, A. Z.; Rossof, A. H. Methods for Treating Leukopenia with Tertiary Amines or Quaternary Ammonium Salts. U.S. Patent 5,466,509, 1987.

Jordan, O. D. Br. Patent 2,081,298, 1982.

Lyle, F. R. U.S. Patent 6,973,257, 1995; *Chem. Abstr.* **1995,** *123,* 2870.
Matsumura, S.; Inadda, H.; Hara, S. (Fuji Film Co.). Jpn. Kokai Tokkyo Koho 87 81,417, 1987.
Merck & Co. Substituted Benzenesulfonamide. German Patent DE 2556122, June 24, 1976.
Norman, L. O. U.S. Patent 4,379,752, 1983.
Tolman, G. L. Trace-Labeled Conjugates of Metallothionein and Target-Seeking Biologically Active Molecules. U.S. Patent 4,732,864, 1988.

Government Publications

Publications of the U.S. government and those of state and local governments can be pamphlets, brochures, books, maps, journals, or almost anything else that can be printed. They may have authors or editors, who may be individuals, offices, or committees, or the author may not be identified. They are published by specific agencies, but they are usually (though not always) available through the Government Printing Office rather than the issuing agency. To enable others to find the publication, the American Library Association suggests that you include as much information as possible in the citation. The following are examples of the most commonly cited types of references.

Publications of Federal Government Agencies

Recommended Format

Author 1; Author 2; etc. Chapter Title. *Document Title;* Government Publication Number; Publishing Agency: Place of Publication, Year; Pagination.

The format resembles that of a serial publication in book format. Include as much information as possible.

The Biologic and Economic Assessment of Maleic Hydrazide; Technical Bulletin No. 1634; U.S. Department of Agriculture, U.S. Government Printing Office: Washington, DC, 1979.
Chemistry Quality Assurance Handbook; Food Safety Inspection Service, U.S. Department of Agriculture, U.S. Government Printing Office: Washington, DC, 1985; Vol. II, Part 5.
Diet and Dental Health: A Study of Relationships; DHHS Publication No. PHS 82-1675; U.S. National Center for Health Statistics, U.S. Government Printing Office: Washington, DC, 1982.
Energy Alternatives and the Environment: 1979; EPA-600/9-80-009; U.S. Environmental Protection Agency, Office of Research and Development,

Office of Environmental Engineering and Technology, U.S. Government Printing Office: Washington, DC, 1979.

Ethylene Dibromide; Position Document 4; U.S. Environmental Protection Agency, Office of Pesticide Programs, U.S. Government Printing Office: Washington, DC, 1983.

Haller, M. H. *Handling, Transportation, Storage and Marketing of Peaches;* USDA Bibliographical Bulletin No. 21; U.S. Department of Agriculture, U.S. Government Printing Office: Washington, DC, 1952.

Haytowitz, D. B.; Matthews, R. H. Vegetables and Vegetable Products—Raw, Processed, Prepared. *Composition of Foods;* USDA Agriculture Handbook No. 8-11; U.S. Department of Agriculture, U.S. Government Printing Office: Washington, DC, revised 1984.

Heller, S. R.; Milne, G. W. A. *EPA/NIH Mass Spectral Data Base;* U.S. Government Printing Office: Washington, DC, 1978.

Interdepartmental Task Force on PCBs. *PCBs and the Environment;* COM 72.10419; U.S. Government Printing Office: Washington, DC, 1972.

National Handbook of Recommended Methods for Water Data Acquisition; Office of Water Data Coordination, U.S. Geological Survey: Reston, VA, 1977; Chapter 5.

Osteen, C. D.; Szmedra, P. I. *Agricultural Pesticide Use: Trends and Policy Issues;* Agricultural Economic Report No. 622; U.S. Department of Agriculture, Economic Research Service, U.S. Government Printing Office: Washington, DC, 1989.

Pesticides Analytical Manual; U.S. Department of Health, Education, and Welfare, Food and Drug Administration, U.S. Government Printing Office: Washington, DC, 1982; Vol. 1.

Reactor Safety Study: An Assessment of Accident Risks in U.S. Commercial Power Plants; NUREG 75/014; Nuclear Regulatory Commission: Washington, DC, 1975.

Sherma, J.; Beroza, M. *Manual of Analytical Quality Control for Pesticides and Related Compounds;* EPA-600/1-79/008; U.S. Environmental Protection Agency, U.S. Government Printing Office: Washington, DC, 1979.

Simpson, R. Analysis of Chloramphenicol in Tissue. *USDA Chemistry Laboratory Guidebook;* U.S. Government Printing Office: Washington, DC, 1985; Part 5-14, Section 5.022.

Author Name Field

Include all author names. With multiple authors, separate the names from one another by semicolons. Always end the author field with a period. List the names in inverted form: last name first, then first initial or name, middle initial or name, and qualifiers (Jr., II). Some publications list the first 10 authors followed by a semicolon and et al.

Chapter Title Field

Chapter titles are set in roman type and end with a period.

Book Title Field

Treat the formal title of the document as the title of a book. These titles are set in italic type and are separated from the next component of the reference by a semicolon.

Government Publication Number Field

The government publication number, also called an agency report number, is very important because it is unique to the publication and because some indexing services provide access by these numbers. These numbers (or number–letter combinations) are usually printed somewhere on the cover or title page of the document and are sometimes identified as a "report/ accession number". Treat a report number the same as a series number; that is, it follows the book title, ends with a semicolon, and is set in roman type.

Publishing Agency Field

The publishing agency field may take on added complexity in government publications. Often, the office or agency issuing the report as well as the Government Printing Office must be cited. The order is department or agency, administration or office, and finally U.S. Government Printing Office, all separated by commas and set in roman type. The field ends with a colon.

Place of Publication Field

For the U.S. Government Printing Office, it is always Washington, DC. The field ends with a comma preceding the date of publication.

Year of Publication Field

The year of publication is set in roman type and ends with a semicolon if it is followed by pagination information. It ends with a period if it is the last field.

Pagination Field

The page numbers are set in roman type and end with a period, unless miscellaneous material is appended to the reference.

Alternative Format

Government agency references can also be given with CASSI abbreviations. In that case, the format is the same as for periodicals.

> Thompson, C. R.; Van Atta, G. R.; Bickoff, E. M.; Walter, E. D.; Livingston, A. L.; Guggloz, J. *Tech. Bull.—U.S. Dep. Agric.* **1957,** *No. 1161,* 63–70.

Federal Register

The *Federal Register* is a periodical and is treated as such in citations.

Environmental Pollutants. *Fed. Regist.* **1987,** *52* (74), 12866–12874.
Use of Lasers. *Fed. Regist.* **1995,** *60* (46), 9745–9821.

Code of Federal Regulations

Cyromazine Proposed Tolerance. *Code of Federal Regulations,* Part 562, Title 21, 1985; *Fed. Regist.* **1984,** *49* (83), 18120.
FIFRA Good Laboratory Practice Standards. *Code of Federal Regulations,* Section 160, Title 40, 1982; *Fed. Regist.* **1983,** *48,* 53946 ff.
Tolerances for Pesticide Chemicals in or on Raw Agricultural Commodities. *Code of Federal Regulations,* Part 180.414, Title 40, 1985.

U.S. Code

Patent Claims Issued Improperly. *U.S. Code,* Section 101, Title 35, 1996.

U.S. Laws

Treat the name of the law as a chapter title (roman; terminated with a period). No publisher name is needed. The number and date of the law are separated by a comma. If additional publication information is given, it is preceded by a semicolon.

Federal Insecticide, Fungicide, and Rodenticide Act. Public Law 92-516, 1972; *Code of Federal Regulations,* Section 136, Title 7, 1990.
Federal Food, Drug, and Cosmetic Act. Public Law 95-532, 1938.
Patent Law Improvements Act of 1984. Public Law 98-6222 (H.R. 6268), 1984.
Toxic Substances Control Act. Public Law 94-469, 1976.

State and Local Government Publications

Recommended Format

Author 1; Author 2; etc. Chapter Title. *Document Title;* Publication Number or Type; Publishing Agency: Place of Publication, Date; Pagination.

Air Quality Aspects of the Development of Offshore Oil and Gas Resources; California Air Resources Board: Sacramento, CA, 1982.
Tofflemire, T. J.; Quinn, S. O. *PCBs in the Upper Hudson River;* Technical Report No. 56; New York State Department of Environmental Conservation: Albany, NY, 1977.

Turner, B.; Powell, S.; Miller, N.; Melvin, J. *A Field Study of Fog and Dry Deposition as Sources of Inadvertent Pesticide Residues on Row Crops;* Report of the Environmental Hazard Assessment Program; California Department of Food and Agriculture: Sacramento, CA, Nov 1989.

Technical Reports and Bulletins

Technical reports and bulletins come in many forms. Examples of some of these have already been presented. Many are in-house publications, and some are government publications. Others are reports of work in progress. The publication itself may include a phrase alluding to its status as a technical report or technical bulletin, but it may also simply be called a report or bulletin. Include whatever information is available, following the format shown for the word "Report", "Report No.", etc. Document titles are set in italic type.

Recommended Format

Author 1; Author 2; etc. *Title of Report or Bulletin;* Technical Report or Bulletin Number; Publisher: Place of Publication, Date; Pagination.

Halbleib, J. A., Sr.; Vandevender, W. H. *TIGER: A One-Dimensional Multilayer Electron–Photon Monte Carlo Transport Code;* Report SLA-73-1026; Sandia National Laboratories: Albuquerque, NM, March 1974.

Morgan, M. G. *Technological Uncertainty in Policy Analysis;* Final Report to the National Science Foundation on Grant PRA-7913070; Carnegie Mellon University: Pittsburgh, PA, 1982.

Reuzel, P. G. J. *Chronic (29-Month) Inhalation Toxicity and Carcinogenicity Study of Methyl Bromide in Rats;* Report No. V86.469/221044; Netherlands Organization for Applied Scientific Research, Division for Nutrition and Food Research TNO: Zeist, The Netherlands, 1987.

Schneider, A. B. *Expert Systems in Analytical Chemistry;* Technical Report No. 1234-56; ABC Co.: New York, 1985.

World Health Organization. *Energy and Protein Requirements;* WHO Technical Report Series No. 522; Geneva, Switzerland, 1973; p 55.

Wintersteen, W.; Hartzler, R. *Pesticides Used in Iowa Crop Production in 1985;* Iowa State Cooperative Extension Service Bulletin Pa-1288; Iowa State University: Ames, IA, Jan 1987.

Unpublished Materials

Material in any stage preceding actual publication falls under this general classification, as do personal communications and work not destined for publication.

Recommended Format for Material Intended for Publication

Author 1; Author 2; etc. Title of Unpublished Work. *Journal Abbreviation,* phrase indicating stage of publication.

Various phrases indicating the stage of publication are acceptable in these references.

- For material accepted for publication, use the phrase “in press”.

  Tang, D.; Jankowiak, R.; Small, G. J.; Tiede, D. M. *Chem. Phys.,* in press.

- For material intended for publication but not yet accepted, use “unpublished work”, “submitted for publication”, or “to be submitted for publication”.

  Chatterjee, K.; Visconti, A.; Mirocha, C. J. Deepoxy T-2 Tetrol: A Metabolite of TO2 Toxin Found in Cow Urine. *J. Agric. Food Chem.,* submitted for publication, 1996.

  Nokinara, K. Duke University, Durham, NC. Unpublished work, 1996.

Recommended Format for Material Not Intended for Publication

Author. Affiliation, City, State. Phrase describing the material, Year.

Henscher, L. X. University of Minnesota, Minneapolis, MN. Personal communication, 1994.

Heltman, L. R. DuPont. Private communication, 1995.

Wagner, R. L. University of Utah, Salt Lake City, UT. Unpublished work, 1995.

Messages sent by electronic mail are considered personal communications and are referred to as such.

Electronic Sources

Computer Programs

References to computer programs must be treated on a case-by-case basis. Five common presentations of computer programs are possible:

1. book format, with the name of the program as the title
2. technical report format
3. CASSI format
4. free style, as a simple listing of program title and author of program
5. thesis style

Book Format

Recommended Format

Author 1; Author 2; etc. *Program Title,* version or edition; Publisher: Place of Publication, Year.

The recommended format is the same as that for a book citation, except that there are no chapters or pages. The name of the computer program, with any descriptors, is considered the title and is set in italic type. If you wish to include additional information about a program that is important for the reader to know, you may add it at the end of the reference with or without parentheses or append it to the title in parentheses before the semicolon.

Binkley, J. S. *GAUSSIAN82;* Department of Chemistry, Carnegie Mellon University: Pittsburgh, PA, 1982.

Main, P. *MULTAN 80: A System of Computer Programs for the Automated Solution of Crystal Structures from X-ray Diffraction Data;* Universities of York and Louvain: York, England, and Louvain, Belgium, 1980.

Motherwell, W. D. S.; Clegg, W. *PLUTO: Program for Plotting Molecular and Crystal Structures;* Cambridge University Press: Cambridge, England, 1978.

Sheldrick, G. M. *SHELX-86: Program for Crystal Structure Solution;* University of Göttingen: Göttingen, Germany, 1986.

Recommended Format for Commercial Software

Program Title, version or edition; comments; Publisher: Place of Publication, Year.

BCI Clustering Package, versions 2.5 and 3.0; Barnard Chemical Information: Sheffield, U.K., 1995.

Mathematica; software for technical computation; Wolfram Research: Champaign, IL, 1996.

Unity Chemical Information Software, version 2.3; Tripos Associates: St. Louis, MO, 1995.

Technical Report Format

Recommended Format

Author. *Title of Report;* Technical Report Number; Publisher: Place of Publication, Year; Pagination (if any).

In a citation to a computer program as a technical report, a report or technical report number is included. As with book format, the name of the computer program is considered the title of the technical report.

Beurskens, P. T.; Bossman, W. P.; Doesburg, H. M.; Gould, R. O.; van der Hark, Th. E. M.; Prick, P. A. J. *DIRDIF: Direct Methods for Difference Structures;* Technical Report 1980/1; Crystallographic Laboratory: Toernooiveld, The Netherlands, 1980.

Johnson, C. K. *ORTEP-II: A Fortran Thermal Ellipsoid Plot Program for Crystal Structure Illustrations;* Report ORNL-5138; National Technical Information Service, U.S. Department of Commerce: Springfield, VA, 1976.

Johnson, C. K. *ORTEP;* Report ORNL-3794; Oak Ridge National Laboratory: Oak Ridge, TN, 1965.

Papelis, C.; Hayes, K. F.; Leckie, J. O. *HYDRAQL: A Program for the Computation of Chemical Equilibrium Composition of Aqueous Batch Systems;* Technical Report No. 306; Department of Civil Engineering, Stanford University: Menlo Park, CA, 1988.

Stewart, J. M.; Machin, P. A.; Dickinson, C. W.; Ammon, H. L.; Heck, H.; Flack, H. *The X-ray 76 System;* Technical Report TR-446; Computer Science Center, University of Maryland: College Park, MD, 1976.

CASSI Format

Because of the broad base from which *Chemical Abstracts* indexes work, computer programs, in the form of technical reports, may be referenced. In such cases, CASSI format would be appropriate.

Johnson, C. K. *Oak Ridge Natl. Lab., [Rep.] ORNL (U.S.)* **1978,** *ORNL-5348.*

Free Style

When only minimal information (e.g., author and program name) is available, present the information as simply as possible.

Programs used in this study included local modifications of Main, Hull, Lessinger, Declerq, and Woolfson's MULTAN78, Jacobson's ALLS, Zalkins's FORDAP, Busing and Levy's ORFEE, and Johnson's ORTEP2.

Lozos, G.; Hoffman, B.; Franz, C. SIMI4A, Chemistry Department, Northwestern University.

SIMI4A: G. Lozos, B. Hoffman, C. Franz, Chemistry Department, Northwestern University, Evanston, IL.

Principal programs used were as follows: REDUCE and UNIQUE, data reduction programs by M. E. Lewonowicz, Cornell University, 1978; BLS78A, an anisotropic block-diagonal least-squares refinement written by K. Hirotsu and E. Arnold, Cornell University, 1980; and PLIPLOT, by G. VanDuyne, Cornell University, 1984.

Thesis Style

Sheldrick, G. M. SHELX-76: Program for Crystal Structure Determination. Cambridge University, 1976.

CD-ROMs

More and more information is being published in CD-ROM (compact disc read-only memory) format in addition to or instead of the traditional printed periodicals and books. The reference style follows that for periodicals and books, as appropriate, and the designation "CD-ROM" is included.

Recommended Format for CD-ROM Periodicals

Author 1; Author 2; Author 3; etc. Title of Article. *Journal Abbreviation* [CD-ROM] **Year,** *Volume,* Inclusive Pagination or other identifying information.

Fleming, S. A.; Jensen, A. W. Substituent Effects on the Photocleavage of Benzyl–Sulfur Bonds. Observation of the "*Meta*" Effect. *J. Org. Chem.* [CD-ROM] **1996,** *61,* 7044.

Recommended Formats for CD-ROM Books

Author 1; Author 2; etc. Chapter Title. In *Book Title,* Edition Number [CD-ROM]; Editor 1, Editor 2, etc., Eds.; Publisher: Place of Publication, Year; Volume Number.

Author 1; Author 2; etc. Chapter Title. *Book Title,* Edition Number [CD-ROM]; Publisher: Place of Publication, Year; Volume Number.

Many books in CD-ROM format are reference works, so they have no authors, editors, or chapter titles.

ACS Directories on Disc, Version 3.0 [CD-ROM]; American Chemical Society: Washington, DC, 1996.

Available Chemicals Directory [CD-ROM]; MDL Information Systems: San Leandro, CA, 1996.

Index Chemicus [CD-ROM]; Institute for Scientific Information: Philadelphia, PA, 1996.

The Merck Index, 12th ed. [CD-ROM]; Chapman & Hall: New York, 1996.

Recommended Format for Conference Proceedings on CD-ROM

Author 1; Author 2; etc. Title of Presentation. *Title of Conference* [CD-ROM]; Publisher: Place of Publication, Year; other identifying information.

Maslen, P.; Faeder, J.; Parson, R. Electronic Structure of Solvated Molecular Anions. *1st Electronic Computational Chemistry Conference* [CD-ROM]; ARInternet Corp.: Landover, MD, 1995; Paper 26.

Internet Sources

The Internet includes World Wide Web (WWW), file transfer protocol (FTP), Telnet, and Gopher sites. These sites can contain almost any kind of information, including databases. Each site has its own electronic address, called a uniform resource locator (URL). Some sites are accessible by anyone, but others are accessible only by subscription through various providers.

Recommended Format

Author (if any). Title of Site. URL (accessed date), other identifying information.

For a subscription-only site, the URL need not be given.

ACS Publications Division Home Page. http://pubs.acs.org (accessed Feb 1997).

ACSWeb. http://www.acs.org (accessed Nov 1996).

Agency for Toxic Substances and Disease Registry, Hazardous Substance Release/Health Effects Database (HazDat). http://atsdr1.atsdr.cdc.gov:8080/atsdrhome.html (accessed May 1996).

Beilstein Online; Beilstein Institute: Frankfurt, Germany (accessed March 1996).

Cambridge Structural Database; Cambridge University: Cambridge, England (accessed Aug 1996).

CAS Home Page. http://cas.org (accessed Oct 1996).

ChemCenter Home Page. http://www.chemcenter.org (accessed Dec 1996).

Computational Chemistry List Archive Gopher Site. infomeister.osc.edu (accessed Nov 1996).

International Union of Pure and Applied Chemistry Home Page. http://alpha.qmw.ac.uk/~ugca00/iupac.html/ (accessed Aug 1996).

Library of Congress Home Page. http://lcweb.loc.gov (accessed Dec 1996).

National Center for Atmospheric Research, University Corporation for Atmospheric Research Programs. http://www.ucar.edu (accessed Dec 1996).

Northern Illinois University Chemistry Gopher Site. hackberry.chem.niu.edu (accessed Nov 1996).

Periodic Table of the Elements. gopher://gopher.tc.umn.edu (accessed Feb 1996), path: Libraries/Reference Works.

STN Beilstein database, Feb 11, 1995.

U.S. Environmental Protection Agency Home Page. http://www.epa.gov (accessed July 1996).

U.S. Geological Survey, Federal Geographic Data Committee, National Geospatial Data Clearinghouse. http://fgdc.er.usgs.gov (accessed June 1996).

Use the title found on the electronic site itself; add the words "Home Page", "Gopher Site", or "database" for clarification when needed. For example, ACSWeb does not need further clarification, but CAS, Library of Congress, and Northern Illinois University do. The date of access is important because the Internet is constantly being changed and updated.

If the URL does not fit on one line, it can be broken, but only according to the following guidelines:

- Break after an ampersand, a slash, or a period, but keep two slashes together.
- Do not add a hyphen to the end of the line.

- Do not break after a hyphen to avoid confusion as to the hyphen's purpose.

Online Periodicals

Online journals (e-journals) and magazines (e-zines or webzines) are proliferating.

Recommended Format

Author 1; Author 2; Author 3; etc. Title of Article. *Journal Abbreviation* [Online] **Year,** *Volume,* Inclusive Pagination or other identifying information.

Craw, J. S.; Hinchliffe, A. *Electron. J. Theor. Chem.* [Online] **1995,** *1,* 8–10.
Tunon, I.; Martins-Costa, M. T. C.; Millot, C.; Ruiz-Lopez, M. F. *J. Mol. Model.* [Online] **1995,** *1,* 196–201.

Electronic Lists and Newsgroups

Recommended Format

Mailing List or Newsgroup Name, other information, electronic address.

Chemical Information List Server, comments in archived messages of July 1995, CHMINF-L@iubvm.ucs.indiana.edu.
CHEM-MOD, discussion of the modeling aspects of computational chemistry, mailbase@mailbase.ac.uk.
Hyperchem Users Group, hyperchem@hyper.com.
ORGCHEM, for exchange of information related to organic chemistry, orgchem@extreme.chem.rpi.edu; also FTP server at extreme.chem.rpi.edu.

Electronic Mail Messages

Whether the message was personal and sent only to you, or whether it was posted in a newsgroup, it is not, in fact, published. Cite these the same as any other personal communication. Include the year and the professional affiliation of the author.

Collating References

Collate all references at the end of the manuscript in numerical order if cited by number and in alphabetical order if cited by author. Do not include items in the reference list that are not cited in the manuscript. Check the publication for which you are writing. Some publications do not allow multiple references to be listed as one numbered entry; they prefer that each numbered entry include only one unique reference.

To collate references according to the author–date style, use the following format.

1. Alphabetize in order of the first authors' surnames.
2. When the same first author is common to multiple references,
   - Group the single-author references first. List them chronologically. To distinguish among references having the same year, add a lowercase letter (a, b, c, etc.) to the year.
   - Group the two-author references next. List them chronologically. To distinguish among references having the same year, add a lowercase letter (a, b, c, etc.) to the year.
   - Group all multiple-author (three or more) references last. List them chronologically. To distinguish among references having the same year, add a lowercase letter (a, b, c, etc.) to the year.

Hamilton, F. J. *Biochemistry* **1995,** *34,* 78–86.
Hamilton, F. J. *J. Agric. Food Chem.* **1996a,** *44,* 1622–1633.
Hamilton, F. J. *J. Org. Chem.* **1996b,** *61,* 298–306.
Hamilton, F. J.; Salvo, P. A. *J. Agric. Food Chem.* **1997,** *45,* 918–924.
Hurd, R. *J. Magn. Reson.* **1990,** *87,* 422.
Mills, M. S.; Thurman, E. M. *Anal. Chem.* **1992,** *64,* 1985–1990.
O'Connor, D. J. *Environ. Eng. ASCE* **1988,** *114,* 507–522.
Rahwan, R. G., Witiak, D. T., Eds. *Calcium Regulation by Calcium Antagonists;* ACS Symposium Series 201; American Chemical Society: Washington, DC, 1982.
Scarponi, T. M.; Moreno, S. P. *Biochemistry* **1994,** *33,* 345–360.
Scarponi, T. M.; Adams, J. S. *J. Pharm. Sci.* **1996,** *85,* 703–712.
Serpone, N.; Pellizetti, E. *Photocatalysis: Fundamentals and Applications;* Wiley & Sons: New York, 1989.
Tewey, L. P.; Rodriguez, R. E.; Jennes, A. C. *J. Agric. Food Chem.* **1994,** *42,* 1879–1886.
Tewey, L. P.; Rodriguez, R. E.; Fortunato, B. D.; Jennes, A. C. *Ind. Eng. Chem. Res.* **1995a,** *34,* 465–472.
Tewey, L. P.; Hiroshi, C. Y.; Allen, P. R.; Lowe, D. L. *Biochemistry* **1995b,** *34,* 11689–11699.
Tewey, L. P.; Rolland, H. J.; Harwood, C. C. *J. Org. Chem.* **1995c,** *60,* 3548–3556.
Tewey, L. P.; Allen, P. R.; Levy, M. S. *J. Am. Chem. Soc.* **1996,** *118,* 2520.

Do not use the Latin terms ibid. (in the same place) or idem (the same) because the actual reference source cannot be searched on electronic databases.

Supplement: 1000+ Journals Most Commonly Cited

Acc. Chem. Res.
Acta Chem. Scand.
Acta Cienc. Indica, Chem.
Acta Crystallogr.
Acta Crystallogr., Sect. A
Acta Crystallogr., Sect. B
Acta Crystallogr., Sect. C
Acta Crystallogr., Sect. D
Acta Endocrinol.
Acta Metall. Mater.
Acta Pharmacol. Toxicol.
Acta Physiol. Scand.
Acta Phys. Pol., A
Acta Phys. Pol., B
Acta Pol. Pharm.
Acta Polym.
Adv. Biosci. (Oxford)
Adv. Cancer Res.
Adv. Carbohydr. Chem. Biochem.
Adv. Catal.
Adv. Chem. Phys.
Adv. Chem. Ser.
Adv. Colloid Interface Sci.
Adv. Cryog. Eng.
Adv. Drug Delivery Rev.
Adv. Environ. Sci. Technol.
Adv. Enzymol. Relat. Areas Mol. Biol.
Adv. Enzymol. Relat. Subj. Biochem.
Adv. Exp. Med. Biol.
Adv. Filtr. Sep. Technol.
Adv. Heterocycl. Chem.
Adv. Inorg. Biochem.
Adv. Inorg. Chem.
Adv. Inorg. Chem. Radiochem.
Adv. Instrum. Control
Adv. Magn. Reson.
Adv. Mater. Opt. Electron.
Adv. Mater. (Weinheim, Ger.)
Adv. Organomet. Chem.
Adv. Org. Geochem.
Adv. Pharmacol.
Adv. Photochem.
Adv. Phys.
Adv. Phys. Org. Chem.
Adv. Polym. Sci.
Adv. Protein Chem.
Adv. Quantum Chem.
Adv. Space Res.
Aerosol Sci. Technol.
Agents Actions
Agric. Biol. Chem.
AIChE J.
AIChE Symp. Ser.
AIDS Res. Hum. Retroviruses
Alcohol.: Clin. Exp. Res.
Aldrichimica Acta
Am. Ceram. Soc. Bull.
Am. Concr. Inst., SP
Am. Ind. Hyg. Assoc. J.
Am. J. Clin. Nutr.
Am. J. Enol. Vitic.
Am. J. Hum. Genet.
Am. J. Pathol.
Am. J. Physiol.
Am. J. Respir. Cell Mol. Biol.
Am. J. Sci.
Am. J. Vet. Res.
Am. Lab.
Am. Mineral.
Am. Rev. Respir. Dis.
Anal. Biochem.
Anal. Chem.
Anal. Chim. Acta
Anal. Lett.
Anal. Proc.
Anal. Sci.

NOTE: Some journals of the same name are published in more than one city. Authors should check the journal name carefully and include the city to prevent misunderstanding.

Analusis
Analyst (Cambridge, U.K.)
Angew. Chem.
Angew. Chem., Int. Ed. Engl.
Angew. Makromol. Chem.
Ann. Chim.
Ann. Intern. Med.
Ann. Neurol.
Ann. N.Y. Acad. Sci.
Ann. Occup. Hyg.
Ann. Phys. (Leipzig)
Annu. Rep. Med. Chem.
Annu. Rep. NMR Spectrosc.
Annu. Rev. Biochem.
Annu. Rev. Biophys. Bioeng.
Annu. Rev. Biophys. Biomol. Struct.
Annu. Rev. Biophys. Biophys. Chem.
Annu. Rev. Cell Biol.
Annu. Rev. Immunol.
Annu. Rev. Mater. Sci.
Annu. Rev. Microbiol.
Annu. Rev. Neurosci.
Annu. Rev. Pharmacol. Toxicol.
Annu. Rev. Phys. Chem.
Annu. Rev. Physiol.
Annu. Tech. Conf.—Soc. Plast. Eng.
An. Quim.
Anti-Cancer Drug Des.
Anticancer Res.
Antimicrob. Agents Chemother.
Antiviral Chem. Chemother.
Antiviral Res.
Appl. Biochem. Biotechnol.
Appl. Catal., A
Appl. Catal., B
Appl. Environ. Microbiol.
Appl. Geochem.
Appl. Magn. Reson.
Appl. Microbiol.
Appl. Microbiol. Biotechnol.
Appl. Occup. Environ. Hyg.
Appl. Opt.
Appl. Organomet. Chem.
Appl. Phys. A
Appl. Phys. B
Appl. Phys. (Berlin)
Appl. Phys. Lett.
Appl. Radiat. Isot.
Appl. Spectrosc.
Appl. Supercond.
Appl. Surf. Sci.
Aquaculture
Aquat. Toxicol.
Arch. Biochem. Biophys.
Arch. Environ. Contam. Toxicol.
Arch. Environ. Health
Arch. Microbiol.
Arch. Pharmacol.
Arch. Pharm. (Weinheim, Ger.)
Arch. Toxicol.
Arch. Virol.
Ark. Kemi
Arzneim.-Forsch.
Asian J. Chem.
ASTM Spec. Tech. Publ.
Astron. Astrophys.
Astron. J.
Astrophys. J.
At. Energ.
Atmos. Environ.
Aust. J. Chem.
Aust. J. Soil Res.
Bandaoti Xuebao
Basic Life Sci.
Ber. Bunsen-Ges. Phys. Chem.
Ber. Dtsch. Chem. Ges.
Biochem. Biophys. Res. Commun.
Biochem. Cell Biol.
Biochemistry
Biochem. J.
Biochem. Mol. Biol. Int.
Biochem. Pharmacol.
Biochem. Soc. Trans.
Biochem. Syst. Ecol.
Biochim. Biophys. Acta
Biochimie
Bioconjugate Chem.
Biodegradation
Bioelectrochem. Bioenerg.
BioEssays
Biofizika
Biokhimiya
Biol. Chem. Hoppe-Seyler
Biol. Mass Spectrom.
Biol. Membr.
Biol. Pharm. Bull.

Biol. Reprod.
Biol. Trace Elem. Res.
Biomaterials
BioMed. Chem. Lett.
Biomed. Chromatogr.
Biomed. Environ. Mass Spectrom.
Biomed. Lett.
Biomed. Res. Trace Elem.
Bioorg. Chem.
Bioorg. Khim.
Bioorg. Med. Chem.
Bioorg. Med. Chem. Lett.
Biophys. Chem.
Biophys. J.
Biopolymers
Bioprocess Eng.
Bioresour. Technol.
Biosci. Biotechnol. Biochem.
Biosens. Bioelectron.
BioTechniques
Biotechnol. Bioeng.
Biotechnol. Lett.
Biotechnology
Biotechnol. Prog.
Biotechnol. Tech.
Blood
Brain Res.
Brain Res. Bull.
Braz. J. Phys.
Br. J. Cancer
Br. J. Clin. Pharmacol.
Br. J. Haematol.
Br. J. Nutr.
Br. J. Pharmacol.
Br. Med. J.
Br. Polym. J.
Bull. Am. Phys. Soc.
Bull. Chem. Soc. Jpn.
Bull. Electrochem.
Bull. Environ. Contam. Toxicol.
Bull. Korean Chem. Soc.
Bull. Mater. Sci.
Bull. Soc. Chim. Belg.
Bull. Soc. Chim. Fr.
Bunseki Kagaku
Byull. Eksp. Biol. Med.
Cailiao Baohu
Cailiao Kexue Yu Gongyi
Calcif. Tissue Int.
Calorim. Anal. Therm.
Cancer
Cancer Chemother. Pharmacol.
Cancer Chemother. Rep.
Cancer Lett.
Cancer Res.
Can. J. Biochem.
Can. J. Chem.
Can. J. Chem. Eng.
Can. J. Fish. Aquat. Sci.
Can. J. Microbiol.
Can. J. Phys.
Can. J. Physiol. Pharmacol.
Can. Mineral.
Carbohydr. Polym.
Carbohydr. Res.
Carbon
Carcinogenesis
Cardiovasc. Res.
Carlsberg Res. Commun.
Catal. Lett.
Catal. Rev.
Catal. Rev.—Sci. Eng.
Catal. Today
Cell
Cell Calcium
Cell Growth Differ.
Cell. Immunol.
Cell. Mol. Biol.
Cell. Signalling
Cell Tissue Res.
Cellul. Chem. Technol.
Cem. Concr. Res.
Ceram. Eng. Sci. Proc.
Ceram. Trans.
Cereal Chem.
Challenges Mod. Med.
Chem. Abstr.
Chem. Anal. (Warsaw)
Chem.-Anlagen Verfahren
Chem. Ber.
Chem. Biol.
Chem.-Biol. Interact.
Chem. Br.
Chem. Commun.
Chem. Eng. Commun.
Chem. Eng. J.

Chem. Eng. News
Chem. Eng. (N.Y.)
Chem. Eng. Prog.
Chem. Eng. Res. Des.
Chem. Eng. Sci.
Chem. Eng. Technol.
Chem.—Eur. J.
Chem. Express
Chem. Geol.
Chem. Health Saf.
Chem. Ind. (London)
Chem.-Ing.-Tech.
Chem. Lett.
Chem. Listy
Chem. Mater.
Chemom. Intell. Lab. Syst.
Chemosphere
Chemotherapy
Chem. Pap.
Chem. Pharm. Bull.
Chem. Phys.
Chem. Phys. Lett.
Chem. Phys. Lipids
Chem. Phys. Processes Combust.
Chem. Res. Toxicol.
Chem. Rev.
Chem. Scr.
Chem. Soc. Rev.
CHEMTECH
Chem.-Ztg.
Chimia
Chim. Ind.
Chin. Chem. Lett.
Chin. J. Chem.
Chin. J. Nucl. Phys.
Chin. J. Phys. (Taipei)
Chin. J. Polym. Sci.
Chin. Phys. Lett.
Chin. Sci. Bull.
Chirality
Chromatographia
Circ. Res.
Circulation
Clay Miner.
Clays Clay Miner.
Clin. Calcium
Clin. Chem.
Clin. Chim. Acta
Clin. Exp. Immunol.
Clin. Exp. Pharmacol. Physiol.
Clin. Immunol. Immunopathol.
Clin. Invest.
Clin. Pharmacol. Ther.
Clin. Sci.
Cold Spring Harbor Symp. Quant. Biol.
Collect. Czech. Chem. Commun.
Colloid Interface Sci.
Colloid Polym. Sci.
Colloids Surf.
Colloids Surf., A
Colloids Surf., B
Combust. Flame
Combust. Sci. Technol.
Comments Inorg. Chem.
Commun. ACM
Commun. Soil Sci. Plant Anal.
Comp. Biochem. Physiol., A
Comp. Biochem. Physiol., B
Comp. Biochem. Physiol., C
Compos. Sci. Technol.
Comput. Appl. Biosci.
Comput. Chem.
Comput. Chem. Eng.
Comput. J.
Comput. Phys. Commun.
Coord. Chem. Rev.
Corrosion
Corros. Sci.
C. R. Acad. Sci.
C. R. Hebd. Seances Acad. Sci.
Crit. Rev. Anal. Chem.
Crit. Rev. Biochem.
Crit. Rev. Biochem. Mol. Biol.
Crit. Rev. Food Sci. Nutr.
Crit. Rev. Toxicol.
Croat. Chem. Acta
Crop Sci.
Cryobiology
Cryogenics
Cryst. Res. Technol.
Cryst. Struct. Commun.
Cuihua Xuebao
Curr. Biol.
Curr. Genet.
Curr. Microbiol.
Curr. Opin. Cell Biol.

Curr. Opin. Struct. Biol.
Cytogenet. Cell Genet.
Denki Kagaku oyobi Kogyo Butsuri Kagaku
Desalination
Dev. Biol.
Dev. Brain. Res.
Development
Dev. Food Sci.
Dev. Plant Soil Sci.
Diabetes
Diamond Relat. Mater.
Diandu Yu Huanbao
Diffus. Defect Data, Pt. A
Diffus. Defect Data, Pt. B
Dig. Dis. Sci.
Discuss. Faraday Soc.
DNA Cell Biol.
Dokl. Akad. Nauk
Dokl. Akad. Nauk Resp. Uzb.
Dokl. Bulg. Akad. Nauk.
Dopov. Akad. Nauk Ukr.
Drug Dev. Ind. Pharm.
Drug Dev. Res.
Drug Metab. Dispos.
Drug Metab. Rev.
Drugs
Drugs Exp. Clin. Res.
Drugs Future
Earth Planet. Sci. Lett.
Econ. Geol.
Ecotoxicol. Environ. Saf.
Electroanalysis
Electrochim. Acta
Electron. Lett.
Electrophoresis
Elektrokhimiya
EMBO J.
Endocrine
Endocrinology
Energy Fuels
Environ. Eng.
Environ. Geol. Water Sci.
Environ. Health Perspect.
Environ. Lett.
Environ. Mol. Mutagen.
Environ. Pollut.
Environ. Prog.
Environ. Res.
Environ. Sci. Pollut. Control Ser.
Environ. Sci. Technol.
Environ. Technol.
Environ. Technol. Lett.
Environ. Toxicol. Chem.
Enzyme Microb. Technol.
Eur. Biophys. J.
Eur. J. Biochem.
Eur. J. Cancer
Eur. J. Cell Biol.
Eur. J. Clin. Chem. Clin. Biochem.
Eur. J. Clin. Pharmacol.
Eur. J. Endocrinol.
Eur. J. Immunol.
Eur. J. Med. Chem.
Eur. J. Mineral.
Eur. J. Pharmacol.
Eur. J. Pharmacol., Mol. Pharmacol. Sect.
Eur. J. Solid State Inorg. Chem.
Europhys. Lett.
Eur. Polym. J.
Exp. Cell Res.
Experientia
Exp. Eye Res.
Exp. Fluids
Exp. Neurol.
Faraday Discuss.
Faraday Discuss. Chem. Soc.
Faraday Symp. Chem. Soc.
Farmaco
FASEB J.
FEBS Lett.
Fed. Proc.
Fed. Regist.
FEMS Immunol. Med. Microbiol.
FEMS Microbiol. Lett.
Fenxi Ceshi Xuebao
Fenxi Huaxue
Fenxi Shiyanshi
Ferroelectrics
Fertil. Steril.
Fire Technol.
Fish. Sci.
Fitoterapia
Fiz. Khim. Obrab. Mater.
Fiz. Met. Metalloved.

Flavour Fragrance J.
Fluid Phase Equilib.
Food Agric. Immunol.
Food Chem.
Food Chem. Toxicol.
Food Cosmet. Toxicol.
Food Hydrocolloids
Food Rev. Int.
Food Sci. Technol. (London)
Food Technol.
Fragrance J.
Free Radical Biol. Med.
Free Radical Res. Commun.
Fresenius Environ. Bull.
Fresenius' J. Anal. Chem.
Fuel
Fuel Process. Technol.
Fuel Sci. Technol. Int.
Fundam. Appl. Toxicol.
Fusion Eng. Des.
Fusion Technol.
Galvanotechnik
Gangtie Yanjiu Xuebao
Gaodeng Xuexiao Huaxue Xuebao
Gaofenzi Cailiao Kexue Yu Gongcheng
Gaofenzi Xuebao
Gaoneng Wuli Yu Hewuli
Gastroenterology
Gaussian
Gazz. Chim. Ital.
Gen. Comp. Endocrinol.
Gene
Genes Dev.
Genetics
Genome
Genomics
Gen. Pharmacol.
Geochim. Cosmochim. Acta
Geokhimiya
Geol. Geofiz.
Geophys. Res. Lett.
Glycobiology
Glycoconjugate J.
Gongneng Gaofenzi Xuebao
Ground Water
Guangxue Xuebao
Health Phys.
Hecheng Xiangjiao Gongye
Hejishu
Helv. Chim. Acta
Helv. Phys. Acta
Hepatology
Heteroat. Chem.
Heterocycles
Histochemistry
Hoppe-Seyler's Z. Physiol. Chem.
Horm. Metab. Res.
HortScience
Hum. Genet.
Hum. Immunol.
Hum. Mol. Genet.
Hum. Mutat.
Hum. Reprod.
Hydrocarbon Process.
Hydrometallurgy
Hyperfine Interact.
IAHS Publ.
IEEE Electron Device Lett.
IEEE J. Quantum Electron.
IEEE Trans. Electron Devices
IEEE Trans. Magn.
IEEE Trans. Nucl. Sci.
Igaku no Ayumi
Igaku to Yakugaku
Immunity
Immunogenetics
Immunol. Lett.
Immunology
Immunol. Today
Ind. Eng. Chem.
Ind. Eng. Chem. Fundam.
Ind. Eng. Chem. Process Des. Dev.
Ind. Eng. Chem. Prod. Res. Dev.
Ind. Eng. Chem. Res.
Indian Drugs
Indian J. Chem., Sect. A
Indian J. Chem., Sect. B
Indian J. Exp. Biol.
Indian J. Pure Appl. Phys.
Indian J. Technol.
Infect. Immun.
Inorg. Chem.
Inorg. Chim. Acta
Inorg. Nucl. Chem. Lett.
Inorg. Synth.
Insect Biochem. Mol. Biol.

Int. Arch. Allergy Immunol.
Int. Arch. Occup. Environ. Health
Int. DATA Ser., Sel. Data Mixtures, Ser. A
Int. Immunol.
Int. J. Biochem.
Int. J. Biol. Macromol.
Int. J. Cancer
Int. J. Chem. Kinet.
Int. J. Environ. Anal. Chem.
Int. J. Environ. Stud.
Int. J. Exp. Pathol.
Int. J. Heat Mass Transfer
Int. J. Hydrogen Energy
Int. J. Mass Spectrom. Ion Phys.
Int. J. Mass Spectrom. Ion Processes
Int. J. Multiphase Flow
Int. J. Numer. Methods Fluids
Int. J. Oncol.
Int. J. Pept. Protein Res.
Int. J. Pharm.
Int. J. Polym. Mater.
Int. J. Quantum Chem.
Int. J. Radiat. Biol.
Int. J. Radiat. Oncol., Biol., Phys.
Int. J. Sports Med.
Int. J. Thermophys.
Int. Rev. Phys. Chem.
Inzh.-Fiz. Zh.
ISIJ Int.
Isr. J. Chem.
Izv. Akad. Nauk, Ser. Fiz.
Izv. Akad. Nauk, Ser. Khim.
Izv. Vyssh. Uchebn. Zaved., Chern. Metall.
Izv. Vyssh. Uchebn. Zaved., Fiz.
J. Adhes.
J. Adhes. Sci. Technol.
J. Aerosol Sci.
J. Agric. Food Chem.
J. Air Pollut. Control Assoc.
J. Air Waste Manage. Assoc.
J. Alloys Compd.
JAMA, J. Am. Med. Assoc.
J. Am. Ceram. Soc.
J. Am. Chem. Soc.
J. Am. Oil Chem. Soc.
J. Am. Soc. Hortic. Sci.
J. Am. Soc. Mass Spectrom.
J. Am. Stat. Assoc.
J.—Am. Water Works Assoc.
J. Anal. Appl. Pyrolysis
J. Anal. At. Spectrom.
J. Anim. Sci.
J. Antibiot.
J. Antimicrob. Chemother.
J. AOAC Int.
J. Appl. Bacteriol.
J. Appl. Chem.
J. Appl. Crystallogr.
J. Appl. Electrochem.
J. Appl. Phys.
J. Appl. Physiol.
J. Appl. Polym. Sci.
J.—Assoc. Off. Anal. Chem.
J. Atmos. Chem.
J. Atmos. Sci.
J. Bacteriol.
J. Biochem. Biophys. Methods
J. Biochem. (Tokyo)
J. Bioenerg. Biomembr.
J. Biol. Chem.
J. Biomed. Mater. Res.
J. Biomol. NMR
J. Biomol. Struct. Dyn.
J. Biotechnol.
J. Bone Miner. Res.
J. Can. Pet. Technol.
J. Carbohydr. Chem.
J. Cardiovasc. Pharmacol.
J. Catal.
J. Cell. Biochem.
J. Cell Biol.
J. Cell. Physiol.
J. Cell Sci.
J. Ceram. Soc. Jpn.
J. Chem. Crystallogr.
J. Chem. Doc.
J. Chem. Ecol.
J. Chem. Educ.
J. Chem. Eng. Data
J. Chem. Eng. Jpn.
J. Chem. Inf. Comput. Sci.
J. Chemom.
J. Chem. Phys.
J. Chem. Res., Miniprint

J. Chem. Res., Synop.
J. Chem. Soc. A
J. Chem. Soc. B
J. Chem. Soc. C
J. Chem. Soc., Chem. Commun.
J. Chem. Soc. D
J. Chem. Soc., Dalton Trans.
J. Chem. Soc., Faraday Trans.
J. Chem. Soc., Faraday Trans. 1
J. Chem. Soc., Faraday Trans. 2
J. Chem. Soc., Perkin Trans. 1
J. Chem. Soc., Perkin Trans. 2
J. Chem. Technol. Biotechnol.
J. Chem. Thermodyn.
J. Chim. Phys. Phys.-Chim. Biol.
J. Chin. Chem. Soc.
J. Chromatogr.
J. Chromatogr., A
J. Chromatogr., B
J. Chromatogr. Sci.
J. Clin. Endocrinol. Metab.
J. Clin. Invest.
J. Clin. Microbiol.
J. Clin. Oncol.
J. Clin. Pharmacol.
J. Colloid Interface Sci.
J. Colloid Sci.
J. Comp. Neurol.
J. Comput.-Aided Mater. Des.
J. Comput.-Aided Mol. Des.
J. Comput. Chem.
J. Comput. Phys.
J. Contam. Hydrol.
J. Controlled Release
J. Coord. Chem.
J. Crystallogr. Spectrosc. Res.
J. Cryst. Growth
J. Cryst. Mol. Struct.
J. Dairy Res.
J. Dairy Sci.
J. Dent. Res.
J. Dispersion Sci. Technol.
J. Econ. Entomol.
J. Electroanal. Chem.
J. Electroanal. Chem. Interfacial Electrochem.
J. Electrochem. Soc.
J. Electron. Mater.
J. Electron Spectrosc. Relat. Phenom.
J. Endocrinol.
J. Environ. Eng. (N.Y.)
J. Environ. Health
J. Environ. Polym. Degrad.
J. Environ. Qual.
J. Environ. Sci. Health, Part A, B, or C
J. Essent. Oil Res.
JETP Lett. (Engl. Transl.)
J. Eur. Ceram. Soc.
J. Exp. Biol.
J. Exp. Bot.
J. Exp. Med.
J. Ferment. Bioeng.
J. Fish. Res. Board Can.
J. Fluid Mech.
J. Fluids Eng.
J. Fluoresc.
J. Fluorine Chem.
J. Food Biochem.
J. Food Prot.
J. Food Sci.
J. Gen. Chem. USSR
J. Gen. Microbiol.
J. Gen. Physiol.
J. Gen. Virol.
J. Geochem. Explor.
J. Geophys. Res.
J. Great Lakes Res.
J. Hazard. Mater.
J. Heat Transfer
J. Heterocycl. Chem.
J. High Resolut. Chromatogr.
J. Histochem. Cytochem.
J. Hydrol.
J. Hypertens.
Jiegou Huaxue
Jikken Igaku
J. Immunol.
J. Immunol. Methods
J. Inclusion Phenom.
J. Inclusion Phenom. Mol. Recognit. Chem.
J. Indian Chem. Soc.
J. Infect. Dis.
J. Inorg. Biochem.
J. Inorg. Nucl. Chem.
J. Insect Physiol.

Jinshu Xuebao
J. Inst. Chem. (India)
J. Invest. Dermatol.
J. Jpn. Pet. Inst.
J. Korean Chem. Soc.
J. Korean Phys. Soc.
J. Labelled Compd. Radiopharm.
J. Less-Common Met.
J. Leukocyte Biol.
J. Lipid Mediators
J. Lipid Res.
J. Liq. Chromatogr.
J. Low Temp. Phys.
J. Lumin.
J. Macromol. Sci., Chem.
J. Macromol. Sci., Phys.
J. Macromol. Sci., Pure Appl. Chem.
J. Magn. Magn. Mater.
J. Magn. Reson., Ser. A
J. Magn. Reson., Ser. B
J. Mater. Chem.
J. Mater. Res.
J. Mater. Sci.
J. Mater. Sci. Lett.
J. Mater. Sci.: Mater. Med.
J. Math. Chem.
J. Math. Phys.
J. Med. Chem.
J. Membr. Biol.
J. Membr. Sci.
J. Microcolumn Sep.
J. Mol. Biol.
J. Mol. Catal.
J. Mol. Cell. Cardiol.
J. Mol. Evol.
J. Mol. Graphics
J. Mol. Liq.
J. Mol. Recognit.
J. Mol. Spectrosc.
J. Mol. Struct.
J. Muscle Res. Cell Motil.
J. Natl. Cancer Inst.
J. Nat. Prod.
J. Neurochem.
J. Neuroimmunol.
J. Neurophysiol.
J. Neurosci.
J. Neurosci. Res.
J. Non-Cryst. Solids
J. Non-Newtonian Fluid Mech.
J. Nucl. Mater.
J. Nucl. Med.
J. Nucl. Sci. Technol.
J. Nutr.
J. Occup. Med.
J. Occup. Med. Toxicol.
J. Oceanogr.
J. Opt. Soc. Am.
J. Opt. Soc. Am. A
J. Opt. Soc. Am. B
J. Organomet. Chem.
J. Org. Chem.
J. Pharmacokinet. Biopharm.
J. Pharmacol. Exp. Ther.
J. Pharm. Biomed. Anal.
J. Pharm. Pharmacol.
J. Pharm. Sci.
J. Phase Equilib.
J. Photochem.
J. Photochem. Photobiol., A
J. Photochem. Photobiol., B
J. Photopolym. Sci. Technol.
J. Phys. A: Gen. Phys.
J. Phys. A: Math. Gen.
J. Phys. A: Math., Nucl. Gen.
J. Phys. B: At. Mol. Opt. Phys.
J. Phys. C: Solid State Phys.
J. Phys. Chem.
J. Phys. Chem. Ref. Data
J. Phys. Chem. Solids
J. Phys.: Condens. Matter
J. Phys. D: Appl. Phys.
J. Phys. E: Sci. Instrum.
J. Phys. II
J. Physiol. (London)
J. Physiol. Pharmacol.
J. Phys. IV
J. Phys., Lett.
J. Phys. Org. Chem.
J. Phys. (Paris)
J. Phys. Radium
J. Phys. Soc. Jpn.
J. Planar Chromatogr.-Mod. TLC
J. Plant Nutr.
J. Plant Physiol.
Jpn. J. Appl. Phys., Part 1

Jpn. J. Appl. Phys., Part 2
Jpn. J. Cancer Res.
Jpn. J. Pharmacol.
Jpn. J. Toxicol. Environ. Health
J. Polym. Sci.
J. Polym. Sci., Part A: Polym. Chem.
J. Polym. Sci., Part B: Polym. Phys.
J. Polym. Sci., Polym. Chem. Ed.
J. Polym. Sci., Polym. Lett. Ed.
J. Polym. Sci., Polym. Phys. Ed.
J. Polym. Sci., Polym. Symp.
J. Power Sources
J. Prakt. Chem./Chem.-Ztg.
J. Protein Chem.
J. Quant. Spectrosc. Radiat. Transfer
J. Radioanal. Nucl. Chem.
J. Raman Spectrosc.
J. Reprod. Fertil.
J. Res. Natl. Bur. Stand. (U.S.)
J. Rheol.
J. Sci. Food Agric.
J. Sci. Instrum.
J. Serb. Chem. Soc.
J. Soc. Ind. Appl. Math.
J. Soil Sci.
J. Sol.-Gel Sci. Technol.
J. Solid State Chem.
J. Solution Chem.
J. Stat. Phys.
J. Steroid Biochem.
J. Steroid Biochem. Mol. Biol.
J. Struct. Biol.
J. Supercond.
J. Supercrit. Fluids
J. Surg. Res.
J. Text. Assoc.
J. Text. Inst.
J. Texture Stud.
J. Theor. Biol.
J. Therm. Anal.
J. Toxicol. Environ. Health
Justus Liebigs Ann. Chem.
J. Vac. Sci. Technol., A
J. Vac. Sci. Technol., B
J. Virol.
J. Virol. Methods
J.—Water Pollut. Control Fed.
Kagaku Kogaku
Kagaku Kogaku Ronbunshu
Kagaku (Kyoto)
Kagaku to Seibutsu
Kankyo Kagaku
Kautsch. Gummi Kunstst.
Key Eng. Mater.
Khim.-Farm. Zh.
Khim. Fiz.
Khim. Prir. Soedin.
Kidney Int.
Kidorui
Kinet. Katal.
Kobunshi Ronbunshu
Kolloidn. Zh.
Kolloid-Z.
Kolloid Z. Z. Polym.
Koord. Khim.
Kristallografiya
Lab. Invest.
Lancet
Langmuir
Laser Inst. Am. [Publ.]
LC-GC
Lebensm.-Wiss. Technol.
Lett. Appl. Microbiol.
Liebigs Ann. Chem.
Life Sci.
Light Met.
Limnol. Oceanogr.
Lipids
Liq. Cryst.
Macromol. Chem.
Macromol. Chem. Phys.
Macromolecules
Macromol. Rapid Commun.
Macromol. Symp.
Macromol. Theory Simul.
Magn. Reson. Chem.
Makromol. Chem.
Makromol. Chem., Macromol. Symp.
Makromol. Chem., Rapid Commun.
Mamm. Genome
Mar. Chem.
Mar. Ecol.: Prog. Ser.
Mar. Environ. Res.
Mar. Pollut. Bull.

Mass Spectrom. Rev.
Mater. Chem. Phys.
Materia
Mater. Lett.
Mater. Res. Bull.
Mater. Res. Soc. Symp. Proc.
Mater. Sci. Eng., A
Mater. Sci. Eng., B
Mater. Sci. Forum
Mater. Sci. Technol.
Mater. Trans., JIM
Meas. Sci. Technol.
Mech. Dev.
Med. Chem. Res.
Med. Res. Rev.
Med. Sci. Res.
Med. Tr. Prom. Ekol.
Mendeleev Commun.
Metab., Clin. Exp.
Metall. Mater. Trans. A
Metall. Mater. Trans. B
Metally
Meteoritics
Methods Enzymol.
Methods Mol. Biol.
Methods Neurosci.
Met. Ions Biol. Syst.
Microbeam Anal.
Microbiol. Immunol.
Microbiology
Microbiol. Rev.
Microchem. J.
Microelectron. Eng.
Microporous Mater.
Mikrochim. Acta
Milchwissenschaft
Miner. Eng.
Modell., Meas. Control, A
Modell., Meas. Control, B
Modell., Meas. Control, C
Mod. Phys. Lett. A
Mod. Phys. Lett. B
Mol. Biochem. Parasitol.
Mol. Biol. Cell
Mol. Biol. (Moscow)
Mol. Cell. Biochem.
Mol. Cell. Biol.
Mol. Cell. Endocrinol.
Mol. Cells
Mol. Cryst. Liq. Cryst. Sci. Technol., Sect. A
Mol. Cryst. Liq. Cryst. Sci. Technol., Sect. B
Mol. Endocrinol.
Mol. Gen. Genet.
Mol. Immunol.
Mol. Med. (Tokyo)
Mol. Microbiol.
Mol. Pharmacol.
Mol. Phys.
Mol. Plant-Microbe Interact.
Mol. Reprod. Dev.
Mol. Simul.
Monatsh. Chem.
Mon. Not. R. Astron. Soc.
MRS Bull.
Mutat. Res.
Nahrung
Naihuo Cailiao
Nanostruct. Mater.
Nat. Genet.
Nat. Prod. Rep.
Nat. Struct. Biol.
Nature (London)
Naturwissenschaften
N. Engl. J. Med.
Neorg. Mater.
Neurochem. Int.
Neurochem. Res.
Neuroendocrinology
Neuron
Neuropharmacology
NeuroReport
Neuroscience
Neurosci. Lett.
New J. Chem.
New Phytol.
Nippon Ishikai Zasshi
Nippon Kagaku Kaishi
Nippon Kikai Gakkai Ronbunshu, B-hen
Nippon Kinzoku Gakkaishi
Nippon Oyo Jiki Gakkaishi
Nippon Suisan Gakkaishi

Nouv. J. Chim.
Nucleic Acids Res.
Nucl. Eng. Des.
Nucl. Fusion
Nucl. Instrum. Methods
Nucl. Instrum. Methods Phys. Res., Sect. A
Nucl. Instrum. Methods Phys. Res., Sect. B
Nucl. Med. Biol.
Nucleosides Nucleotides
Nucl. Phys. A
Nucl. Phys. B
Nucl. Phys. B, Proc. Suppl.
Nucl. Technol.
Nucl. Tracks Radiat. Meas.
Nuovo Cimento Soc. Ital. Fis., A
Nutr. Rep. Int.
Nutr. Res. (N.Y.)
Occup. Environ. Med.
Occup. Hazards
Occup. Health Saf.
Off. Methods Anal.
Oil Gas J.
Oncogene
Opt. Commun.
Opt. Eng.
Opt. Lett.
Opt. Spektrosk.
Organohalogen Compd.
Organometallics
Org. Geochem.
Org. Magn. Reson.
Org. Mass Spectrom.
Org. Prep. Proced. Int.
Org. Process Res. Dev.
Org. React.
Org. Synth.
Ozone: Sci. Eng.
PCR Methods Appl.
Pediatr. Res.
Pept. Chem.
Peptides
Perfum. Flavor.
Pestic. Biochem. Physiol.
Pestic. Sci.
Pfluegers Arch.
Pharmacol., Biochem. Behav.
Pharmacol. Commun.
Pharmacology
Pharmacol. Rev.
Pharmacol. Ther.
Pharmacol. Toxicol.
Pharm. Acta Helv.
Pharmazie
Pharm. Res.
Pharm. Weekbl.
Phase Transitions
Philos. Mag. A
Philos. Mag. B
Philos. Trans. R. Soc. London, Ser. A
Phosphorus Relat. Group V Elem.
Phosphorus Sulfur Relat. Elem.
Phosphorus, Sulfur Silicon Relat. Elem.
Photobiochem. Photobiophys.
Photochem. Photobiol.
Photogr. Sci. Eng.
Photosynth. Res.
Phys. Chem.
Phys. Chem. Glasses
Phys. Chem. Liq.
Phys. Chem. Miner.
Phys. Fluids
Physica A
Physica B
Physica C
Physica D
Physiol. Behav.
Physiol. Plant.
Physiol. Rev.
Phys. Lett. A
Phys. Lett. B
Phys. Plasmas
Phys. Rep.
Phys. Rev. A: At., Mol., Opt. Phys.
Phys. Rev. B: Condens. Matter
Phys. Rev. B: Solid State
Phys. Rev. C: Nucl. Phys.
Phys. Rev. D: Part. Fields
Phys. Rev. E: Stat. Phys., Plasmas, Fluids, Relat. Interdiscip. Top.
Phys. Rev. Lett.
Phys. Scr.
Phys. Scr., T
Phys. Status Solidi A
Phys. Status Solidi B

Phys. Today
Phys. Z.
Phytochem. Anal.
Phytochemistry
Pis'ma Zh. Eksp. Teor. Fiz.
Pis'ma Zh. Tekh. Fiz.
Planet. Space Sci.
Planta
Planta Med.
Plant Cell
Plant Cell Physiol.
Plant Cell Rep.
Plant Foods Hum. Nutr.
Plant J.
Plant Mol. Biol.
Plant Physiol.
Plant Physiol. Biochem.
Plant Sci.
Plant Soil
Plasma Phys. Controlled Fusion
Plast., Rubber Compos. Process. Appl.
Pol. J. Chem.
Polycyclic Aromat. Compd.
Polyhedron
Polym. Adv. Technol.
Polym. Bull.
Polym. Commun.
Polym. Degrad. Stab.
Polym. Eng. Sci.
Polymer
Polym. Int.
Polym. J.
Polym. Mater. Sci. Eng.
Polym. Prepr. (Am. Chem. Soc., Div. Polym. Chem.)
Polym. Sci. U.S.S.R.
Poroshk. Metall. (Kiev)
Poult. Sci.
Poverkhnost
Powder Technol.
Prepr. Pap.—Am. Chem. Soc., Div. Fuel Chem.
Proc. Am. Assoc. Cancer Res.
Proc. Cambridge Philos. Soc.
Proc. Chem. Soc., London
Proc. Natl. Acad. Sci. U.S.A.
Proc. Phys. Soc., London
Proc. R. Soc. London
Proc. R. Soc. London, Ser. A or B
Proc. Soc. Exp. Biol. Med.
Proc. SPIE-Int. Soc. Opt. Eng.
Prog. Biophys. Mol. Biol.
Prog. Biotechnol.
Prog. Colloid Polym. Sci.
Prog. Energy Combust. Sci.
Prog. Inorg. Chem.
Prog. Lipid Res.
Prog. Nucleic Acid Res. Mol. Biol.
Prog. Nucl. Magn. Reson. Spectrosc.
Prog. Phys. Org. Chem.
Prog. Polym. Sci.
Prog. React. Kinet.
Prog. Solid State Chem.
Prog. Surf. Sci.
Prog. Theor. Phys.
Protein Eng.
Protein Sci.
Proteins: Struct., Funct., Genet.
Przem. Chem.
Psychopharmacology (Berlin)
Pure Appl. Chem.
QCPE
QCPE Bull.
Q. Rev. Biophys.
Quant. Struct.-Act. Relat.
Quim. Anal. (Barcelona)
Quim. Nova
Radiat. Eff. Defects Solids
Radiat. Meas.
Radiat. Phys. Chem.
Radiat. Prot. Dosim.
Radiat. Res.
Radiocarbon
Radiochim. Acta
Radiokhimiya
Rapid Commun. Mass Spectrom.
React. Kinet. Catal. Lett.
React. Polym.
Recl. Trav. Chim. Pays-Bas
Regul. Pept.
Regul. Toxicol. Pharmacol.
Rep. Prog. Phys.
Res. Chem. Intermed.
Res. Commun. Chem. Pathol. Pharmacol.
Res. Commun. Mol. Pathol. Pharmacol.
Res. Discl.

Residue Rev.
Res. J. Water Pollut. Control Fed.
Rev. Chim. (Bucharest)
Rev. Chim. Miner.
Rev. Environ. Contam. Toxicol.
Rev. Mod. Phys.
Rev. Roum. Chim.
Rev. Sci. Instrum.
Rheol. Acta
Rinaho Byori
RNA
Rubber Chem. Technol.
Russ. Chem. Rev. (Engl. Transl.)
Russ. J. Inorg. Chem. (Engl. Transl.)
Russ. J. Phys. Chem. (Engl. Transl.)
Sae Mulli
Saishin Igaku
Scand. J. Immunol.
Sci. Am.
Sci. China, Ser. A
Sci. China, Ser. B
Science (Washington, D.C.)
Sci. Tech. Froid
Sci. Total Environ.
Semicond. Sci. Technol.
Sens. Actuators, A
Sens. Actuators, B
Sep. Sci. Technol.
Soc. Automot. Eng., [Spec. Publ.] SP
Soc. Automot. Eng. Tech. Pap. Ser.
Soc. Neurosci. Abstr.
Soc. Pet. Eng. J.
Soil Biol. Biochem.
Soil Sci.
Soil Sci. Soc. Am. J.
Soil Sci. Soc. Am. Proc.
Sol. Energy Mater.
Sol. Energy Mater. Sol. Cells
Solid State Commun.
Solid-State Electron.
Solid State Ionics
Solvent Extr. Ion Exch.
Sov. Phys. Crystallogr.
Sov. Phys. JETP
Sov. Phys. Solid State
Spec. Publ.—R. Soc. Chem.
Spectrochim. Acta, Part A
Spectrochim. Acta, Part B
Spectrosc. Lett.
Stahl
Steroids
Struct. Biol.
Struct. Bonding (Berlin)
Struct. Chem.
Structure
Stud. Environ. Sci.
Stud. Phys. Theor. Chem.
Stud. Surf. Sci. Catal.
Supercond. Sci. Technol.
Superlattices Microstruct.
Supramol. Chem.
Surf. Coat. Technol.
Surf. Interface Anal.
Surf. Sci.
Surf. Sci. Lett.
Surf. Sci. Rep.
Svar. Proizvod.
Symp. (Int.) Combust., [Proc.]
Synapse
Synlett
Synth. Commun.
Synthesis
Synth. Met.
Synth. React. Inorg. Met.-Org. Chem.
Talanta
Tappi J.
Technometrics
Tellus
Tetrahedron
Tetrahedron: Asymmetry
Tetrahedron Comput. Methodol.
Tetrahedron Lett.
Tetsu to Hagane
THEOCHEM
Theor. Appl. Genet.
Theor. Chim. Acta
Thermochim. Acta
Thin Solid Films
Thromb. Haemostasis
Thromb. Res.
Top. Curr. Chem.
Top. Stereochem.
Toxicol. Appl. Pharmacol.
Toxicol. Environ. Chem.
Toxicol. in Vitro
Toxicol. Lett.

Toxicologist
Toxicology
Toxicon
Trans. Faraday Soc.
Trans. Inst. Chem. Eng.
Transition Met. Chem. (London)
Transition Met. Chem. (N.Y.)
Transplantation
Transplant. Proc.
Trans. Soc. Rheol.
Trends Anal. Chem.
Trends Biochem. Sci.
Trends Biotechnol.
Trends Food Sci. Technol.
Trends Genet.
Trends Neurosci.
Trends Pharmacol. Sci.
Trends Polym. Sci.
Tsvetn. Met.
Ukr. Biokhim. Zh.
Ukr. Khim. Zh. (Russ. Ed.)
Ultramicroscopy
Usp. Khim.
Vaccine
Vacuum
VDI-Ber.
Vib. Spectrosc.
Virology
Vysokomol. Soedin.
Water, Air, Soil Pollut.
Water Chem. Nucl. React. Syst.
Water Environ. Res.
Water Res.
Water Resour. Res.
Water Sci. Technol.
Wear
Weed Res.
Weed Sci.
Xenobiotica
Xiangjiao Gongye
Yad. Fiz.
Yakubutsu Dotai
Yakugaku Zasshi
Yakuri to Chiryo
Yeast
Yingyong Huaxue
Yukagaku
Z. Anorg. Allg. Chem.
Z. Anorg. Chem.
Zavod. Lab.
Z. Chem.
Z. Elektrochem.
Zeolites
Zh. Anal. Khim.
Zh. Eksp. Teor. Fiz.
Zh. Fiz. Khim.
Zh. Neorg. Khim.
Zh. Obshch. Khim.
Zh. Org. Khim.
Zh. Prikl. Khim.
Zh. Strukt. Khim.
Zh. Tekh. Fiz.
Z. Kristallogr.
Z. Lebensm.-Unters.-Forsch.
Z. Metallkd.
Z. Naturforsch., A: Phys. Sci.
Z. Naturforsch., B: Chem. Sci.
Z. Naturforsch., C: Biosci.
Z. Phys. A: Hadrons Nucl.
Z. Phys. B: Condens. Matter
Z. Phys. C: Part. Fields
Z. Phys. Chem. (Leipzig)
Z. Phys. Chem. (Munich)
Z. Phys. D: At., Mol. Clusters
Z. Physiol. Chem.

CHAPTER 7

Names and Numbers for Chemical Compounds

The use of proper chemical nomenclature is essential for effective scientific communication. More than one million new substances are reported each year, each of which must be identified clearly, unambiguously, and completely in the primary literature. Chemical compounds are named according to the rules established by the International Union of Pure and Applied Chemistry (IUPAC), the International Union of Biochemistry and Molecular Biology (IUBMB) [formerly the International Union of Biochemistry (IUB)], the Chemical Abstracts Service (CAS), the Committee on Nomenclature of the American Chemical Society, and other authorities as appropriate. For those rules, refer to the bibliography at the end of this chapter. This chapter gives the editorial conventions and style points for chemical compound names that are presumed to be correct.

Components of Chemical Names

The names of chemical compounds may consist of one or more words, and they may include locants, descriptors, and syllabic portions. Locants and descriptors can be numerals, element symbols, small capital letters, Greek letters, Latin letters, italic words and letters, and combinations of these. Treat the word or syllabic portions of chemical names just like other common nouns: use roman type, keep them lowercase in text, capitalize them at the beginnings of sentences and in titles, and hyphenate them only when they do not fit completely on one line.

Locants and Descriptors

◆ Numerals used as locants can occur at the beginning of or within a chemical name. They are set off with hyphens.

6-aminobenzothiazole
di-2-propenylcyanamide
3′-methylphthalanilic acid
6-hydroxy-2-naphthalenesulfonic acid
5,7-dihydroxy-3-(4-hydroxyphenyl)-4*H*-1-benzopyran-4-one
4a,8a-dihydronaphthalene

◆ Use italic type for chemical element symbols that denote attachment to an atom or a site of ligation.

N-acetyl group
N,*N*′-bis(3-aminopropyl)-1,4-butanediamine
bis[(ethylthio)acetato-*O*,*S*]platinum
B,*B*′-di-3-pinanyldiborane
N-ethylaniline
glycinato-*N*
S-methyl benzenethiosulfonate
P-phenylphosphinimidic acid
O,*O*,*S*-triethyl phosphorodithioate

◆ When element symbols are used with a type of reaction as a noun or adjective, use roman type for the symbol and hyphenate it to the word that follows it.

N-oxidation
N-acetylation
O-substituted
S-methylation
N-oxidized
N-acetylated
O-substitution
S-methylated

◆ Use italic type for the capital H that denotes indicated or added hydrogen.

3*H*-fluorene
phosphinin-2(1*H*)-one
2*H*-indene
2*H*-pyran-3(4*H*)-thione
1*H*-1,3-diazepine

◆ Use Greek letters, not the spelled-out forms, in chemical names to denote position or stereochemistry. Use a hyphen to separate them from the chemical name.

α-amino acid, *not* alpha amino acid
β-naphthol, *not* beta naphthol
5α,10β,15α,20α-tetraphenylporphyrin

◆ Use italic type for positional, stereochemical, configurational, and descriptive structural prefixes when they appear with the chemical name or formula. Use a hyphen to separate them from the chemical name. Accepted prefixes include

abeo	gem	retro
ac	hexahedro	ribo
altro	hexaprismo	s
amphi	hypho	S
anti	icosahedro	S*
antiprismo	klado	sec
ar	l	sn
arachno	m	sym
as	M	syn
asym	mer	t
c	meso	tert
catena	n	tetrahedro
cis	nido	threo
cisoid	o	trans
closo	octahedro	transoid
cyclo	p	triangulo
d	P	triprismo
dodecahedro	pentaprismo	uns
E	quadro	vic
endo	r	xylo
erythro	R	Z
exo	R*	
fac	rel	

Do not capitalize prefixes that are shown here as lowercase, even at the beginning of a sentence or in a title, and never use lowercase for those that are written in capital letters. Enclose the prefixes *R*, *R**, *S*, *S**, *E*, and *Z* in parentheses.

anti-bicyclo[3.2.1]octan-8-amine
ar-chlorotoluene
as-trichlorobenzene
catena-triphosphoric acid
cis-$[PtCl_2(NH_3)_2]$
cis-diamminedichloroplatinum
cyclo-hexasulfur, *c*-S_6
(*E*,*E*)-2,4-hexadienoic acid
erythro-2,3-dibromosuccinic acid
exo-chloro-*p*-menthane
(*E*,*Z*)-1,3-di-1-propenylnaphthalene
m-ethylpropylbenzene
meso-tartaric acid
o-dibromobenzene
p-aminoacetanilide
s-triazine
(*S*)-2,3-dihydroxypropanoic acid
5-*sec*-butylnonane
sym-dibromoethane
tert-pentyl bromide, *t*-$C_5H_{11}Br$
threo-2,3-dihydroxy-1,4-dimercaptobutane
trans-2,3-dimethylacrylic acid
uns-dichloroacetone
vic-triazine

◆ Use small capital letters D and L to indicate absolute configuration with amino acids and carbohydrates.

L-galactosamine
2-(difluoromethyl)-DL-ornithine
2-*O*-β-D-glucopyranosyl-α-D-glucose

◆ Use plus and minus signs enclosed in parentheses as stereochemical descriptors.

(+)-dihydrocinchonine
(−)-3-(3,4-dihydroxyphenyl)-L-alanine
(±)-2-allylcyclohexanone

◆ When the structural prefixes cyclo, iso, spiro, and neo are integral parts of chemical names, close them up to the rest of the name (without hyphens) and do not italicize them.

cyclohexane isopropyl alcohol neopentane

However, italicize and hyphenate cyclo as a nonintegral structural descriptor.

cyclo-triphosphoric acid *cyclo*-octasulfur

◆ Use numerals separated by periods within square brackets in names of bridged and spiro alicyclic compounds.

bicyclo[3.2.0]heptane
bicyclo[4.4.0]decane
spiro[4.5]decane
1-methylspiro[3.5]non-5-ene

◆ Use italic letters within square brackets in names of polycyclic aromatic compounds.

dibenz[*a,j*]anthracene
dicyclobuta[*de,ij*]naphthalene
1*H*-benzo[*de*]naphthacene
dibenzo[*c,g*]phenanthrene
indeno[1,2-*a*]indene

Syllabic Portion of Chemical Names

Multiplying affixes are integral parts of the chemical name; they are set in roman type and are always closed up to the rest of the name (without hyphens). Use hyphens only to set off intervening locants or descriptors. Use enclosing marks (parentheses, brackets, or braces) to ensure clarity or to observe other recommended nomenclature conventions. Multiplying prefixes include the following:

- hemi, mono, di, tri, tetra, penta, hexa, hepta, octa, ennea, nona, deca, deka, undeca, dodeca, etc.
- semi, uni, sesqui, bi, ter, quadri, quater, quinque, sexi, septi, octi, novi, deci, etc.
- bis, tris, tetrakis, pentakis, hexakis, heptakis, octakis, nonakis, decakis, etc.

3,4′-bi-2-naphthol
2,2′-bipyridine
bis(benzene)chromium(0)
1,4-bis(3-bromo-1-oxopropyl)piperazine

1,3-bis(diethylamino)propane
dichloride
hemihydrate
2,4,6,8-nonanetetrone
3,4,5,6-tetrabromo-*o*-cresol
triamine
tri-*sec*-butylamine
tris(ethylenediamine)cadmium dihydroxide
di-*tert*-butyl malonate
1,2-ethanediol
hexachlorobenzene
pentachloroethane
tetrakis(hydroxymethyl)methane
2,3,5-tris(aziridin-1-yl)-*p*-benzoquinone
triethyl phosphate

Capitalization of Chemical Names

Chemical names are not capitalized unless they are the first word of a sentence or are part of a title or heading. Then, the first letter of the syllabic portion is capitalized, not the locant, stereoisomer descriptor, or positional prefix. Table 1 presents examples of simple chemical names and their capitalization. Table 2 presents chemical names that include locants and descriptors.

◆ Some reaction names are preceded by element symbols; they may be used as nouns or adjectives. When they are the first word of a sentence or appear in titles and headings, capitalize the first letter of the word. Do not italicize the element symbol.

N-Oxidation of the starting compounds yielded compounds **3–10**.

N-Benzoylated amines undergo hydroxylation when incubated with yeast.

Preparation of S-Methylated Derivatives

O-Substituted Structural and Functional Analogs

Punctuation in Chemical Names

◆ Use commas between numeral locants, chemical element symbol locants, and Greek locants, with no space after the comma. When a single locant consists of a numeral and a Greek letter together with no space or punctuation, the numeral precedes the Greek letter. When the Greek letter precedes the numeral, they indicate two different locants and should be separated by a comma. For example, α,2 denotes two locants; 1α is viewed as one locant.

1,2-dinitrobutane
N,*N*-dimethylacetamide
β,4-dichlorocyclohexanepropionic acid
(6α,11β,16α)-6-fluoro-16-methylpregna-1,4-diene
2,3,3a,4-tetrahydro-1*H*-indole

Table 1. Examples of Multiword Chemical Names

In Text	*At Beginning of Sentence*	*In Titles, Headings*
Acids		
benzoic acid	Benzoic acid	Benzoic Acid
ethanethioic *S*-acid	Ethanethioic *S*-acid	Ethanethioic *S*-Acid
hydrochloric acid	Hydrochloric acid	Hydrochloric Acid
Alcohols		
ethyl alcohol	Ethyl alcohol	Ethyl Alcohol
ethylene glycol	Ethylene glycol	Ethylene Glycol
Ketones		
di-2-naphthyl ketone	Di-2-naphthyl ketone	Di-2-naphthyl Ketone
methyl phenyl ketone	Methyl phenyl ketone	Methyl Phenyl Ketone
Ethers		
di-*sec*-butyl ether	Di-*sec*-butyl ether	Di-*sec*-butyl Ether
methyl propyl ether	Methyl propyl ether	Methyl Propyl Ether
Anhydrides		
acetic anhydride	Acetic anhydride	Acetic Anhydride
phthalic anhydride	Phthalic anhydride	Phthalic Anhydride
Esters		
methyl acetate	Methyl acetate	Methyl Acetate
phenyl thiocyanate	Phenyl thiocyanate	Phenyl Thiocyanate
propyl benzoate	Propyl benzoate	Propyl Benzoate
Polymer Names		
1,2-polybutadiene	1,2-Polybutadiene	1,2-Polybutadiene
poly(butyl methacrylate)	Poly(butyl methacrylate)	Poly(butyl methacrylate)
poly(*N,N*-dimethylacrylamide	Poly(*N,N*-dimethylacrylamide)	Poly(*N,N*-dimethylacrylamide)
poly(ethylene glycol)	Poly(ethylene glycol)	Poly(ethylene glycol)
Other Organic Compounds		
aniline hydrochloride	Aniline hydrochloride	Aniline Hydrochloride
benzyl hydroperoxide	Benzyl hydroperoxide	Benzyl Hydroperoxide
butyl chloride	Butyl chloride	Butyl Chloride
tert-butyl fluoride	*tert*-Butyl fluoride	*tert*-Butyl Fluoride
dicyclohexyl peroxide	Dicyclohexyl peroxide	Dicyclohexyl Peroxide
diethyl sulfide	Diethyl sulfide	Diethyl Sulfide
methyl iodide	Methyl iodide	Methyl Iodide
2-naphthoyl bromide	2-Naphthoyl bromide	2-Naphthoyl Bromide
sodium *S*-phenyl thiosulfite	Sodium *S*-phenyl thiosulfite	Sodium *S*-Phenyl Thiosulfite
Inorganic and Coordination Compounds		
ammonium hydroxide	Ammonium hydroxide	Ammonium Hydroxide
calcium sulfate	Calcium sulfate	Calcium Sulfate
bis(diethyl phosphato)zinc	Bis(diethyl phosphato)zinc	Bis(diethyl phosphato)zinc
(dimethyl sulfoxide)-cadmium sulfate	(Dimethyl sulfoxide)-cadmium sulfate	(Dimethyl sulfoxide)-cadmium Sulfate
magnesium oxide	Magnesium oxide	Magnesium Oxide
sodium cyanide	Sodium cyanide	Sodium Cyanide
sulfur dioxide	Sulfur dioxide	Sulfur Dioxide

Table 2. Locants and Descriptors in Chemical Names

In Text	*At Beginning of Sentence*	*In Titles, Headings*
Numeral Locants		
adenosine 5′-triphosphate	Adenosine 5′-triphosphate	Adenosine 5′-Triphosphate
1,3-bis(bromomethyl)benzene	1,3-Bis(bromomethyl)benzene	1,3-Bis(bromomethyl)benzene
1-bromo-3-chloropropane	1-Bromo-3-chloropropane	1-Bromo-3-chloropropane
2-benzoylbenzoic acid	2-Benzoylbenzoic acid	2-Benzoylbenzoic Acid
2-(2-chloroethyl)pentanoic acid	2-(2-Chloroethyl)pentanoic acid	2-(2-Chloroethyl)pentanoic Acid
7-(4-chlorophenyl)-1-naphthol	7-(4-Chlorophenyl)-1-naphthol	7-(4-Chlorophenyl)-1-naphthol
1,2-dicyanobutane	1,2-Dicyanobutane	1,2-Dicyanobutane
4a,8a-dihydronaphthalene	4a,8a-Dihydronaphthalene	4a,8a-Dihydronaphthalene
Element Symbol Locants		
(2,3-butanedione dioximato-*O,O*′)copper	(2,3-Butanedione dioximato-*O,O*′)copper	(2,3-Butanedione dioximato-*O,O*′)copper
N,2-dihydroxybenzamide	*N*,2-Dihydroxybenzamide	*N*,2-Dihydroxybenzamide
N,N′-dimethylurea	*N,N*′-Dimethylurea	*N,N*′-Dimethylurea
N-ethylaniline	*N*-Ethylaniline	*N*-Ethylaniline
3*H*-fluorene	3*H*-Fluorene	3*H*-Fluorene
S-methyl benzenethiosulfonate	*S*-Methyl benzenethiosulfonate	*S*-Methyl Benzenethiosulfonate
O,S,S-triethyl phosphorodithioate	*O,S,S*-Triethyl phosphorodithioate	*O,S,S*-Triethyl Phosphorodithioate
Greek Letter Locants and Descriptors		
β-chloro-1-naphthalenebutanol	β-Chloro-1-naphthalenebutanol	β-Chloro-1-naphthalenebutanol
ω,ω′-dibromopolybutadiene	ω,ω′-Dibromopolybutadiene	ω,ω′-Dibromopolybutadiene
β,4-dichlorocyclohexanepropionic acid	β,4-Dichlorocyclohexanepropionic acid	β,4-Dichlorocyclohexanepropionic Acid
β-endorphin	β-Endorphin	β-Endorphin
α-hydroxy-β-aminobutyric acid	α-Hydroxy-β-aminobutyric acid	α-Hydroxy-β-aminobutyric Acid
1α-hydroxycholecalciferol	1α-Hydroxycholecalciferol	1α-Hydroxycholecalciferol
17α-hydroxy-5β-pregnane	17α-Hydroxy-5β-pregnane	17α-Hydroxy-5β-pregnane
α-methylbenzeneacetic acid	α-Methylbenzeneacetic acid	α-Methylbenzeneacetic Acid
α_1-sitosterol	α_1-Sitosterol	α_1-Sitosterol
tris(β-chloroethyl)amine	Tris(β-chloroethyl)amine	Tris(β-chloroethyl)amine
Small Capital Letter Descriptors		
L-methionine	L-Methionine	L-Methionine
D-serine	D-Serine	D-Serine
DL-alanine	DL-Alanine	DL-Alanine
β-D-arabinose	β-D-Arabinose	β-D-Arabinose
D-1,2,4-butanetriol	D-1,2,4-Butanetriol	D-1,2,4-Butanetriol
D_S-threonine	D_S-Threonine	D_S-Threonine

Continued on next page

Table 2. Locants and Descriptors in Chemical Names—Continued

In Text	*At Beginning of Sentence*	*In Titles, Headings*
Positional and Structural Descriptors		
p-benzenediacetic acid	*p*-Benzenediacetic acid	*p*-Benzenediacetic Acid
7-bromo-*p*-cymene	7-Bromo-*p*-cymene	7-Bromo-*p*-cymene
sec-butyl alcohol	*sec*-Butyl alcohol	*sec*-Butyl Alcohol
n-butyl iodide	*n*-Butyl iodide	*n*-Butyl Iodide
p-*tert*-butylphenol	*p*-*tert*-Butylphenol	*p*-*tert*-Butylphenol
4-chloro-*m*-cresol	4-Chloro-*m*-cresol	4-Chloro-*m*-cresol
2-(*o*-chlorophenyl)-1-naphthol	2-(*o*-Chlorophenyl)-1-naphthol	2-(*o*-Chlorophenyl)-1-naphthol
o-dibromobenzene	*o*-Dibromobenzene	*o*-Dibromobenzene
sym-dibromoethane	*sym*-Dibromoethane	*sym*-Dibromoethane
m-hydroxybenzyl alcohol	*m*-Hydroxybenzyl alcohol	*m*-Hydroxybenzyl Alcohol
tert-pentyl isovalerate	*tert*-Pentyl isovalerate	*tert*-Pentyl Isovalerate
1-(*trans*-1-propenyl)-3-(*cis*-1-propenyl)naphthalene	1-(*trans*-1-Propenyl)-3-(*cis*-1-propenyl)naphthalene	1-(*trans*-1-Propenyl)-3-(*cis*-1-propenyl)naphthalene
s-triazine	*s*-Triazine	*s*-Triazine
Stereochemical Descriptors		
(+)-*erythro*-2-amino-3-methylpentanoic acid	(+)-*erythro*-2-Amino-3-methylpentanoic acid	(+)-*erythro*-2-Amino-3-methylpentanoic Acid
dl-2-aminopropanoic acid	*dl*-2-Aminopropanoic acid	*dl*-2-Aminopropanoic Acid
anti-bicyclo[3.2.1]octan-8-amine	*anti*-Bicyclo[3.2.1]octan-8-amine	*anti*-Bicyclo[3.2.1]octan-8-amine
exo-bicyclo[2.2.2]oct-5-en-2-ol	*exo*-Bicyclo[2.2.2]oct-5-en-2-ol	*exo*-Bicyclo[2.2.2]oct-5-en-2-ol
(1*R**,3*S**)-1-bromo-3-chlorocyclohexane	(1*R**,3*S**)-1-Bromo-3-chlorocyclohexane	(1*R**,3*S**)-1-Bromo-3-chlorocyclohexane
rel-(1*R*,3*R*)-1-bromo-3-chlorocyclohexane	*rel*-(1*R*,3*R*)-1-Bromo-3-chlorocyclohexane	*rel*-(1*R*,3*R*)-1-Bromo-3-chlorocyclohexane
d-camphor	*d*-Camphor	*d*-Camphor
endo-2-chlorobicyclo[2.2.1]-heptane	*endo*-2-Chlorobicyclo[2.2.1]-heptane	*endo*-2-Chlorobicyclo[2.2.1]-heptane
(*Z*)-5-chloro-4-pentenoic acid	(*Z*)-5-Chloro-4-pentenoic acid	(*Z*)-5-Chloro-4-pentenoic Acid
L-*threo*-2,3-dichlorobutyric acid	L-*threo*-2,3-Dichlorobutyric acid	L-*threo*-2,3-Dichlorobutyric Acid
cis-1,2-dichloroethene	*cis*-1,2-Dichloroethene	*cis*-1,2-Dichloroethene
(*S*)-2,3-dihydroxypropanoic acid	(*S*)-2,3-Dihydroxypropanoic acid	(*S*)-2,3-Dihydroxypropanoic Acid
exo-5,6-dimethyl-*endo*-bicyclo[2.2.2]octan-2-ol	*exo*-5,6-Dimethyl-*endo*-bicyclo[2.2.2]octan-2-ol	*exo*-5,6-Dimethyl-*endo*-bicyclo[2.2.2]octan-2-ol
(*E*)-diphenyldiazene	(*E*)-Diphenyldiazene	(*E*)-Diphenyldiazene
sn-glycerol 1-(dihydrogen phosphate)	*sn*-Glycerol 1-(dihydrogen phosphate)	*sn*-Glycerol 1-(Dihydrogen phosphate)
erythro-β-hydroxyaspartic acid	*erythro*-β-Hydroxyaspartic acid	*erythro*-β-Hydroxyaspartic Acid
syn-7-methylbicyclo[2.2.1]-heptene	*syn*-7-Methylbicyclo[2.2.1]-heptene	*syn*-7-Methylbicyclo[2.2.1]-heptene
trans-*cisoid*-*trans*-perhydrophenanthrene	*trans*-*cisoid*-*trans*-Perhydrophenanthrene	*trans*-*cisoid*-*trans*-Perhydrophenanthrene
meso-tartaric acid	*meso*-Tartaric acid	*meso*-Tartaric Acid
(1*Z*,4*E*)-1,2,4,5-tetrachloro-1,4-pentadiene	(1*Z*,4*E*)-1,2,4,5-Tetrachloro-1,4-pentadiene	(1*Z*,4*E*)-1,2,4,5-Tetrachloro-1,4-pentadiene

◆ Use hyphens to separate locants and configurational descriptors from each other and from the syllabic portion of the name.

D-arabinose
2-benzoylbenzoic acid
1,4-bis(2-ethylhexyl) sulfosuccinate
trans-2-bromocyclopentanol
3-chloro-4-methylbenzoic acid
cis-dichloroethylene
(*E*)-2-(3,7-dimethyl-2,6-octadienyl)-1,4-benzenediol
(1,4-dioxaspiro[4.5]dec-2-ylmethyl) guanidine
4-*O*-β-D-galactopyranosyl-D-fructose
N-hydroxy-*N*-nitrosobenzeneamine
α-ketoglutaric acid
N-methylmethanamine
tetrahydro-3,4-dipiperonyl-2-furanol

◆ Do not use hyphens to separate the syllables of a chemical name unless the name is too long to fit on one line.

End-of-Line Hyphenation of Chemical Names

The following is a list of prefixes, suffixes, roots, and some complete words hyphenated as they would be at the end of a line. To hyphenate a chemical name such as

5-(2-chloroethyl)-9-(diaminomethyl)-2-anthracenol

look up each syllable that is to be hyphenated in the list. Also, chemical names can be broken after hyphens that are integral in their names. Follow other standard rules for hyphenation of regular words; for example, try to leave at least three characters on each line. Thus, the example given could be hyphenated as follows:

5-(2-chlo-ro-eth-yl)-9-(di-ami-no-meth-yl)-2-anth-ra-cenol

Most desk dictionaries contain the names of common chemicals; they also give end-of-line hyphenation.

ace-naph-tho
ace-tal
acet-al-de-hyde
acet-amide
acet-am-i-do
acet-amin-o-phen
acet-an-i-lide
ace-tate
ac-et-azol-amide
ace-tic
ace-to
ace-to-ace-tic
ace-tone
ace-to-ni-trile
ace-tyl
acet-y-late
acet-y-lene
acro-le-in
ac-ryl-am-ide
ac-ry-late
acryl-ic
ac-ry-lo
ad-i-po-yl
al-kyl
al-lyl
ami-di-no
amide
ami-do
amine
ami-no
am-mine
am-mo-nio
am-mo-ni-um

an-thra
an-thra-cene
an-thra-ce-no
an-thryl
ar-se-nate
ar-si-no
aryl
azi-do
azi-no
azo
benz-ami-do
ben-zene
benz-hy-dryl
ben-zo-yl
ben-zyl
ben-zyl-i-dene
bi-cy-clo
bo-ryl
bro-mide
bro-mo
bu-tane
bu-ten-yl
bu-tyl
bu-tyl-ene
bu-tyl-i-dene
car-ba-mate
car-bam-ic
car-ba-mide
carb-an-ion
car-ba-ryl
car-ba-zole
car-bi-nol
car-bol-ic
car-bon-ate
car-bon-ic
car-bo-ni-um
car-bon-yl
car-box-a-mi-do
car-boxy
car-box-yl
car-byl-a-mi-no
chlo-ride
chlo-ro
chlo-ro-syl
chlo-ryl
cu-mene
cy-a-nate
cy-a-nide
cy-a-na-to
cy-a-no
cy-clo
cy-clo-hex-ane
cy-clo-hex-yl
di-azo
di-bo-ran-yl
di-car-bon-yl
di-im-ino
di-oxy
di-oyl
diyl
do-de-cyl
ep-oxy
eth-ane
eth-a-no
eth-a-nol
eth-a-no-yl
eth-en-yl
eth-yl
eth-yl-ene
eth-yl-i-dene
eth-yn-yl
fluo-res-cence
fluo-ride
fluo-ro
form-al-de-hyde
form-ami-do
for-mic
form-imi-do-yl
for-myl
fu-ran
ger-myl
gua-ni-di-no
gua-nyl
halo
hep-tane
hep-tyl
hex-ane
hex-yl
hy-dra-zide
hy-dra-zine
hy-dra-zi-no
hy-dra-zo
hy-dra-zo-ic
hy-dric
hy-dride
hy-dri-od-ic
hy-dro
hy-dro-chlo-ric
hy-dro-chlo-ride
hy-dro-chlo-ro
hy-drox-ide
hy-droxy
hy-drox-yl
imi-da-zole
imide
imi-do
imi-do-yl
imi-no
in-da-mine
in-da-zole
in-dene
in-de-no
in-dole
io-date
io-dide
iodo
io-do-syl
io-dyl
iso-cy-a-na-to
iso-cy-a-nate
iso-cy-a-nide
iso-pro-pen-yl
iso-pro-pyl
mer-cap-to
mer-cu-ric
meth-an-ami-do
meth-ane
meth-ano
meth-yl
meth-yl-ate
meth-yl-ene
meth-yl-i-dene
mono
mono-ac-id
mono-amine
naph-tha-lene
naph-tho
naph-thyl
neo-pen-tyl
ni-trate
ni-tric
ni-trile
ni-trilo
ni-trite
ni-tro
ni-troso
oc-tane
oc-tyl
ox-idase
ox-ide

ox-ido
ox-ime
oxo
ox-o-nio
oxy
palm-i-toyl
pen-tane
pen-tyl
pen-tyl-i-dene
per-chlo-rate
per-chlo-ride
per-chlo-ryl
per-man-ga-nate
per-ox-idase
per-ox-ide
per-oxy
phen-ac-e-tin
phen-an-threne
phen-an-thro
phen-an-thryl
phen-a-zine
phe-no
phe-nol
phe-none
phen-ox-ide
phen-oxy
phen-yl
phen-yl-ene
phos-phate
phos-phide
phos-phine
phos-phi-no
phos-phin-yl
phos-phite
phos-pho
phos-pho-nio
phos-pho-no
phos-phor-anyl
phos-pho-li-pase
phos-pho-lip-id
phos-pho-ni-um
phos-pho-ric
phos-pho-rus
phos-pho-ryl
plum-byl
pro-pane
pro-pen-yl
pro-pen-yl-ene
pro-pyl
pro-pyl-ene
pro-pyl-i-dene
pu-rine
py-ran
pyr-a-zine
pyr-a-zole
pyr-i-dine
pyr-id-a-zine
pyr-role
quin-o-line
qui-none
sel-e-nate
se-le-nic
sel-e-nide
sel-e-nite
se-le-no
si-lane
sil-anyl
sil-ox-anyl
sil-ox-yl
si-lyl
spi-ro
stan-nic
stan-nite
stan-nous
stan-nyl
stib-ino
sty-rene
sty-ryl
sul-fa-mo-yl
sul-fate
sul-fe-no
sul-fe-nyl
sul-fide
sul-fi-do
sul-fi-no
sul-fi-nyl
sul-fite
sul-fo
sul-fon-ami-do
sul-fo-nate
sul-fone
sul-fon-ic
sul-fo-nio
sul-fo-nyl
sulf-ox-ide
sul-fu-ric
sul-fu-rous
sul-fu-ryl
tet-ra
thio
thio-nyl
thio-phene
thi-oxo
thi-oyl
tol-u-ene
tol-u-ide
tol-yl
tri-a-zine
tri-a-zole
tri-yl
urea
ure-ide
ure-id-o
uric
vi-nyl
vi-nyl-i-dene
xan-thene
xan-tho
xy-lene
xy-li-dine
xy-lyl
xy-li-din-yl
yl-i-dene

Polymers

Polymer names are often one or two words in parentheses following the prefix “poly”. “Poly” is a syllabic prefix, not a descriptor, and thus is set in roman type.

Here is a short list of correctly formatted names of frequently cited polymers. (These names are not necessarily IUPAC or CA index preferences.)

nylon-6
nylon-6,6
polyacrylamide
poly(acrylic acid)
polyacrylonitrile
polyamide
poly(aryl sulfone)
polybutadiene
1,2-polybutadiene
1,4-polybutadiene
poly(butyl acrylate)
poly(butyl methacrylate)
poly(*n*-butyl methacrylate)
poly(butylene terephthalate)
polycarbonate
polychloroprene
poly(*N*,*N*-dimethylacrylamide)
poly(dimethylsiloxane)
polyester
polyether
poly(ether imide)
poly(ether ketone)
poly(ether sulfone)
poly(ethyl acrylate)
poly(ethyl methacrylate)
polyethylene
poly(ethylene adipate)
poly(ethylene glycol)
poly(ethylene oxide)
poly(ethylene terephthalate)
polyformaldehyde
poly(*N*,*N*′-hexamethyleneadipamide)
polyimidazole
polyimide
poly(isobutyl methacrylate)
polyisobutylene
polyisoprene
poly(methacrylic acid)
poly(methyl acrylate)
poly(methyl methacrylate)
poly(methylene)
poly(oxyethylene)
poly(oxymethylene)
poly(oxy-1,4-phenylene)
poly(phenylene ether)
poly(phenylene oxide)
poly(phenylene sulfide)
polypropylene
poly(propylene glycol)
polystyrene
polysulfide
polysulfone
poly(tetrafluoroethylene)
poly(tetramethylene oxide)
polythiazole
poly(thiocarbonate)
polyurethane
poly(vinyl acetate)
poly(vinyl alcohol)
poly(vinyl butyral)
poly(vinyl chloride)
poly(vinyl ether)
poly(vinyl trichloroacetate)
poly(vinylidene chloride)
poly(vinylpyrrolidone)
povidone

◆ In text, keep polymer names lowercase. As the first word of a sentence and in titles or headings, capitalize only the first letter ("P").

Poly(vinyl chloride) is a less useful polymer than poly(ethylene glycol).

Reactions of Poly(methyl methacrylate)

New Uses for Poly(ethylene terephthalate)

◆ In copolymer nomenclature, descriptive lowercase italic infixes may be used. These include alt, blend, block (or b), co, cross, graft (or g), inter, per, stat, and ran. At the beginning of a sentence, in titles, and in headings, capitalize only the "p" of the examples that follow.

polybutadiene-*graft*-[polystyrene:poly(methyl methacrylate)]
poly(*cross*-butadiene)
poly[*cross*-(ethyl acrylate)]-*inter*-polybutadiene
poly(ethylene-*alt*-carbon monoxide)
polyisoprene-*blend*-polystyrene
poly[(methyl methacrylate)-*b*-(styrene-*co*-butadiene)]
poly[(methyl methacrylate)-*co*-styrene]
polystyrene-*block*-polybutadiene
poly(styrene-*co*-butadiene)
poly(styrene-*g*-acrylonitrile)
poly(vinyl trichloroacetate)-*cross*-polystyrene

Amino Acids

◆ Always capitalize the three-letter and one-letter abbreviations for amino acids.

Amino Acid	*Three-Letter Abbreviation*	*One-Letter Abbreviation*
alanine	Ala	A
arginine	Arg	R
asparagine	Asn	N
aspartic acid	Asp	D
cysteine	Cys	C
glutamic acid	Glu	E
glutamine	Gln	Q
glycine	Gly	G
histidine	His	H
isoleucine	Ile	I
leucine	Leu	L
lysine	Lys	K
methionine	Met	M
phenylalanine	Phe	F
proline	Pro	P
serine	Ser	S
threonine	Thr	T
tryptophan	Trp	W
tyrosine	Tyr	Y
valine	Val	V

◆ In sequences of amino acids, separate the three-letter abbreviations with hyphens.

Pro-Gln-Ile-Ala

- Do not use abbreviations for individual amino acids in running text.

  Selective labeling of arginine and serine made it possible to monitor the kinetics of folding of the individual residues.

Chemical Abstracts Service Registry Numbers

David W. Weisgerber
Chemical Abstracts Service

Chemical substances, their syntheses, the determination of their properties, and their applications are the core of chemistry and the main occupation of chemists. In their communications, chemists represent chemical substances by structural diagrams, names, molecular formulas, codes, and identification numbers. One of the most frequently used identification numbers is the Chemical Abstracts Service (CAS) Registry Number. Today, CAS Registry Numbers are often used to identify chemical substances in handbooks, indexes, databases, and inventories, and even on many commercial product labels.

The CAS Chemical Registry System is a computer-based system that uniquely identifies chemical substances on the basis of their molecular structures. Begun originally in 1965 to support indexing for *Chemical Abstracts*, the Registry System now serves not only as a support system for identifying substances within CAS operations but also as an international resource for chemical substance identification by scientists, industry, and regulatory bodies.

The CAS Chemical Registry database is the largest collection of information on naturally occurring and synthetic chemical substances in the world, including organic compounds, inorganic compounds, polymers, coordination compounds, alloys, biosequences, and minerals. It contains CAS Registry Numbers, structures, and names for more than 16 million substances, covering the chemical literature from 1957 to the present, in addition to substances registered from special collections, for governmental and industrial organizations, and for individual requesters. CAS Registry Numbers may be obtained for a fee through CAS Client Services.

CAS Registry Numbers are assigned in sequential order as substances are entered into the CAS Registry database for the first time; the numbers have no chemical significance. Registry Numbers link the molecular structure diagram, systematic CA index name, synonyms, molecular formula, and other identifying information for each substance. Because CAS Registry Numbers are independent of the many different systems of chemical nomenclature, they can bridge them and link often unrecognized synonymous names.

A format was developed using hyphens to make the numbers easier to read and to recognize. A CAS Registry Number includes up to nine digits that are separated into three parts by hyphens. The first part, starting from the left, has up to six digits, the second part has two digits, and the final part is a single check digit to verify the validity of the total number (e.g., 7732-18-5 for water).

Within the Registry System, each substance is assigned a separate CAS Registry Number. For example, each salt of an acid receives a distinct number, individual stereoisomers receive distinct numbers, and an ion receives a number different from that of the neutral compound.

The CAS Registry began initially as an internal indexing and processing aid at CAS, but it has developed into an international resource for chemical substance identification. The CAS Registry Numbers generated by the system have become a standard for substance identification throughout the industrial world, bridging the many differences in systematic, generic, proprietary, and trivial substance names.

CAS Registry Numbers are included in the printed *Chemical Abstracts* chemical substance and formula indexes and in the CAS databases. The full set of CAS Registry database information—structures, names, formulas, and ring data—is available for search and display through STN International. CAS Registry information is also available in CAS databases offered by other online system vendors.

In addition to their inclusion in the CAS databases, CAS Registry Numbers are used in many public and private databases. Currently, the cluster of databases on STN International containing Registry Numbers includes 7 CAS databases and 53 non-CAS databases. Among the latter are such diverse files as AIDSLINE, BIOSIS, Design Institute for Physical Property Data File (DIPPR), MEDLINE, Natural Products Alert (NAPRALERT), Pharmaceutical News Index (PNI), and Plastics Materials Selection Database (PLASPEC).

Many handbooks, guides, and other reference works include CAS Registry Numbers and provide special indexes that allow the reader to find the proper place in the text without first having to identify the full name of the substance. The reader benefits because the full name may differ from handbook to handbook.

Among these reference works with CAS Registry Numbers are the *CRC Handbook of Chemistry and Physics*, the *Kirk-Othmer Encyclopedia of Chemical Technology*, *The Merck Index*, the *USP Dictionary of USAN and International Drug*

Names, the National Institute of Occupational Safety and Health (NIOSH) Registry of Toxic Effects of Chemical Substances, and the proposed and recommended International Nonproprietary Names lists published by the World Health Organization.

CAS Registry Numbers are also widely used as standard identifiers for chemical substances in many of the commercial chemical inventories of governmental regulatory agencies, such as the Toxic Substances Control Act (TSCA) Inventory in the United States, the Ministry of International Trade and Industry (MITI) Inventory in Japan, the European Inventory of Existing Commercial Chemical Substances (EINECS), and the Canadian Domestic and Non-Domestic Substances Lists (DSL/NDSL).

Whenever a chemical substance is sold, transported, imported, exported, reported to a regulatory agency, or disposed of, a CAS Registry Number is probably involved.

Bibliography

The International Union of Pure and Applied Chemistry (IUPAC) Nomenclature Documents Home Page on the World Wide Web can be found at http://www.chem.qmw.ac.uk/iupac/. This site contains information on IUPAC itself, the names and publishers of references on chemical nomenclature, and many of the recommendations.

Organic Chemistry

IUPAC. *Nomenclature of Organic Chemistry: Sections A, B, C, D, E, F, and H;* Rigaudy, J., Klesney, S. P., Eds.; Pergamon: Elmsford, NY, 1979 (commonly referred to as the "blue book").

A Guide to IUPAC Nomenclature of Organic Compounds, Recommendations 1993; Panico, R., Powell, W. H., Richer, C., Eds.; Blackwell Scientific Publications: Oxford, England, 1993.

Ring Systems Handbook; American Chemical Society: Columbus, OH, 1988 (and supplements).

Inorganic Chemistry

IUPAC. *Nomenclature of Inorganic Chemistry, Recommendations 1990;* Blackwell Scientific Publications: Oxford, England, 1990 (commonly referred to as the "red book").

Block, B. P.; Powell, W. H.; Fernelius, W. C. *Inorganic Chemical Nomenclature: Principles and Practice;* American Chemical Society: Washington, DC, 1990.

General Chemistry

Chemical Abstracts Index Guide; American Chemical Society: Columbus, OH; Appendix IV (updated periodically).

Compendium of Chemical Nomenclature; Gold, V., Loening, K. L., McNaught, A. D., Sehmi, P., Eds.; Blackwell Scientific Publications: Oxford, England, 1993.

The CRC Handbook of Chemistry and Physics, 77th ed.; Lide, D. R., Ed.; CRC Press: Boca Raton, FL, 1996.

Kirk-Othmer Encyclopedia of Chemical Technology, 4th ed.; Wiley & Sons: New York, 1993; Vol. 5.

The Merck Index: An Encyclopedia of Chemicals, Drugs, and Biologicals, 12th ed.; Budavari, S., Ed.; Merck & Company: Whitehouse Station, NJ, 1996.

Biochemistry

Biochemical Nomenclature and Related Documents, 2nd ed.; Liébecq, C., Ed.; Portland Press: London, 1992 (commonly referred to as the "white book").

Enzyme Nomenclature; Academic: New York, 1992.

Drug Names

The USP Dictionary of USAN and International Drug Names; U.S. Pharmacopeial Convention: Rockville, MD, 1995 (updated annually).

Polymer Chemistry

IUPAC. *Compendium of Macromolecular Nomenclature;* Blackwell Scientific Publications: Oxford, England, 1991.

Advice on *Chemical Abstracts* nomenclature is available from the Manager of Nomenclature Services, Department 64, Chemical Abstracts Service, P.O. Box 3012, Columbus, OH 43210. A name generation service is available through CAS Client Services, Chemical Abstracts Service, P.O. Box 3343, Columbus OH 43210; e-mail address, answers@cas.org; and URL, http://www.cas.org/Support/client.html.

Acknowledgment

We thank the ACS Nomenclature Committee, chaired by John A. Secrist, for reading this chapter and advising us on its contents.

How Many Words?

One word:

ethylenediamine
methanethiolate
methylcellulose
tetraethyllead

Two words:

diethyl ether
diethyl sulfide
dimethyl sulfoxide
ethylene glycol
ethylene oxide

Three words:

lithium aluminum hydride
sodium dodecyl sulfate

Which Words?

Correct forms for alcohols are

1-butanol *or* butyl alcohol
2-butanol *or* *sec*-butyl alcohol
2-methyl-1-propanol *or* isobutyl alcohol, *not* isobutanol
2-methyl-2-propanol *or* *tert*-butyl alcohol
1-propanol *or* propyl alcohol
2-propanol *or* isopropyl alcohol, *not* isopropanol

Incorrect: any combination of n, sec, or tert with the words butanol or propanol.

CHAPTER *8*

Conventions in Chemistry

This chapter presents a quick reference guide for the use of typefaces (roman, italic, and bold), Greek letters, superscripts and subscripts, and special symbols that are commonly used in chemistry.

Detailed recommendations from the International Union of Pure and Applied Chemistry (IUPAC) are given in the book titled *Quantities, Units and Symbols in Physical Chemistry*, 2nd edition, nicknamed the "green book", published by Blackwell Science, Oxford, England, 1993. Updates are published as articles in the journal *Pure and Applied Chemistry.*

Detailed recommendations from the International Organization for Standardization (ISO) are given in the ISO Standards Handbook 2, *Quantities and Units,* published by ISO, Geneva, Switzerland, 1993.

Subatomic Particles and Quanta

◆ Use lowercase Latin or Greek letters for abbreviations for subatomic particles.

alpha particle	α	neutrino	ν_e
beta particle	β	neutron	n
deuteron	d	photon	γ
electron	e	pion	π
helion	h	proton	p
muon	$\mu^{\pm}$	triton	t

◆ Indicate electric charges with the appropriate superscript (+, –, or 0).

n^0 e^+ e^- $\pi^{\pm}$

If the symbols p and e are used without indication of charge, they refer to positive proton and negative electron, respectively.

Electronic Configuration

◆ Denote electron shells with the uppercase roman letters K, L, M, and N.

◆ Name electron subshells and atomic orbitals with the lowercase roman letters s, p, d, and f. Write principal energy levels 1–7 on the line and to the left of the letter; give the number of electrons in the orbital as a superscript to the right of the letter. Specify orbital axes with italic subscripts.

7s electron	$5f^2$ ions	5f orbital
6d orbital	sp^3 hybrid orbital	$f^{n-3}ds^2$ configuration
$3d^4 4s 4p^2$ configuration	$p_x p_y p_z$	$d_{xz} d_{yz} d_{xy}$
d_{z^2}	$d_{x^2-y^2}$	

The ground state of boron is $1s^2 2s^2 2p_x^{\ 1} 2p_y^{\ 0} 2p_z^{\ 0}$.
The valence-shell configuration of nitrogen is $2s^2 2p_x^{\ 1} 2p_y^{\ 1} 2p_z^{\ 1}$.
The electronic configuration of potassium is $1s^2 2s^2 2p^6 3s^2 3p^6 4s^1$.
The valence-electron configuration is described by $5d^{10} 6s^1$.

◆ Use Greek letters for some bonding orbitals and the bonds they generate.

π bond	σ orbital	σ* orbital

◆ Name the *electronic states of atoms* with the uppercase roman letters S, P, D, F, G, H, I, and K, corresponding to quantum numbers *l* = 0–7. Use the corresponding lowercase letters to indicate the orbital angular momentum of a single electron. The left superscript is the spin multiplicity; the right subscript is the total angular momentum quantum number *J*.

$^2S_0$	$^4P_{1/2}$	$^7F_0$	$^7D_1$
$^2s_0$	$^4p_{1/2}$	$^7f_0$	$^7d_1$
$^8F_{1/2}$	$^8G_{1/2}$	$^2P_{3/2}$	
$^8f_{1/2}$	$^8g_{1/2}$	$^2p_{3/2}$	

◆ Name the *electronic states of molecules* with the uppercase roman letters A, B, E, and T; the ground state is X. Use the corresponding lowercase letters for one-electron orbitals. A tilde (~) is added for polyatomic molecules. The subscripts describe the symmetry of the orbital.

$\tilde{A}$	$^2A_{1g}$	A_{2g}	$^3B_1$
$\tilde{a}$	$^2a_{1g}$	a_{2g}	$^3b_1$
E_g	E_{2g}	T_{2g}	
e_g	e_{2g}	t_{2g}	

Chemical Elements and Formulas

◆ Write the names of the chemical elements in roman type and treat them as common nouns.

calcium	californium	carbon
einsteinium	francium	hydrogen
helium	oxygen	seaborgium
uranium		

◆ Write the symbols for the chemical elements in roman type with an initial capital letter.

Ca	Cf	C	Es
Fr	H	He	O
Sg	U		

The complete list of chemical elements and symbols is given in Table 1.

◆ Even when symbols are used, the element's name is pronounced. Therefore, choose the article (a or an) preceding the element symbol to accommodate the pronunciation of the element name. (This usage does not apply to isotopes, as described in the section on isotopes.)

a Au electrode (pronounced "a gold electrode")
a N-containing compound (pronounced "a nitrogen-containing compound")
a He–Ne laser (pronounced "a helium–neon laser")

◆ Write the names of chemical compounds in roman type and treat them as common nouns. (Names for chemical compounds are discussed further in Chapter 7.)

benzaldehyde	ferric nitrate	mercuric sulfate
calcium carbonate	hydrochloric acid	methyl salicylate
chlorobenzene	isopropyl iodide	phenol
ethanol	magnesium sulfate	sodium hydroxide

◆ Use roman type for the symbols for chemical compounds.

$BaSO_4$	C_2H_5OH	C_6H_5OH
$CaCO_3$	CH_3COOH	C_6H_5Cl
$Fe(NO_3)_3$	H_3PO_4	HCl
$HgSO_4$	NaOH	$Ni_3P_2O_8$
P_2S_5	VF_5	$Zn(C_2H_3O_2)_2$

◆ You may use both chemical symbols and element names in text, but it is best to use one or the other consistently. Do not mix symbols and words within a name.

NaCl *or* sodium chloride, *not* Na chloride

Table 1. Element Symbols, Atomic Numbers, and Atomic Weights

Element	*Symbol*	*Atomic Number*	*Atomic Weight*[a]
actinium	Ac	89	(227.03)
aluminum	Al	13	26.98
americium	Am	95	(243.06)
antimony	Sb	51	121.76
argon	Ar	18	39.95
arsenic	As	33	74.92
astatine	At	85	(209.98)
barium	Ba	56	137.33
berkelium	Bk	97	(247.07)
beryllium	Be	4	9.01
bismuth	Bi	83	208.98
boron	B	5	10.81
bromine	Br	35	79.90
cadmium	Cd	48	112.41
calcium	Ca	20	40.08
californium	Cf	98	(251.08)
carbon	C	6	12.01
cerium	Ce	58	140.12
cesium	Cs	55	132.91
chlorine	Cl	17	35.45
chromium	Cr	24	51.99
cobalt	Co	27	58.93
copper	Cu	29	63.55
curium	Cm	96	(247.07)
dysprosium	Dy	66	162.50
einsteinium	Es	99	(252.08)
erbium	Er	68	167.26
europium	Eu	63	151.97
fermium	Fm	100	(257.09)
fluorine	F	9	18.99
francium	Fr	87	(223.01)
gadolinium	Gd	64	157.25
gallium	Ga	31	69.72
germanium	Ge	32	72.61
gold	Au	79	196.97
hafnium	Hf	72	178.49
hahnium[b]	Ha	105	(262.11)
hassium[b]	Hs	108	
helium	He	2	4.00
holmium	Ho	67	164.93
hydrogen	H	1	1.01
indium	In	49	114.82
iodine	I	53	126.90
iridium	Ir	77	192.22

Continued on next page

Table 1. Element Symbols, Atomic Numbers, and Atomic Weights—Continued

Element	*Symbol*	*Atomic Number*	*Atomic Weight*[a]
iron	Fe	26	55.85
krypton	Kr	36	83.80
lanthanum	La	57	138.91
lawrencium	Lr	103	(262.11)
lead	Pb	82	207.2
lithium	Li	3	6.94
lutetium	Lu	71	174.97
magnesium	Mg	12	24.31
manganese	Mn	25	54.94
meitnerium[b]	Mt	109	
mendelevium	Md	101	(258.10)
mercury	Hg	80	200.59
molybdenum	Mo	42	95.94
neodymium	Nd	60	144.24
neon	Ne	10	20.18
neptunium	Np	93	(237.05)
nickel	Ni	28	58.69
nielsbohrium[b]	Ns	107	(262.12)
niobium	Nb	41	92.91
nitrogen	N	7	14.01
nobelium	No	102	(259.10)
osmium	Os	76	190.2
oxygen	O	8	15.99
palladium	Pd	46	106.42
phosphorus	P	15	30.97
platinum	Pt	78	195.08
plutonium	Pu	94	(244.06)
polonium	Po	84	(208.98)
potassium	K	19	39.1
praseodymium	Pr	59	140.91
promethium	Pm	61	(144.91)
protactinium	Pa	91	(231.04)
radium	Ra	88	(226.03)
radon	Rn	86	(222.02)
rhenium	Re	75	186.21
rhodium	Rh	45	102.91
rubidium	Rb	37	85.47
ruthenium	Ru	44	101.07
rutherfordium[b]	Rf	104	(261.11)
samarium	Sm	62	150.36
scandium	Sc	21	44.96
seaborgium[b]	Sg	106	(263.12)
selenium	Se	34	78.96

Continued on next page

Table 1. Element Symbols, Atomic Numbers, and Atomic Weights—Continued

Element	*Symbol*	*Atomic Number*	*Atomic Weight*[a]
silicon	Si	14	28.09
silver	Ag	47	107.87
sodium	Na	11	22.99
strontium	Sr	38	87.62
sulfur	S	16	32.07
tantalum	Ta	73	180.95
technetium	Tc	43	(97.91)
tellurium	Te	52	127.60
terbium	Tb	65	158.93
thallium	Tl	81	204.38
thorium	Th	90	(232.04)
thulium	Tm	69	168.93
tin	Sn	50	118.71
titanium	Ti	22	47.88
tungsten	W	74	183.85
uranium	U	92	238.03
vanadium	V	23	50.94
xenon	Xe	54	131.29
ytterbium	Yb	70	173.04
yttrium	Y	39	88.91
zinc	Zn	30	65.39
zirconium	Zr	40	91.22

NOTE: This table is mostly based on the guidelines of the Inorganic Chemistry Division, Commission on Atomic Weights and Isotopic Abundances, "Atomic Weights of the Elements 1993", which appeared in *Pure Appl. Chem.* **1994,** *66* (12), 2423–2444.

[a] Values in parentheses are the relative atomic masses of the longest-lived isotopes.

[b] Names recommended by the ACS Committee on Nomenclature.

◆ You may use common abbreviations for organic groups in formulas and structures, but not in text. These (and only these) abbreviations need not be defined.

Ac	acetyl	Bz	benzoyl
Ar	aryl	Et	ethyl
Bu	butyl	Me	methyl
i-Bu	isobutyl	Ph	phenyl
n-Bu	*n*-butyl	Pr	propyl
sec-Bu	*sec*-butyl	*i*-Pr	isopropyl
t-Bu	*tert*-butyl	R, R′	alkyl

◆ Use square brackets in formulas for *coordination entities.*

$[Cr(C_6H_6)_2]$ $K[PtCl_3(C_2H_4)]$

◆ In the formula for an *addition compound,* use a centered dot, closed up on each side. (Although the IUPAC books show a space on each side, this spacing would wreak havoc with many typesetting systems.)

$BH_3{\cdot}NH_3$ $Ni(NO_3)_2{\cdot}2Ni(OH)_2$

Water of hydration follows a centered dot, closed up on each side.

$Na_2SO_4{\cdot}10H_2O$ $Zn(NO_3)_2{\cdot}H_2O$

◆ Use either a slash or an en dash between components of a mixture, but not a colon.

dissolved in 5:1 glycerin/water
dissolved in 5:1 glycerin–water

the metal/ligand (1:1) reaction mixture
the metal–ligand (1:1) reaction mixture
the metal–ligand (1/1) reaction mixture

the methane/oxygen/argon (1/50/450) matrix
the methane/oxygen/argon (1:50:450) matrix

Reference to the Periodic Table

◆ Always use lowercase for the word "group", even with a specific number.

group 15 elements group IVB elements

◆ Always use lowercase for the words "periodic table".

The elements in group 8d of the periodic table are Fe, Ru, and Os.

Atoms and Molecules

Nuclide descriptors are specified with superscripts and subscripts to the element symbol, as follows.

Use the Left Superscript for Mass Number

◆ The mass number of an atom is usually shown only for isotopes or in discussions of isotopes.

^{32}S ^{12}C ^{35}Cl

Use the Left Subscript for Atomic Number

◆ The atomic number of an atom is usually used only in discussions of nuclear chemistry.

$_{6}C$ $_{16}S$

Use the Right Superscript for Ionic Charge

◆ The charge number is followed by the sign of the ionic charge. When the charge number is 1, only the sign is used.

Na^{+} NO_3^{-} Ca^{2+}

◆ Stagger the subscript and superscript; do not align them. The subscript comes first.

$PO_4{}^{3-}$

◆ Do not use multiple plus or minus signs, and do not circle the charge.

Hg^{2+}, *not* Hg^{++}

Use the Right Asterisk for Excited Electronic State

He* NO*

Use the Right Superscript for Oxidation Number

◆ You may use superscript roman numerals for oxidation numbers. In formulas, do not use numbers on the line to avoid confusion with the symbols for iodine or vanadium.

Co^{III}	O^{-II}	Ni^{0}
$Mn^{III/IV}$	$Pb^{IV}O_2$	$(NH_3)_2Pt^{II}$
$Fe^{II}Cl_2$	$Mn^{IV}O_2$	$Ru^{II}–Ru^{III}$
Ru^{II}/Ru^{III}	$Mn^{III}–Mn^{IV}$	Mn^{III}/Mn^{IV}

◆ Stagger the subscript and superscript; do not align them. The subscript follows the superscript.

$Pb^{II}{}_2$

◆ You may also write oxidation numbers on the line in parentheses closed up to the element name or symbol.

copper(II)	Cu(II)	iron(II)
Fe(II)	cobalt(III)	Co(III)
manganese(IV) oxide	ferrate(VI) ion	diammineplatinum(II)

potassium tetracyanonickelate(0)	iron(II) chloride
Mn(III)–Mn(IV) complex	Mn(III)/Mn(IV) complex

Use the Right Subscript for Number of Atoms

◆ With an element symbol, use a subscript to indicate the number of atoms, whether in formulas or in narrative text.

C_6	Fe_3	NH_4
Al_2O_3	$FeSi_2$	H_2S
$C_6H_5CH_3$	$(CH_3)_4C$	

The C_{60} fullerene molecule is shaped like a soccer ball.

◆ With an element name, follow the usual conventions for numbers in text.

Molecules composed of 60 carbon atoms are shaped like soccer balls.

In this reaction, three hydrogen atoms are lost.

Atom in a Specific Position

◆ Use either words or symbols and numbers on the line to refer to an atom in a specific position.

at the carbon in the 6-position *or* at C6 *or* at C-6

the atom in the β-position *or* the β atom

Isotopes

◆ Specify the isotope of an element by a mass number written as a left superscript to the element symbol.

^{29}Si	^{13}C	^{32}S
^{15}N	^{235}U	

◆ Alternatively, indicate an isotope by using the spelled-out element name hyphenated to its mass number.

uranium-235 carbon-14

◆ In either case, the isotope name or symbol is pronounced first, then the number. Thus, ^{14}C is pronounced "c fourteen". Consequently, choose the article (a or an) preceding the isotope to accommodate the pronunciation of the element name or symbol, not the number.

a carbon-14 isotope	a ^{14}C isotope (pronounced "c fourteen")
a nitrogen-15 isotope	an ^{15}N isotope (pronounced "en fifteen")
a hydrogen-3 isotope	an ^{3}H isotope (pronounced "aitch three")

◆ Use the symbols D for deuterium and T for tritium when no other nuclides are present.

D_2O	CD_2H_2	HD
CH_3OT	$(T_2N)_2CO$	$HDSO_4$
D_2S	CH_2TOH	

An *isotopically unmodified* compound is one whose isotopic nuclides are present in the proportions that occur in nature. An *isotopically modified* compound has a nuclide composition that deviates measurably from that occurring in nature.

An *isotopically substituted* compound has a composition such that all of the molecules of the compound have only the indicated nuclides at the designated positions. To indicate isotopic substitution in formulas, the nuclide symbols are incorporated into the formulas. To indicate isotopic substitution in spelled-out compound names, the number and symbol (and locants if needed) are placed in parentheses closed up to the name.

$N^{15}NF_2$	$^{24}NaCl$	$H_2N^{14}CONH_2$
$^{14}CH_4$	$^{238}UCl_3$	$^{32}PO_4^{3-}$
$Mo(^{12}CO)_6$	$Na_2{}^{35}S$	
(^{15}N)ammonia	(^{14}C)glucose	(1,3-$^3H_2$)benzene
2,4-diamino(^{18}O)phenol	chloro(^{3}H)benzene	

An *isotopically labeled* compound is a mixture of an isotopically unmodified compound with an analogous isotopically substituted compound or compounds. Isotopically labeled compounds may be specifically labeled or selectively labeled. To indicate isotopic labeling, the number and symbol (and locants if needed) are enclosed in square brackets closed up to the compound name or formula.

specifically labeled:	$[^{14}C]H_4$	$CH_2[^2H_2]$	$CH_3CH_2[^{18}O]H$
selectively labeled:	$[^{15}N]NaNO_2$	$[^{15}N]NH_4Cl$	$[^{36}Cl]SOCl_2$

[6,7-^{15}N]adenosine	[^{15}N]alanine	[^{15}N]ammonium chloride
[1,3-$^3H_2$]benzene	chloro[^{3}H]benzene	[^{57}Co]cyanocobalamin
2,4-diamino[^{18}O]phenol	[2,8-^{3}H]inosine	[2-^{14}C]leucine

◆ When the isotope position is specified by a group name that is part of the parent compound, italicize the group name.

[*methyl*-^{14}C]toluene

◆ Isotopically labeled compounds may also be described by inserting the symbol in brackets into the name of the compound.

hydrogen [^{36}Cl]chloride [^{35}S]sulfuric [^{2}H]acid

◆ Do not use the left superscript within an abbreviation.

[^{32}P]CMP, *not* CM^{32}P

◆ To indicate *general labeling*, use the symbol G in the names of selectively labeled compounds in which all positions of the designated element are labeled, but not necessarily in the same isotopic ratio.

D-[G-^{14}C]glucose

◆ To indicate *uniform labeling*, use the symbol U in the names of selectively labeled compounds in which all positions of the designated element are labeled in the same isotopic ratio.

D-[U-^{14}C]glucose

◆ When it is unknown or irrelevant whether the compound is isotopically labeled or isotopically substituted, simply hyphenate the isotope symbol to the compound name and do not use square brackets or parentheses.

^{14}C-glucose ^{15}N-adenosine ^{3}H-benzene

The *Boughton system*, used in *Chemical Abstracts*, does not distinguish between labeling and substitution. The isotopic variation is shown by the symbol for the isotope (with a subscript numeral to indicate the number of isotopic atoms) placed after the name or relevant portion of the name; locants are cited if necessary. The locants and symbols are in italics, except subscripts and Greek letters, and hyphens are used to separate them.

acetamide-*1*-^{13}*C*-^{15}*N*   benzeneacetic-*carboxy*,α-^{14}*C*$_2$ acid
toluene-*methyl*-^{14}*C*   4-(2-propenyl-*3*-^{13}*C*-oxy)benzoic acid
acetic-^{17}*O*$_2$ acid   benzoic-^{18}*O* acid

◆ In this system, deuterium and tritium are represented by italic lowercase letters *d* and *t*, respectively.

ethane-*1*-*d*-*2*-*t*   methane-d_4 acetic-t_3 acid-*t*
urea-t_4 ammonia-*d*-*t*   methan-*t*-ol
tri(silyl-d_3)phosphine   alanine-*N*,*N*,*1*-d_3
1-(ethyl-*2*,*2*,*2*-d_3)-4-(methyl-d_3)benzene

Radicals

◆ In the formula of a free radical, indicate the unshared electron by a superscript or centered dot closed up to the element symbol or formula.

$\bullet CH_3$ $C_6H_5\bullet$ $HO^{\bullet}$
$H^{\bullet}$ $\bullet SnH_3$ $(SiH_3)^{\bullet}$

$\bullet SH$ $\bullet NH_2$ $Br^{\bullet}$
$Br\bullet$

◆ Charged radical cations and anions are often indicated by the symbol, formula, or structure with a superscript dot followed by a plus or minus sign. However, in mass spectrometry, the reverse is used. Therefore, use the order of dots and signs for charges that is appropriate for the context.

$R^{\bullet -}$ $R_2^{\bullet +}$ $R^{\bullet 2-}$
$C_6H_5NO^{\bullet 3-}$ $HCO^{\bullet +}$ $(Ag_2)^{\bullet +}$
$(SO_2)^{\bullet -}$

mass spectrometry: $R^{+\bullet}$ $C_3H_6^{+\bullet}$

Bonds

◆ For linear formulas in text, do not show single bonds unless the bonds are the subject of the discussion.

H_2SO_4 $C_6H_5CH_3$ CH_3COOH
$CH_3CHOHCH_3$ $C_6H_5COOCOCH_3$

◆ When necessary for the discussion, indicate bonds by en dashes.

the $–CH_2–$ segment the C–H distances the C–C–C angle

◆ When necessary, show double and triple bonds in linear formulas.

$CH_3C{\equiv}CH$ $CH_2{=}CH_2$

R–C(=O)–OH (O shown double-bonded below C) *is better as* RCOOH *or* RCO_2H *or* RC(=O)OH

◆ Use three centered dots to indicate association of an unspecified type (e.g., hydrogen bonding, bond formation, or bond breaking).

C···Pt F···H–NH_3 Mg^{2+}···O– Ni···Al
H_2O···π aromatic hydrogen bonding

Crystallography

Planes and Directions in Crystals

◆ Miller indices of a crystal face or a single net plane are enclosed in parentheses. (123) or (*hkl*) is a plane or set of planes that describe crystal faces; $(h_1h_2h_3)$ is a single net plane.

◆ Laue indices are not enclosed. 123 or *hkl* is the Bragg reflection from the set of net planes (123) or (*hkl*), respectively.

◆ Indices of a set of all symmetrically equivalent crystal faces or net planes are enclosed in braces. {*hkl*} is a form.

◆ Indices of a zone axis or lattice direction are enclosed in square brackets. [123] or [*uvw*] is a direction.

◆ Indices of a set of symmetrically equivalent lattice directions are enclosed in angle brackets. <*uvw*> represents all crystallographically equivalent directions of the type [*uvw*].

$1\bar{2}0$ 1,10,1 11,0,1 1,–2,0 reflections
the (111) face the (120) face the [001] axis
the [010] direction the [101] direction
*h*00 diffraction lines the *hk*0 zone
the 002 reflection the 00*l* class of reflections

◆ When indices are used with spelled-out element names, separate the name of the element and the index with a space.

the silver (110) surface copper (111) rhenium (010)
on silicon (111) surfaces a gold (111) substrate

◆ However, when indices are used with element symbols, close up the element symbol to the index.

Cu(111) Rh(010) GaAs(100)
Si(400) Au(210) CdTe(100)
the Ag(110) surface an iodine-modified Ag(111) electrode

Types of Crystal Lattices

fcc	face-centered cubic
bcc	body-centered cubic
hcp	hexagonal close-packed
ccp	cubic close-packed

Symmetry Operations and Structural Point Groups

◆ Use italic type for the letters in symmetry operations and structural point groups. The symbols (Schoenflies) are as follows: *E*, identity; *C*, cyclic; *D*, dihedral; *T*, tetrahedral; *O*, octahedral; *I*, icosahedral; *S*, rotation–reflection; and σ, mirror plane. Align subscripts and superscripts.

C_1 C_i C_s C_2
C_2^3 C_2^4 C_∞ C_{2v}
C_{3v} C_{4v} $C_{\infty v}$

C_{2h}	C_{3h}	D_{2d} (V_d)	D_{3d}
D_{4d}	D_{2h} (V_h)	D_{3h}	D_{4h}
$D_{\infty h}$	T_d	O_h (K_h)	
I_h	S_3	S_4	$2S_6$
S_8	σ	$2\sigma_v$	$3\sigma_v$
$4\sigma_v$	$3\sigma_d$	σ_h	

Crystallographic Point Groups

◆ Use arabic numbers or combinations of numbers and the italic letter *m* to designate the 32 crystallographic point groups (Hermann–Mauguin). The number is the degree of the rotation, and *m* stands for mirror plane. Use an overbar to indicate rotation inversion.

1	*m*	2/*m*
*mm*2	*mmm*	4*mm*
32	622	6/*mmm*
$\bar{4}3m$	$m\bar{3}m$	

Space Groups

◆ Designate space groups by a combination of unit cell type and point group symbol, modified to include screw axes and glide planes (Hermann–Mauguin); 230 space groups are possible. Use italic type for conventional types of unit cells (or Bravais lattices): *P*, primitive; *I*, body-centered; *A*, A-face-centered; *B*, B-face-centered; *C*, C-face-centered; *F*, all faces centered; and *R*, rhombohedral.

Pnma	*C*2/*c*	*Pbcn*
$I4_1/a$	*Fd*3*m*	*Pnn*2
$P4_32_12$	$Fm\bar{3}m$	*R*3*m*
$Cmc2_1$	*P*1	*Fdd*2
*Aba*2	*I*4/*mmm*	

Chirality

◆ Use italic type for certain chirality symbols and symmetry site terms.

A	anticlockwise
C	clockwise
CU	cube
DD	dodecahedron
OC	octahedron
TP	trigonal phase
TPR	trigonal prism
TPY	trigonal pyramid

These symbols are often combined with coordination numbers and position designations for stereochemical descriptors (e.g., *OC*-6-11′).

◆ In chemical names, use (*R*) and (*S*) as prefixes to designate absolute configuration.

(*R*)-hydroxyphenylacetic acid (*S*)-2,3-dihydroxypropanoic acid
(1*S*,2*S*,4*R*)-trichloro-1,2,4-trimethylcyclohexane

◆ Indicate optical rotation by plus and minus signs in parentheses and hyphenate them to the chemical name.

(+)-glucose (–)-tartaric acid (±)-4-(2-aminopropyl)phenol

◆ Use small capital letters D and L for absolute configuration with amino acids and carbohydrates.

D-glucose	L-alanine	D-valine
L-phenylalanine	L-ascorbic acid	DL-leucine
hydroxy-L-proline	D-allothreonine	L-alloisoleucine
hydroxy-DL-glutamic acid	5-hydroxy-L-lysine	β-D-cellotetraose

◆ Use a hyphen between (+) or (–) and D or L.

(–)-D-glyceraldehyde (+)-L-phosphoglycerol (–)-D-fructose

Concentration

◆ Use square brackets enclosing an element symbol or formula to indicate its concentration in reactions and equations, but not in narrative text.

Correct $[Mg^{2+}] = 3 \times 10^{-2}$ M

Correct The Mg concentration decreased with repeated washings.

Incorrect The [Mg] was found to be greater in the unwashed samples.

◆ Do not use square brackets to indicate concentration with a spelled-out name.

$[Ca^{2+}]$, *not* [calcium]
[NaCl], *not* [sodium chloride]

◆ Do not use italic type for the chemical concentration unit M (molar, moles per cubic decimeter, moles per liter) or the unit N (normal). Use italic type for the unit *m* (molal, moles per kilogram). Use a space between the number and these abbreviations, that is, on each side of these abbreviations.

8 M urea	1 mM EDTA	6 N HCl
2.0 *m* NaOH		

◆ When concentration is given as percentage, use the percent sign closed up to the number.

20% H_2SO_4 90% acetonitrile/10% water

◆ Generally, the negative logarithm of the hydrogen ion concentration is denoted by pH; the negative logarithm of the hydroxide ion concentration is denoted by pOH. Use a space to separate pH or pOH and the number. Use roman type for pH and pOH; always use lowercase for "p"; always capitalize "H" and "OH".

The salt was precipitated at pH 9.0 using NaOH as a precipitating agent.

Solutions were titrated to pH >11.

The UV spectra were measured at pH 6.

A pOH of <12 was acceptable.

Chemical Reactions

◆ Short chemical reactions may be run into text or they may be displayed and numbered, if numbering is needed. Long chemical reactions should be displayed separately from the text. The sequential numbering system used may integrate both chemical and mathematical equations, or separate sequences using different notations may be used for different types of equations (e.g., eqs 1–3 could be used for a set of chemical reactions and eqs I–III could be used for a set of mathematical equations). The use of lettering, rather than numbering, sequences is also acceptable.

$$Cr(CO)_4 + CO \rightarrow Cr(CO)_5 \quad (1)$$

$$NH_3 + HCOOH \rightarrow NH_2CHO + H_2O \quad (2)$$

$$(C_6H_5)_2P\text{–}P(C_6H_5)_2 \rightarrow 2(C_6H_5)_2P^{\bullet} \quad (3)$$

$$Fe(CO)_5 + OCH_3^- \rightarrow Fe(CO)_4(CO_2CH_3)^- \quad (4)$$

◆ Many kinds and combinations of arrows can be used. For example, two full arrows in opposite directions (⇆) indicate a reaction that is proceeding in both directions. Two arrows with half heads in opposite directions (⇌) indicate a reaction in equilibrium. A single arrow with heads on both sides (↔) indicates resonance structures, not a reaction.

◆ Specify the number of each species (molecules, atoms, ions, etc.) of reactants and products by a numeral written on the line and closed up to the symbol.

$$2Al + 6NaOH \rightarrow 2Na_3AlO_3 + 3H_2 \quad (5)$$

◆ To indicate the aqueous, solid, liquid, or gas state, use the appropriate abbreviations on the line, in parentheses, and with no space preceding them.

$$\mathrm{Ag(s) + H^+(aq) + Cl^-(aq) \rightarrow AgCl(s) + ½H_2(g)} \quad (6)$$

$$\mathrm{4FeS(s) + 7O_2(g) \rightarrow 2Fe_2O_3(s) + 4SO_2(g)} \quad (7)$$

◆ Indicate reaction conditions and catalysts over and under the arrow in a smaller type size. The Greek capital letter delta indicates heat; $h\nu$ indicates light, where h is Planck's constant and the Greek letter nu is the photon frequency.

$$\mathrm{PhS^-} \xrightarrow{h\nu} \mathrm{PhS^\bullet + e^-}$$

$$\mathrm{C_3H_8(g) + 5O_2(g)} \xrightarrow{\Delta} \mathrm{3CO_2(g) + 4H_2O(g)}$$

$$\mathrm{RC{\equiv}N + 2H_2} \xrightarrow{\text{Pt or Pd}} \mathrm{RCH_2NH_2}$$

$$\mathrm{2H_2O + CH_3C(CH_3){=}C(CH_3)CH_3} \xrightarrow[\mathrm{H_2SO_4,\Delta}]{\mathrm{KMnO_4}} \mathrm{2CH_3C({=}O)CH_3 + 2H_2}$$

$$\mathrm{(C_2H_5)_2C{=}O + 2CH_3OH} \xrightarrow[125\,^\circ\mathrm{C}]{\mathrm{H^+}} \mathrm{(C_2H_5)_2C(OCH_3)_2 + H_2O}$$

◆ Specify *nuclear reactions* according to the following scheme:

initial nuclide (incoming particle(s) or quanta , outgoing particle(s) or quanta) final nuclide

Examples

$^{14}\mathrm{N}(\alpha,\mathrm{p})^{17}\mathrm{O}$ $\quad$ $^{59}\mathrm{Co}(\mathrm{n},\gamma)^{60}\mathrm{Co}$ $\quad$ $^{23}\mathrm{Na}(\gamma,3\mathrm{n})^{20}\mathrm{Na}$

$^{31}\mathrm{P}(\gamma,\mathrm{pn})^{29}\mathrm{Si}$

◆ Treat *chemical equations* that include structures with rings as illustrations. They are discussed in Chapter 9.

◆ Abbreviate reaction types with capital roman letters and arabic numbers.

S_N1	S_N2	first- and second-order nucleophilic substitution, respectively
E1	E2	first- and second-order elimination, respectively
$S_{RN}1$	$S_{RN}2$	first- and second-order radical nucleophilic substitution

Reporting Analytical Data

There is no best way to present data. A presentation that is suitable for one paper or publication may be unsuitable for another. The following are examples of acceptable presentations of analytical data. These are not necessarily real examples; they may be combinations of data from two or more samples, intended to show various style possibilities. You need not define the abbreviations and symbols in the paper.

Melting and Boiling Points

mp 175.5 °C (lit. 175–176 °C)
mp 225 °C dec
bp 127 °C

Abbreviations: mp, melting point; bp, boiling point; lit., literature value; and dec, decomposition. A full space is used between the number and the unit °C; the degree symbol is closed up to the C.

Specific Rotation

$[\alpha]_D^{20}$ +25.4° (*c* 1.00, $CHCl_3$)

Abbreviations: α, specific rotation; D, the sodium D line or wavelength of light used for the determination; and the superscript number, temperature (°C) at which the determination was made. In parentheses: *c* stands for concentration; the number following *c* is the concentration in grams per 100 mL; and last is the solvent name or formula.

NMR Spectroscopy

^{1}H NMR (400 MHz, CD_3OD, δ): 8.73 (s, 3H, $-OCH_3$), 7.50 (s, 1H, CH), 7.15 (d, *J* = 8.2 Hz, 1H, Ar H), 6–3 (br s, 5H, NH and NH_2).

^{1}H NMR (500 MHz, $CDCl_3$, δ): 1.12 (t, *J* = 7.1 Hz, $-CH_2\mathit{CH_3}$, 3H), 3.34 (q, *J* = 7.1 Hz, $-\mathit{CH_2}CH_3$, 2H), 3.38 (t, *J* = 6.0 Hz, $-\mathit{CH_2}CH_2OH$, 2H), 3.72 (t, *J* = 6.0 Hz, $-CH_2\mathit{CH_2}OH$, 2H), 6.57 (dd, *J* = 8.7 Hz, Ar H, 2H).

^{13}C NMR (DMSO-d_6, δ): 175.4 (C=O), 156.5 (C_4), 147.4 (C_6), 138.3 (C_2), 110.5 (d, *J* = 11.3 Hz, C_5), 52.3 (CH_3), 28.4 and 28.8 (C_7).

^{13}C NMR (DMSO-d_6, δ): 0.43 (2C), 27.56 (4C), 131.8 (1C), 161.9 (2C).

^{13}C NMR ($CDCl_3$, 75.4 MHz): δ 213.50 (s, C-21), 178.27 (s, C-2), 168.69 (s, C-8), 164.61 (d, C-10), 119.67 (d, C-7), 52.45 (t, C-22), 38.95 (q, C-25).

^{13}C NMR (C_6D_{12}, δ): 6.51 (s, $C_5\mathit{Me}_5$), 14.41 (d, *J* = 157 Hz, PMe_3), 28.68 (s, Me), 105.1 (t, *J* = 3.7 Hz, C_5Me_5), 128.52 (s), 135.19 (br s), 212.56 (C=O).

If the experimental conditions have already been described elsewhere in the paper, they need not be repeated.

NMR: 3.81, 2.56, and 2.12 ppm.

Compound **27**: NMR 5.14, 3.90, 2.67, and 1.88 ppm.

Abbreviations: δ, chemical shift in parts per million (ppm) downfield from the standard; *J*, coupling constant in hertz; and the multiplicities s, singlet; d, doublet; t, triplet; q, quartet; and br, broadened. Italicized elements or groups are those that are responsible for the shifts.

IR Spectroscopy

IR (KBr) $\tilde{\nu}_{max}$: 967 (Ti=O), 3270 cm^{-1} (NH).

IR (KBr, thin film) $\tilde{\nu}_{max}$ (cm^{-1}): 3017, 2953 (s, OH), 2855 (s), 2192, 1512, 1360, 1082, 887.

IR (dry film) $\tilde{\nu}_{max}$: 3324 (OH), 2973–2872 (CH, aliphatic), 1706 (C=O, ketone), 1595, 1437, 1289, 1184, 1048, 870, 756, 677 cm^{-1}.

IR: 2000, 2030, 2040, 2050 cm^{-1}.

IR (cm^{-1}): 3130, 3066, 2964, 1654, 1500, 1371.

Compound **6**: IR 2910, 2487, 1972, 1564, 1190 cm^{-1}.

GC–FTIR $\tilde{\nu}_{max}$ (cm^{-1}): 2979 (w), 1400 (m), 1264 (s), 827 (vs).

Abbreviations: $\tilde{\nu}_{max}$ is the wavenumber of maximum absorption peaks in reciprocal centimeters, and the absorptions are w, weak; m, medium; s, strong; vw, very weak; vs, very strong; and br, broad.

Mass Spectrometry

MS *m/z* (relative intensity): 238.2058 (44.8%), 195.1487 (100%), 153.1034 (21.2%).

GC–MS *m/z* (% relative intensity, ion): 202 (9, M + 4), 200 (32, M + 2), 198 (23, M^+), 142 (35, M – 2CO), 321 (95, M – Me), 415 (M^+ – Cl).

HRMS–FAB (*m/z*): $[M + H]^+$ calcd for $C_{21}H_{38}N_4O_6S$, 475.259; found, 475.256.

EIMS (70 eV) *m/z*: M^+ 420 (15), 241 (15), 201 (59), 135 (14), 69 (23).

Abbreviations: *m/z* is the mass-to-charge ratio, M is the molecular weight of the molecule itself, M^+ is the molecular ion, HRMS is high-resolution mass spectrometry, FAB is fast atom bombardment, and EIMS is electron-impact mass spectrometry.

UV–Visible Spectroscopy

UV (hexanes) λ_{max}, nm (ε): 250 (1070).

UV (CH_3OH) λ_{max} (log ε) 210 (3.33), 242 (sh, 3.02), 288 (sh, 2.21), 421 nm (3.16).

Abbreviations: λ_{max} is the wavelength of maximum absorption in nanometers; ε is the extinction coefficient or molar absorptivity (units are not necessary); and sh is the shoulder. The wavenumber, $\bar{\nu}$, in reciprocal micrometers, might also be given.

Quantitative Analysis

Anal. Calcd for $C_{45}H_{28}N_4O_7$: C, 62.47; H, 3.41; N, 6.78. Found: C, 61.80; H, 3.55; N, 6.56.

Anal. ($C_{12}H_{20}N_2O_8S$) C, H, S. N: calcd, 6.99; found, 7.55.

All values are given as percentages.

X-ray Diffraction Spectroscopy

◆ Separate the element and its spectral line by a space.

Cu Kα	Cu Kβ	Mo Kα
Au L_I	Au L_{III}	Cr K
Cu Kα radiation	the Ni K edge	the Co K absorption edge

Citing ASTM, ANSI, and ISO Standards

ASTM is the American Society for Testing and Materials, ANSI is the American National Standards Institute, and ISO is the International Organization for Standardization.

◆ For ASTM standards, separate the letter and the number of the standard by a space. For ANSI standards, close up the letter and the number.

ASTM	D 3137	D 573	D 130	D 1660
	DS 4B	DS 64	DS 48A	E 380-93
	STP 1249	STP 169C	STP 315I	
ANSI	Z358.1-1990	Z88.2-1992		
ISO	14020	9000	71.040	ISO/DIS 14010

as described in ASTM Standard D 1223
according to ISO Standard 11634
as noted in ANSI Standard H35.1M-1993

Symbols for Commonly Used Physical Quantities

Atoms and Molecules

Name	*Symbol*	*SI Unit*
Atomic mass	m_a	kg
Atomic mass constant	m_u	kg
Atomic number	Z	dimensionless
Decay constant	λ	s^{-1}
Electron rest mass	m_e	kg
Electronic term	T_e	m^{-1}
Elementary charge (of a proton)	e	C
g factor, g value	g	dimensionless
Ionization energy	E_i, I	J
Magnetogyric ratio	γ	$s^{-1} \cdot T^{-1}$
Mass number	A	dimensionless
Neutron number	N	dimensionless
Nucleon number	A	dimensionless
Planck constant	h	J·s
Planck constant/2π	$\hbar$	J·s
Proton number	Z	dimensionless
Proton rest mass	m_p	kg
Rotational constants	A, B, C	m^{-1}
Rotational term	F	m^{-1}
Total angular momentum component	m_j, m_J	dimensionless
Total term	T	m^{-1}
Unified atomic mass unit	m_u	kg
Vibrational quantum number	v	dimensionless
Vibrational term	G	m^{-1}

Chemical Kinetics

Name	*Symbol*	*SI Unit*
Arrhenius or activation energy	E_a, E_A	J/mol, J·mol^{-1}
Boltzmann constant	k, k_B	J/K, J·K^{-1}
Energy of activation	E	J/mol, J·mol^{-1}
Half-life	$t_{1/2}$	s
Photochemical yield	ϕ	dimensionless
Quantum yield	ϕ	dimensionless
Rate constant, first order	k	s^{-1}
Rate constant, second order	k	mol^{-1}·s^{-1}
Rate of concentration change of substance B	r_B	mol/(m^3·s), mol·m^{-3}·s^{-1}
Rate of conversion	$\dot{\zeta}$	mol/s, mol·s^{-1}
Rate of reaction	v	mol/(m^3·s), mol·m^{-3}·s^{-1}
Relaxation time	τ	s
Scattering angle	θ	rad
Standard enthalpy of activation	$\Delta^{\ddagger}H^{\circ}$, $\Delta H^{\ddagger}$	J/mol, J·mol^{-1}
Standard entropy of activation	$\Delta^{\ddagger}S^{\circ}$, $\Delta S^{\ddagger}$	J/(mol·K), J·mol^{-1}·K^{-1}
Standard Gibbs energy of activation	$\Delta^{\ddagger}G^{\circ}$, $\Delta G^{\ddagger}$	J/mol, J·mol^{-1}
Standard internal energy of activation	$\Delta^{\ddagger}U^{\circ}$, $\Delta U^{\ddagger}$	J/mol, J·mol^{-1}
Temperature, absolute	T	K
Thermal energy	kT	J
Volume of activation	$\Delta^{\ddagger}V$, $\Delta V^{\ddagger}$	J/mol, J·mol^{-1}

Electricity and Magnetism

Name	*Symbol*	*SI Unit*
Capacitance	C	C/V, $C{\cdot}V^{-1}$, F
Charge density	ρ	C/m^3, $C{\cdot}m^{-3}$
Conductance	G	S
Conductivity	κ	S/m, $S{\cdot}m^{-1}$
Dielectric polarization	$\boldsymbol{P}$	C/m^2, $C{\cdot}m^{-2}$
Electric charge	Q	C
Electric current	I	A
Electric current density	j	A/m^2, $A{\cdot}m^{-2}$
Electric dipole moment	p, $\boldsymbol{\mu}$	C·m
Electric displacement	$\boldsymbol{D}$	C/m^2, $C{\cdot}m^{-2}$
Electric field strength	$\boldsymbol{E}$	V/m, $V{\cdot}m^{-1}$
Electric potential	V, ϕ	V, J/C, $J{\cdot}C^{-1}$
Electric potential difference	U, ΔV, $\Delta\phi$	V
Electric resistance	R	Ω
Electric susceptibility	χ_e	dimensionless
Electromagnetic moment	$\boldsymbol{\mu}$, $\boldsymbol{m}$	$A{\cdot}m^2$, J/T, $J{\cdot}T^{-1}$
Impedance	Z	Ω
Inductance	$\boldsymbol{H}$	A/m, $A{\cdot}m^{-1}$, H
Magnetic field strength	$\boldsymbol{H}$	A/m, $A{\cdot}m^{-1}$, H
Magnetic flux	Φ	Wb
Magnetic flux density	$\boldsymbol{B}$	T
Magnetic induction	$\boldsymbol{B}$	T
Magnetic moment	$\boldsymbol{M}$	A/m, $A{\cdot}m^{-1}$
Magnetic susceptibility	χ	dimensionless
Magnetization	$\boldsymbol{M}$	A/m, $A{\cdot}m^{-1}$
Permeability	μ	H/m, $H{\cdot}m^{-1}$
Permittivity	ε	F/m, $F{\cdot}m^{-1}$
Polarization (of a particle)	α	$m^2{\cdot}C{\cdot}V^{-1}$
Relative permeability	μ_r	dimensionless
Relative permittivity	ε_r	dimensionless
Resistance	R	Ω
Resistivity	ρ	Ω·m
Self-inductance	L	H
Voltage	U, ΔV, $\Delta\phi$	V

Electrochemistry

Name	*Symbol*	*SI Unit*
Charge number of an ion	z	dimensionless
Conductivity	κ	S/m, $S{\cdot}m^{-1}$
Diffusion rate constant	k_d	m/s, $m{\cdot}s^{-1}$
Electric current	I	A
Electric current density	j	A/m^2, $A{\cdot}m^{-2}$
Electric mobility	u	$m^2/(V{\cdot}s)$, $m^2{\cdot}V^{-1}{\cdot}s^{-1}$
Electrode potential	E	V
Electrolytic conductivity	κ	S/m, $S{\cdot}m^{-1}$
Electromotive force (emf)	E	V
Elementary charge	e	C
Faraday constant	F	C/mol, $C{\cdot}mol^{-1}$
Half-wave potential	$E_{1/2}$	V
Ionic strength		
concentration basis	I_c, I	mol/m^3, $mol{\cdot}m^{-3}$
molality basis	I_m, I	mol/kg, $mol{\cdot}kg^{-1}$
Mass-transfer coefficient	k_d	m/s, $m{\cdot}s^{-1}$
Molar conductivity	Λ	$S{\cdot}m^2/mol$, $S{\cdot}m^2{\cdot}mol^{-1}$
Reduction potential	$E°$	V
Standard electrode potential	$E°$	V
Standard electromotive force (emf)	$E°$	V
Surface charge density	σ	C/m^2, $C{\cdot}m^{-2}$
Transport number	t	dimensionless

General Chemistry

Name	*Symbol*	*SI Unit*
Amount concentration	c	mol/m^3, mol·m^{-3}
Amount of substance	n	mol
Atomic weight	A_r	dimensionless
Concentration	c	mol/m^3, mol·m^{-3}
Degree of dissociation	α	dimensionless
Extent of reaction	ζ	mol
Mass fraction	w	dimensionless
Molality	m, b	mol/kg, mol·kg^{-1}
Molar mass	M	kg/mol, kg·mol^{-1}
Molar volume	V_m	m^3/mol, m^3·mol^{-1}
Molarity	M	mol/L, mol·L^{-1}
Mole fraction	x	dimensionless
Molecular weight	M_r	dimensionless
Number concentration	C, n	m^{-3}
Number density of entities	C, n	m^{-3}
Number of entities	N	dimensionless
Partial pressure of substance B	p_B	Pa
Relative atomic mass	A_r	dimensionless
Relative molecular mass	M_r	dimensionless
Stoichiometric coefficient	ν	dimensionless
Surface concentration	Γ	mol/m^2, mol·m^{-2}
Volume fraction	ϕ	dimensionless

Mechanics

Name	*Symbol*	*SI Unit*
Angular momentum	$\boldsymbol{L}$	J·s
Density	ρ	kg/m^3, $kg{\cdot}m^{-3}$
Energy	E	J
Force	$\boldsymbol{F}$	N
Gravitational constant	G	$N{\cdot}m^2/kg^2$, $N{\cdot}m^2{\cdot}kg^{-2}$
Hamiltonian function	H	J
Kinetic energy	E_k, K	J
Lagrange function	L	J
Mass	m	kg
Moment of force	M	N·m
Moment of inertia	I	$kg{\cdot}m^2$
Momentum	$\boldsymbol{p}$	kg·m/s, $kg{\cdot}m{\cdot}s^{-1}$
Potential energy	E_p, V	J
Power	P	W
Pressure	P, p	Pa, N/m^2, $N{\cdot}m^{-2}$
Reduced mass	μ	kg
Relative density	d	dimensionless
Specific volume	v	m^3/kg, $m^3{\cdot}kg^{-1}$
Surface tension	γ	N/m, $N{\cdot}m^{-1}$, J/m^2, $J{\cdot}m^{-2}$
Torque	$\boldsymbol{T}$	N·m
Weight	$\boldsymbol{G}$, $\boldsymbol{W}$	N
Work	w, W	J

NMR Spectroscopy

Name	*Symbol*	*SI Unit*
Bohr magneton	μ_B, β	J/T, $J \cdot T^{-1}$
Bohr radius	a_0	m
Chemical shift, δ scale	δ	dimensionless
Coupling constant		
(indirect) spin–spin	J_{AB}	Hz
direct (dipolar)	D_{AB}	Hz
reduced spin–spin	K_{AB}	$T^2 \cdot J^{-1}$, $N \cdot A^{-2} \cdot m^{-3}$
Delay time	τ	s
Electron spin quantum component	m_s, m_S	dimensionless
Electron spin quantum number	s, S	dimensionless
Hyperfine coupling constant	a, A, $\boldsymbol{T}$	Hz
Larmor angular frequency	ω_L	s^{-1}
Larmor frequency	ν_L	Hz
Magnetogyric ratio	γ	$s^{-1} \cdot T^{-1}$
Nuclear magneton	μ_N	J/T, $J \cdot T^{-1}$
Nuclear spin quantum component	M_I	dimensionless
Nuclear spin quantum number	I	dimensionless
Orbital quantum number	l, L	dimensionless
Orbital quantum number component	m_l, m_L	dimensionless
Principal quantum number	n	dimensionless
Quadrupole moment	Q, Θ	$C \cdot m^2$
Relaxation time		
longitudinal	T_1	s
transverse	T_2	s
Shielding constant	σ	dimensionless

Polymer Chemistry

Name	*Symbol*	*SI Unit*
Bulk modulus	K	Pa
Complex permittivity	ε^*	F/m, $F \cdot m^{-1}$
Crack-tip radius	ρ_c	m
Electrophoretic mobility	μ	$m^2 \cdot V^{-1} \cdot s^{-1}$
Flory–Huggins interaction parameter	χ	
Fracture strain	γ_f, ε_f	dimensionless
Fracture stress	σ_f	Pa
Glass-transition temperature	T_g	K
Modulus of elasticity	E	Pa
Tensile strength	σ	Pa
Viscosity	ν	Pa•s
Volume fraction	V_f	dimensionless
Yield stress	σ_y	Pa
Young's modulus	E	Pa

Radiation

Name	*Symbol*	*SI Unit*
Absorbance	A	dimensionless
Absorption factor	α	dimensionless
Angle of optical rotation	α	dimensionless, rad
Angular frequency	ω	s^{-1}, rad/s, $rad \cdot s^{-1}$
Emissivity, emittance	ε	dimensionless
Frequency	ν	Hz
Linear decadic absorption coefficient	a	m^{-1}
Molar decadic absorption coefficient	ε	m^2/mol, $m^2 \cdot mol^{-1}$
Molar refraction	R_m	m^3/mol, $m^3 \cdot mol^{-1}$
Radiant energy	Q, W	J
Radiant intensity	I	W/sr, $W \cdot sr^{-1}$
Radiant power	P	W
Refractive index	n	dimensionless
Speed of light	c	m/s, $m \cdot s^{-1}$
Stefan–Boltzmann constant	σ	$W/(m^2 \cdot K^4)$, $W \cdot m^{-2} \cdot K^{-4}$
Transmittance	T	dimensionless
Wavelength	λ	m
Wavenumber (in a vacuum)	$\bar{\nu}$	m^{-1}

Space and Time

Name	*Symbol*	*SI Unit*
Acceleration	$\boldsymbol{a}$	m/s^2, $m{\cdot}s^{-2}$
Area	A, S, A_S	m^2
Cartesian space coordinates	x, y, z	m
Characteristic time interval	τ	s
Circular frequency	ω	s^{-1}, rad/s, $rad{\cdot}s^{-1}$
Diameter	d	m
Frequency	ν, f	Hz
Height	h	m
Length	l	m
Position vector	$\boldsymbol{r}$	m
Radius	r	m
Speed	v, u, w, c	m/s, $m{\cdot}s^{-1}$
Thickness, distance	d, δ	m
Time	t	s
Time constant	τ	s
Velocity	$\boldsymbol{v}$, $\boldsymbol{u}$, $\boldsymbol{w}$, $\boldsymbol{c}$	m/s, $m{\cdot}s^{-1}$
Volume	V	m^3

Thermodynamics

Name	*Symbol*	*SI Unit*
Absolute activity	λ	dimensionless
Affinity of a reaction	A	J/mol, $J \cdot mol^{-1}$
Chemical potential	μ	J/mol, $J \cdot mol^{-1}$
Cubic expansion coefficient	α	K^{-1}
Energy	E	J
Enthalpy	H	J
Entropy	S	J/K, $J \cdot K^{-1}$
Fugacity	f	Pa
Gas constant	R	J/(K·mol), $J \cdot K^{-1} \cdot mol^{-1}$
Gibbs energy	G	J
Heat	q, Q	J
Heat capacity, molar	C_m	J/(K·mol), $J \cdot K^{-1} \cdot mol^{-1}$
Heat capacity at constant pressure	C_p	J/K, $J \cdot K^{-1}$
Heat capacity at constant volume	C_v	J/K, $J \cdot K^{-1}$
Helmholtz energy	A	J
Internal energy	U	J
Isothermal compressibility	κ	Pa^{-1}
Joule–Thomson coefficient	μ	K/Pa, $K \cdot Pa^{-1}$
Pressure, osmotic	Π	Pa
Pressure coefficient	β	Pa/K, $Pa \cdot K^{-1}$
Specific heat capacity	c	J/(K·kg), $J \cdot K^{-1} \cdot kg^{-1}$
Surface tension	γ, σ	J/m^2, $J \cdot m^{-2}$, N/m, $N \cdot m^{-1}$
Temperature		
Celsius	t, θ	°C
thermodynamic	T	K
Viscosity	η	Pa·s
Work	w, W	J

◆ Subscripts denote a chemical process or reaction.

ads	adsorption	imm	immersion
at	atomization	mix	mixing
c	combustion	r	reaction in general
dil	dilution	sol	solution
dpl	displacement	sub	sublimation
f	formation	trs	transition
fus	fusion	vap	vaporization

- Certain superscripts are recommended.

$\ddagger$	activated complex, transition state
$'$	apparent
E	excess quantity
id	ideal
∞	infinite dilution
$*$	pure substance
$^\circ$, $^\ominus$	standard state

Transport Properties

Name	*Symbol*	*SI Unit*
Coefficient of heat transfer	h	$W/(m^2{\cdot}K)$, $W{\cdot}m^{-2}{\cdot}K^{-1}$
Diffusion coefficient	D	m^2/s, $m^2{\cdot}s^{-1}$
Flux of a quantity x	J_x, J	varies
Heat flow rate	ϕ	W
Kinematic viscosity	ν	m^2/s, $m^2{\cdot}s^{-1}$
Mass flow rate	q_m	kg/s, $kg{\cdot}s^{-1}$
Mass-transfer coefficient	k_d	m/s, $m{\cdot}s^{-1}$
Thermal conductivity	λ, k	$W/(m{\cdot}K)$, $W{\cdot}m^{-1}{\cdot}K^{-1}$
Thermal diffusion coefficient	D_T	m^2/s, $m^2{\cdot}s^{-1}$
Thermal diffusivity	a	m^2/s, $m^2{\cdot}s^{-1}$
Viscosity	η	$Pa{\cdot}s$
Volume flow rate	q_v, V	m^3/s, $m^3{\cdot}s^{-1}$

Acknowledgments

We thank the ACS Nomenclature Committee, chaired by John A. Secrist, and Thomas Sloan of Chemical Abstracts Service for reading this chapter and advising us on its contents.

CHAPTER 9

Illustrations and Tables

This chapter is a general guide to preparing and submitting illustrations and tables for a scientific paper. Certain style points and other requirements differ from journal to journal or publisher to publisher. For ACS journals, consult recent issues as well as the Guide, Notes, or Instructions for Authors that appear in each journal's first issue of the year. For ACS books, consult the brochure "How To Prepare Your Manuscript for the ACS Symposium Series" or "Instructions for Authors".

This chapter does not contain recommendations for or discussions of software because the products available are constantly changing. The guides for authors published by the individual ACS journals and the ACS Books Department contain the most up-to-date information on ACS's software preferences.

Illustrations

When To Use Illustrations

Illustrations can play a major role in highlighting, clarifying, and summarizing results and data when words are not enough. Appropriate, well-drawn illustrations allow readers to reach their own conclusions. Illustrations such as charts and graphs can substantially increase comprehension of the text and can convey trends, comparisons, and relationships more clearly than text. Line graphs are best to show trends. Bar graphs compare magnitudes. Pie charts show relative portions of a whole. Photographs can provide absolute proof of findings. Chemical structures, reactions, and mathematical expressions can be essential to understanding theories and processes. Schematic drawings best illustrate processes and sequences of events.

Illustrations that are poorly rendered or do not clarify the discussion but merely repeat data already presented in text decrease comprehension and cause confusion. Illustrations are time-consuming and costly for authors to prepare and also for publishers to reproduce, so they should justify the time, effort, and expense.

Line Art

Line art usually consists of only black markings on a white background and no shades of gray (see p 283). If you are drawing it or having it drawn by hand, use black ink on white, high-quality, smooth, opaque paper. If you are generating artwork with a computer program, print it on white, high-quality, smooth, laser-printer paper. Use the highest quality resolution available [a graphics plotter or a 600-dpi (dots per inch) or higher laser printer]. Be sure that your printer has enough toner to generate a crisp, dark image (see pp 284 and 285).

Determine the column width and page length of the printed publication to which the art will be submitted. For best results, prepare your artwork in the actual size at which it will appear. **For ACS journals, try always to design illustrations to fit the width of one column.** Table 1 gives the column widths and page lengths of ACS publications. The values given are

Table 1. Column Dimensions in ACS Publications

	Column Width			*Page Length*		
Publication	*picas*	*inches*	*centi-meters*	*picas*	*inches*	*centi-meters*
Books, trim size						
6 × 9 inches	27	4½	11	42	7	17½
7 × 10 inches	33	5½	13½	51	8½	21½
8 × 11 inches				56	9	23
single column	20	3¼	8¼			
double column	42	7	17½			
Journals and magazines, two-column format				60	10	25½
single column	20	3¼	8¼			
double column	42	7	17½			
Magazines, three-column format				60	10	25½
single column	13	2	5			
double column	27½	4½	11			
triple column	42	7	17½			

Line art and halftone. Top, line art. Bottom, halftone made from a continuous-tone print.

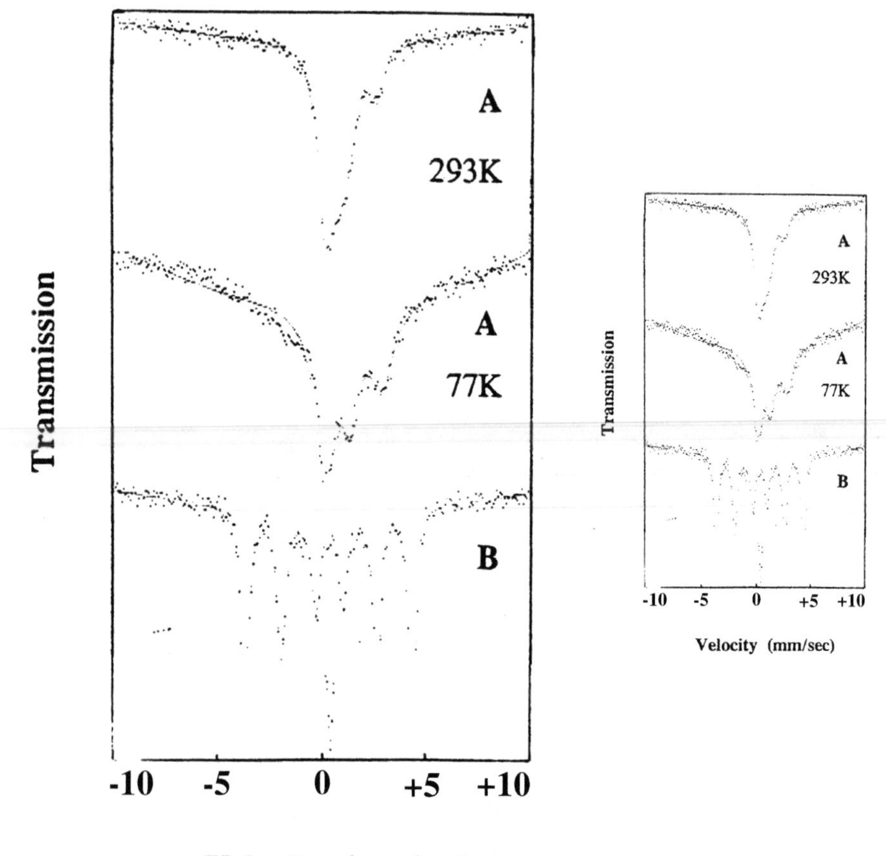

The points in this figure are too faint. Left, original size; right, how this art would reproduce in a publication. When the top figure is reproduced at a smaller size, as it would be in a publication, some points disappear.

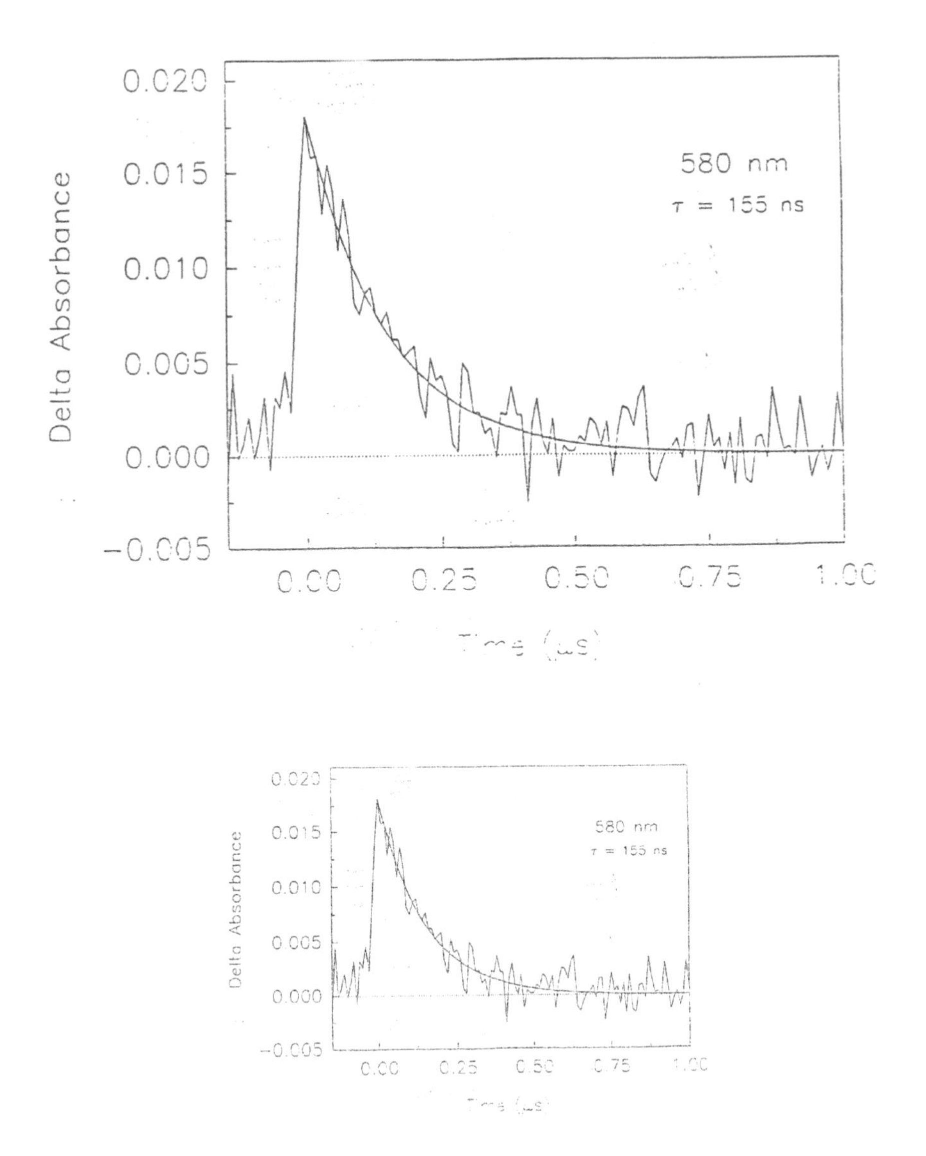

The figure on the top contains type that is faint; it has a "milky" appearance. The art is also dirty. There are many causes of these problems: They occur often with a copy of a copy. Each generation away from the original loses some sharpness and clarity. Also, the original art may have been printed without enough toner. Top, original size; bottom, how this art would reproduce in a publication. It becomes difficult to read.

the maximum space available for the art itself as well as any captions or titles. Therefore, the art must be small enough to leave space for the caption. Furthermore, the width and depth of an illustration should not exceed the needs of that illustration. Be economical; do not waste space. Here are some ways to improve the quality of the art you submit:

◆ Scale the length, width, type, symbols, and lines of the art proportionally; keep the symbols, lines, and type at uniform density; and size the art so that the symbols and type will be legible if the art needs to be reduced to fit the column width. Production problems can be caused by very thin lines, light material, and flaws on the artwork. Make the lines at least 0.5 point wide, select type size 7–10 points (type font Helvetica or Times Roman), and make the symbols at least the size of a lowercase letter "oh" (about 2 mm). (See the box on p 287.) If the type is very large and the symbols are very small, when the art is reduced to fit the column width, the type will be an appropriate size, but the symbols will be so small that they will be unintelligible. If the symbols are large and the type is small, reduction will make the type illegible. Likewise, if the overall size is very large and the type and symbols are small, reduction to fit the column will make the type and symbols unreadable. (See the art on pp 288–290.)

◆ Keep illustrations clear and simple. Keep words to a minimum.

◆ Do not use shades of gray in line art; when it is necessary to distinguish solid areas, such as in a bar graph, use open, solid, hatched, and crosshatched patterns rather than shades of gray (see p 291).

◆ Use simple, common symbols that would not be confused with each other and would be readily available in any publishing house, for example, ○, ●, □, ■, △, ▲, ▽, ▼, ◇, ◆, +, ×, ☆, and ★.

◆ You may combine curves plotted on the same set of axes, but do not put more than four or five curves on one set of axes. Label all curves clearly. Leave sufficient space between curves; they should not overlap so much that the symbols are indistinguishable.

◆ Keep illustrations compact; draw axes only long enough to define the contents. For example, if the highest data point on the curve is 14, then the scale should extend no longer than 15. Furthermore, the origin or lowest point on the axes does not have to be zero. For example, if the lowest data point on the curve is 4, then the axis can start at 3. Put grid marks on the axes to indicate the scale divisions.

◆ Label each axis with the parameter or variable being measured, the units of measure in parentheses, and the scale. Use initial capital letters only,

Type Size and Font

In publishing, type is measured in points; space is measured in picas. There are 72 points to an inch; there are 6 picas to an inch. The size of the type you are reading on this page is 10 points. The column is 27 picas wide, and the text page is 44 picas long.

You can use a pica ruler to measure space, but there is no ruler for measuring type: you must compare it to known type sizes. For example, in 10-point type, no character actually is 10 points high.

The font (or typeface) is the style or design of the letters. There are literally hundreds of fonts, but plain, simple fonts such as Helvetica or Times Roman are best for scientific art.

Type also comes in different weights. Most of the type you are reading is lightface; **this is boldface type**. Type may be *italic* or roman. Generally, you should use 10-point lightface, roman type for illustrations and tables.

This is 12-point Times Roman.
$^{14}C_6H_6$ shows subscripts and superscripts.

This is 12-point Helvetica.
$^{14}C_6H_6$ shows subscripts and superscripts.

This is 14-point Times Roman.
$^{14}C_6H_6$ shows subscripts and superscripts.

This is 14-point Helvetica.
$^{14}C_6H_6$ shows subscripts and superscripts.

not all capitals: Time (min), Reaction Temperature (°C), Thickness (μm). Place all labels outside and parallel to the axes. Letters and numbers on the abscissa and ordinate should read from left to right and from bottom to top, respectively. Do not place arrowheads on the ends of the axis lines. Use only two axes, one horizontal and one vertical, for a set of units; do not box the graph. Use initial capital letters for callouts or labels within the illustration, and use the same size type as for the axis labels.

◆ Leave clear margins at least an inch (2.5 cm) wide on all sides of the artwork to allow the editors space for marking codes and identification and for handling.

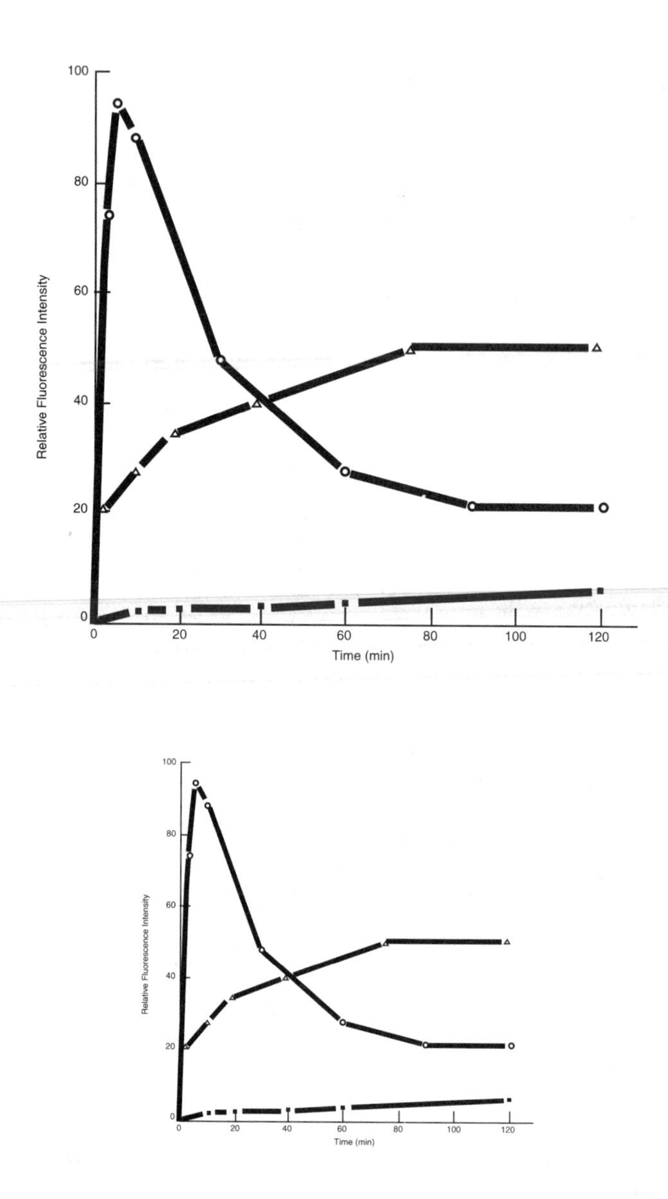

This figure has type and symbols that are too small in proportion to the rest of the figure. The lines on the graph are also too thick. Top, original size; bottom, how this art would reproduce to fit on a 13-pica column. The type becomes unreadable.

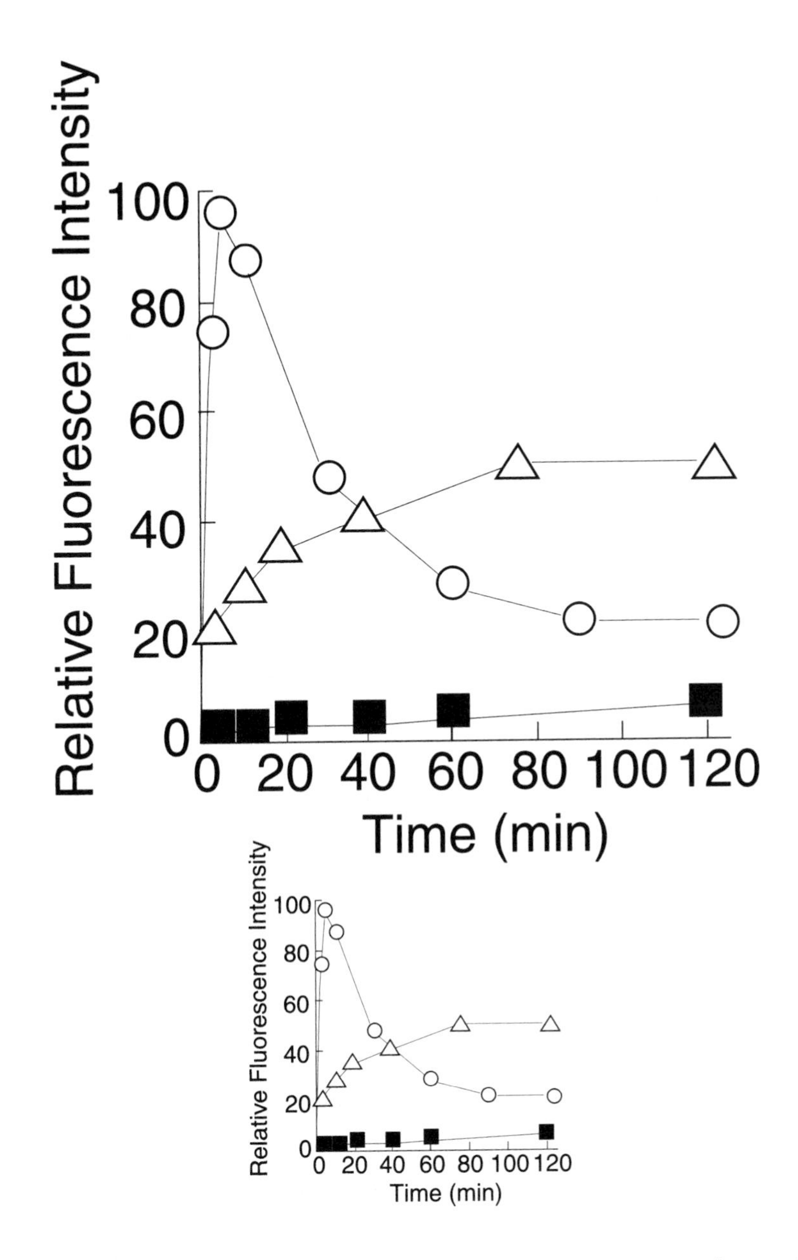

This figure has type and symbols that are too large in proportion to the rest of the figure. The lines on the graph are also too thin. Top, original size; bottom, how small the art would need to be to make the type the right size. The thin lines may also disappear.

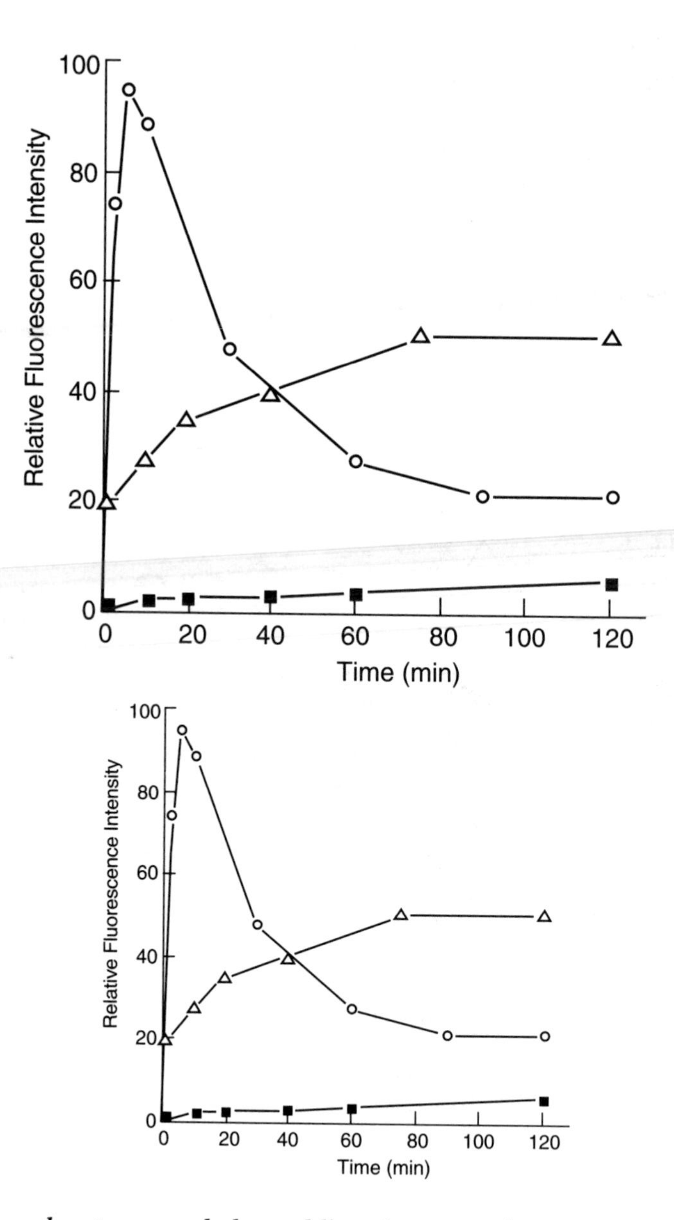

This figure has type, symbols, and lines in proportion to the rest of the figure. Top, original size; bottom, how this art would reproduce in a publication.

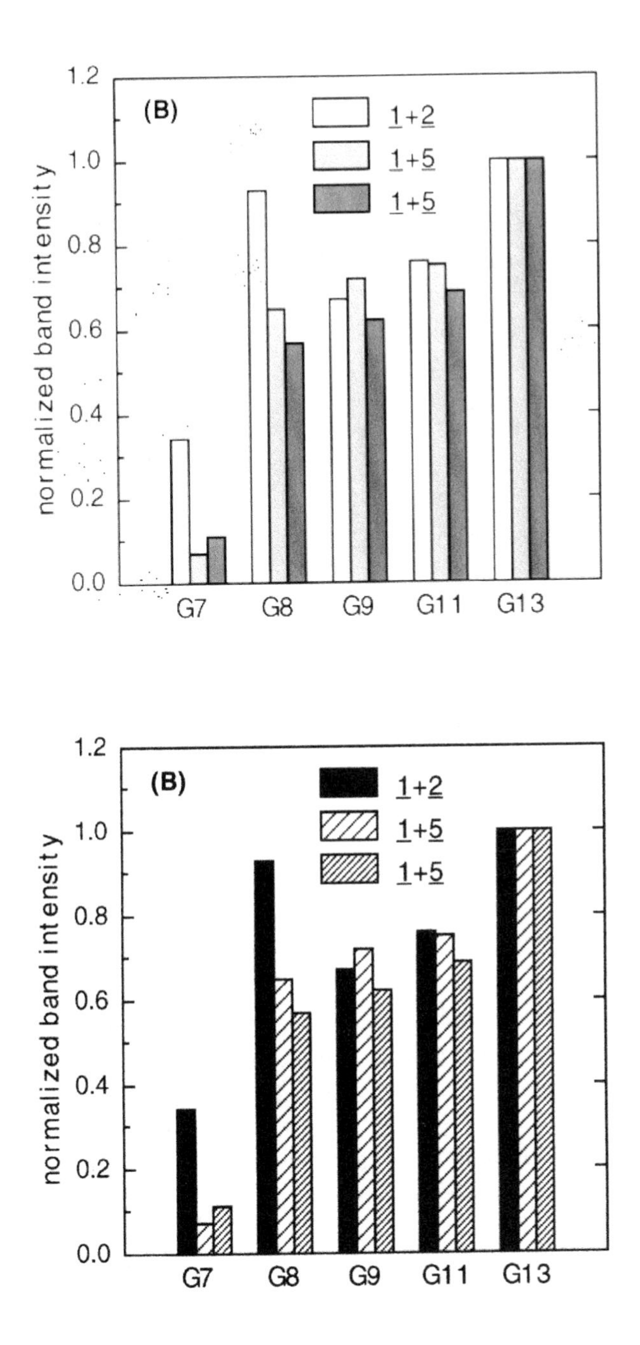

The original of the top figure contained shading, which did not reproduce well. The bottom figure has bars that are hatched. These bars are much clearer and easier to read.

◆ If your art is hand-drawn, have a stat made of the originals if possible. A stat (also called Photostat or PMT) is a photographic print on a special coated paper. By submitting stats, you can avoid losing type that may flake off the original. Also, corrections made with opaquing fluid will not reproduce on a stat.

◆ Computer-generated illustrations are acceptable if they meet the criteria for line weight; type size, font, and density; and overall size. Drawings produced on older dot matrix printers are usually unacceptable by these standards.

Continuous-Tone Prints

All black-and-white photographs contain shades of gray and are called continuous-tone prints (see p 283). Drawings or sketches that contain shades of gray are also continuous-tone prints. They can be reproduced as halftones for publication in two ways: by photographic screening or by digital scanning.

When a photograph is screened, it is photographed by a graphic arts camera that converts shades of gray into dot patterns suitable for printing on a press; the converted print is called a halftone. The various shades of gray are achieved with dot patterns of various densities: the greater the density, the darker the gray. In a newspaper photograph, you can see the dots easily, but the dot patterns are much denser in a book or journal photograph because book and journal printers use finer screens. A digital scanner also converts continuous tones of gray into dot patterns. For the best photographic reproduction, submit only original black-and-white photographs or (if necessary) prints produced on a high-quality laser printer. Negatives, slides, and overhead transparencies cannot be screened and are thus not usable for reproduction. Have them converted to prints.

Photographs always lose some detail in reproduction, so it is best to submit original photographs that have good tonal definition. More tonal definition and contrast can be achieved by placing an object against a light or dark background. Also, most commercial photography laboratories can adjust contrast to optimize tonal definition. When you get your film developed, tell the technician that the photograph will be used in a publication.

Because the continuous-tone process is a method for reproducing shades of gray, dark lines and type often appear lighter or blurred in continuous-tone photographs. Therefore, stats of line art will usually reproduce better. Here are some ways to improve the quality of the photographs you submit:

◆ Handle photographs with care! Printers cannot correct flaws on photographs, nor can they selectively omit them. Most flaws on a continuous-tone print will be reproduced in the published photograph. You can damage continuous-tone prints by using paper clips on them, by writing on the back with a sharp pen, by folding them, by stapling them, or by using tape on them. Depending on the amount and location of the damage, you may have to reshoot the photograph.

◆ Generally, do not submit prescreened illustrations. For reproduction, they must be screened again, and screening an existing screen can yield unwanted and unappealing results. However, some illustrations that require shading to convey information cannot easily be created as continuous-tone prints. Such cases can be reproduced reasonably well if you follow these suggestions: Be sure that the art is at the proper size and will not need reduction to fit a column. Use line screens of 85 lines per inch or lower. Use gray levels in increments of at least 20%. Do not use gray levels below 20% (because they fade to white) or above 70% (because they are often indistinguishable from black). Do not layer words or symbols over shaded areas.

◆ Do not submit color prints to be reproduced in black and white.

In some cases, color is necessary to convey the material or the results. Use color judiciously. Color printing is expensive because color photographs must ’ “color-separated” into four separate plates. Most publications cannot afford these extra costs, so authors must pay for color. Procedures differ for different publications; check with the specific editorial office.

To use a photograph of a person in a publication, you must obtain written permission from the person or the person’s heirs. Likewise, to use a photograph of commercial equipment, you must obtain written permission from the manufacturer.

How To Submit Illustrations

For line art, submit original camera-ready* artwork, printouts, stats, or glossy photographs. For continuous-tone prints, submit the original photographs. Mount stereoviews so that they can be used without modification (see p 294). Be sure that the distances between similar points are between 5.5 and 6.4 cm. Submit all original art with the original manuscript and photocopies of all the art with each copy of the manuscript. If possible, submit artwork that does not have to be reduced to fit a column width. Check Table 1 for column widths.

---

*The term “camera ready” originally meant “ready for the printer’s camera” because the first step in the printing process was the printer’s photograph of the final copy. Currently, most artwork is scanned digitally, and the electronic files are sent to the printer, so that no camera is involved. Nevertheless, the term “camera ready” is still used to indicate art in a final form, ready for the publisher’s production process.

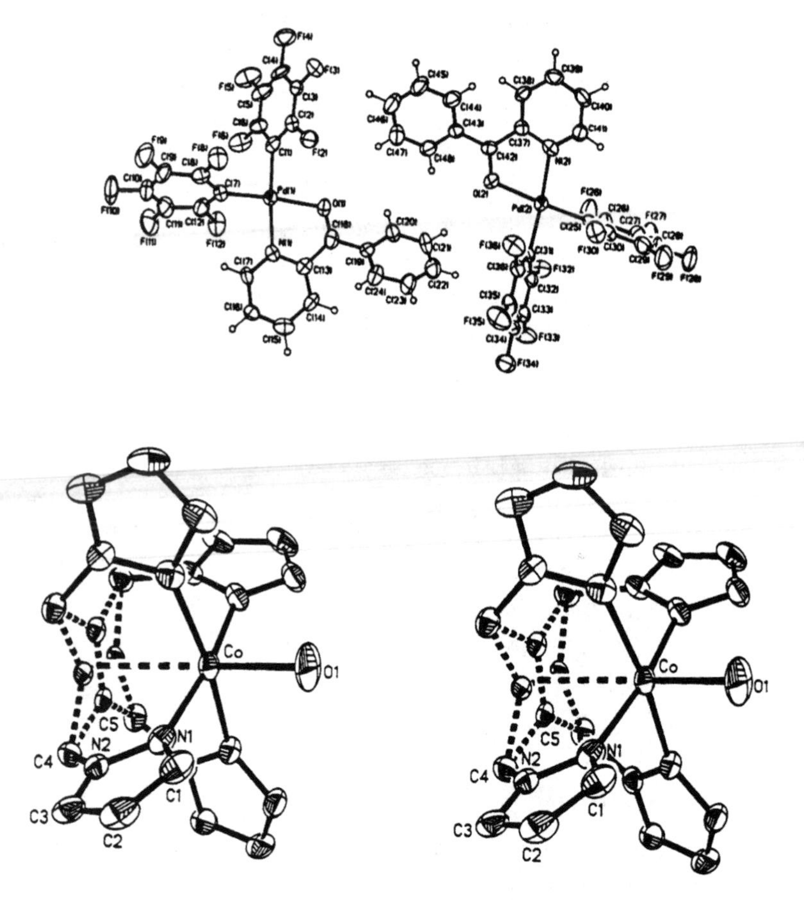

ORTEP is a computer program for producing three-dimensional art. Its art shows proportion and direction with ellipsoids that are divided into quarters. The figure on top is a bad ORTEP drawing. The labels are too small to read, and the differentiations among the quarters are not clearly seen. The bottom figure is a good ORTEP: clear, precise, and readable. It is also a good example of a stereoview.

All artwork (figures, structures, schemes, etc.) should be labeled and handled according to these instructions:

- Place each illustration on a separate sheet; do not incorporate illustrations into the text, but group them at the end of the paper.
- Note the figure, structure, or scheme number and the first author's surname either on the back or on the front of each piece of artwork, about an inch (2.5 cm) clear of the image area. Do not write on the front or back of the image area; write only in the margins. Indicate the top of the illustration with the word "top" if the correct orientation is not obvious.
- The best way to identify photographs is by marking a pressure-sensitive label and then affixing the label to the back of the photograph. If you must mark the back of a photograph, write very lightly and use a very soft pencil. The pressure of a ballpoint pen will indent the photo, the writing will show on the photograph face, and it will be reproduced in processing.
- Do not fold or roll artwork or photographs; protect art with cardboard or heavy paper for transport. Do not staple, clip, or punch holes in photographs or any artwork.

Figures

How To Cite Figures

Number figures sequentially with arabic numerals in order of discussion in the text (Figure 1, Figure 2, etc.). Designate parts of a figure by using a combination of the arabic number and a sequence of consistent labels, usually (but not always) alphabet letters: Figure 1a, Figure 1b; Figure 1A, Figure 1B; Figure 1-I, Figure 1-II. Discuss, or at least cite, all figures sequentially in the text. A multipart figure is considered as one figure for citation purposes.

> The block copolymers may contain a small but detectable fraction of impurities, as shown in Figures 1 and 2.
>
> Figures 3–5 show the production of acid-reactive substances in three different oils.
>
> The deuterium-labeled substrate gave rise to the partial ^{1}H NMR spectra shown in Figure 2a,b.
>
> As seen in Figure 3b–d, the catalytic wave shows a small but distinct decrease upon addition of the nucleophile.
>
> Parts a and b of Figure 4 illustrate that the voltammetric plateau current depends on the number of enzyme monolayers.
>
> Curves c–e of Figure 5 were obtained for various methyl groups in the protein.

Chemical structures and schemes should not be numbered as figures; they should be labeled according to separate sequences. See the section "How To Cite Structures and Schemes" on p 299.

Figure Captions

Every figure must have a caption that includes the figure number and a brief, informative description, preferably in nonsentence format.

> Figure 2. Mass spectrum obtained when laboratory ambient air containing 2.5 ppm of **1** was introduced into the MS system.
>
> Figure 4. Change in carotenoid contents during maturation of three varieties of grapes.
>
> Figure 6. Variable-temperature NMR spectra of **3d** in CD_2Cl_2 solution at 500 MHz.
>
> Figure 7. Reaction rate constants as a function of proton affinity for the reactions shown in eqs 5–7: k_{exp}, experimental; k_c, calculated.
>
> Figure 1. Specificity of bovine muscle LDH antibodies in a sandwich ELISA. Data are the average of three replicates.

If more information is necessary, use complete sentences and standard punctuation. The caption should be understandable without reference to the text and should not include material that is not in the text. Use similar wording for captions of related figures.

If the art contains many symbols and the key to symbols will be large and give the artwork a cluttered appearance, put the key in the caption. If the artwork contains unusual symbols, and these symbols may not be available to the publisher for use in the caption, identify the symbols within the artwork. Make sure that the symbols and abbreviations in the caption agree with those in the figure itself and in the text.

Submit the figure captions separately from the artwork, typed double-spaced on a page at the end of the text. Verify that the number of captions agrees with the number of figures. The figures and captions follow different routes in the production process, so if you place the caption on the art, the editors usually will need to delete it and typeset the caption according to the publication's style.

In ACS publications, credit lines for art reproduced from previously published work appear at the end of the caption in parentheses in one of two formats and follow three possible wordings, depending on the original source:

Format 1

Most publishers: Reprinted with permission from ref XX. Copyright Year Copyright Owner's Name.

Published by ACS: Reprinted from ref XX. Copyright Year American Chemical Society.

Published by the U.S. government: Reprinted from ref XX.

Examples

Reprinted with permission from ref 10. Copyright 1993 American Pharmaceutical Association.

Reprinted from ref 12. Copyright 1995 American Chemical Society.

Reprinted from ref 23.

Format 2

Most publishers: Reprinted with permission from Author Names (Year of Publication). Copyright Year Copyright Owner's Name.

Published by ACS: Reprinted from Author Names (Year of Publication). Copyright Year American Chemical Society.

Published by the U.S. government: Reprinted from Author Names (Year of Publication).

Examples

Reprinted with permission from Camiola and Altieri (1997). Copyright 1997 American Institute of Physics.

Reprinted from Fitzgerald and Cheng (1995). Copyright 1995 American Chemical Society.

Reprinted from Takanishi and Schmidt (1994).

Chemical Structures and Schemes

When To Use Structures and Schemes

Structures and reaction schemes are essential in some papers. Examples are papers that describe a previously unreported synthesis or reaction sequence, those that discuss structure–activity relationships, and those that describe compounds that are not well-known. However, structures are not necessary just because you discuss a compound, especially if you give the systematic name of the compound. Often, a line formula is adequate, such as CH_3COOH, C_6H_6, ClC_6H_5, or 1,4-$Cl_2C_6H_4$. Do not supply art for material that can and should be portrayed on one line. Do not make a formula angular if it can be portrayed linearly. Likewise, do not display reactions if they can be explained clearly in text.

How To Prepare Structures and Schemes

The overall quality and size requirements for structures and schemes are the same as those for other illustrations. Do not draw structures freehand.

◆ Arrange structures in horizontal rows the width of a single column.

◆ The size of the rings and the size of the type should be proportional. The published size of six-membered rings should be approximately ¼ in. (6 mm) in diameter; the published size of five-membered rings should be slightly smaller. The type size should be 5–8 points.

◆ Keep oddly shaped rings and the shapes of bicyclic structures consistent throughout a manuscript. In multiring structures such as steroids, use partial structures that show only the pertinent points.

◆ In three-dimensional drawings, use dashed lines for those in the background that are crossed by lines in the foreground to give a greater three-dimensional effect. Make lines in the foreground heavier.

◆ Center the compound labels (numbers or letters) just below the structures. If you are discussing compounds of similar structure, draw only one parent structure, use a general designation (e.g., R or Ar) at the position(s) where the substituents differ, and specify modifications below the structure.

◆ For structures mentioned in tables, provide labeled structures as separate artwork (in text, charts, etc.) and use only their labels in the table. Do not place structures within tables unless you provide the entire table as camera-ready copy.

◆ Center the reaction arrows vertically on the midline of the structure height. Align the centers of all structures. The midline for all structures is the center of the "tallest" structure on the same line.

◆ Do not waste space, either vertical or horizontal. Use the full column width before starting a new line. A compact presentation is most effective. Avoid using vertical arrows unless it is necessary to portray a cyclic or "square" reaction scheme. Schemes generally read from left to right, just like English sentences. As long as the proper sequence is maintained, it does not matter on which line any given structure appears. If a reaction continues to the next line, keep the arrow or other operator on the top line.

◆ Do not place circles around plus or minus signs.

How To Cite Structures and Schemes

To identify structures that you cite in the text, use boldface numerals (arabic or roman), boldface alphabet letters (capital or lowercase), or a combination of these. For several structures or for a series of structures, use a consistent sequence of labels, for example

1, 2, 3	**I, II, III**	**A, B, C**
a, b, c	**1a, 1b, 1c**	**Ia, Ib, Ic**
1A, 1B, 1C	**IA, IB, IC**	

If you discuss or mention a compound several times in text, refer to it only by its label throughout the text (e.g., "as exhibited by the reaction with **6**"). This principle applies to any chemical species that is the topic of frequent discussion or mention.

It is not necessary to number structures that you will not be specifically citing in the text. However, be sure to indicate clearly the placement of unmentioned structures.

Reactions can be labeled in the same sequence as mathematical equations, with lightface numerals or alphabet letters or combinations of them in parentheses at the right margin. Depending on the complexity of the material, you may use one set of sequential labels for both chemical reactions and mathematical equations, or two separate but different numbering sequences (e.g., eqs 1, 2, 3, ... and reactions I, II, III, ...). Do not confuse structure labels with reaction labels.

Here is a properly drawn reaction scheme. The arrow and the one-line structure are centered from top to bottom in the height of the long structure; the reaction arrow and the plus sign have an equal amount of space on both sides; and the full column width is used.

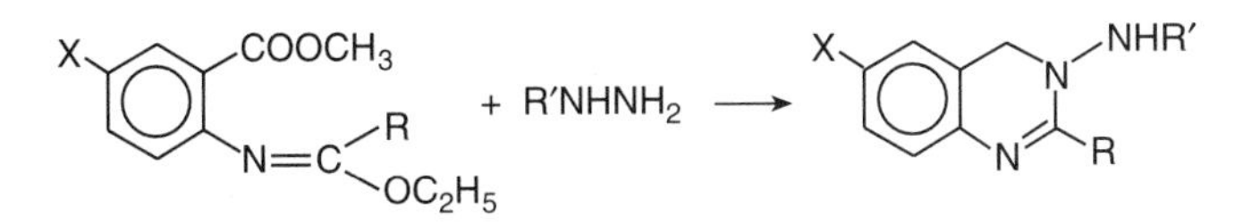

Groups of structures are called charts; groups of reactions are called schemes. Schemes show action; charts do not. Reaction schemes that include several reactions should be numbered separately with arabic numerals and cited as "Scheme 1", "Scheme 2", and so on. Charts are numbered similarly: Chart 1, Chart 2, and so on. Very large charts or schemes may require a double-column width. Charts and schemes may have brief titles describing their contents; they may also have footnotes.

Tables

When To Use Tables

Use tables when the data cannot be presented clearly as narrative, when many precise numbers must be presented, or when more meaningful interrelationships can be conveyed by the tabular format. Tables should supplement, not duplicate, text and figures. Examples of material that is best handled as narrative in text are results of IR absorption and NMR chemical shift studies, unless they are major topics of discussion. In many instances, one table with representative data, rather than several tables, is all that is needed to illustrate a point.

How To Prepare Tables

There are two kinds of tables: informal (or in-text) and formal. An informal table consists of three to five lines and is no more than four columns wide; it cannot exceed the width of a text column. Informal tables may be placed in text following an introductory sentence. They are not given titles or numbers, nor do they contain footnotes.

A formal table should consist of at least three interrelated columns and three rows. If you have only two columns, try writing the material as narrative. If you have three columns, but they do not relate to each other, perhaps the material is really a list of items and not a table at all. If your table has unusual alignment and positioning requirements, perhaps it should really be a figure. It is important to understand these differences because tables are much more expensive to produce than text; the larger the table, the higher the cost. A well-constructed, meaningful table is worth the expense, but anything else is wasteful and does not enhance your paper.

Tables should be simple and concise; arrange all data for optimal use of space. If you have many small tables, consider combining some. Combining is usually possible when the same column is repeated in separate tables or when the same type of material is presented in several small tables. Use consistent wording for all elements of similar or related tables. Be consistent with symbols and abbreviations among tables and between tables and text.

The table width will of course depend on the widths of the individual columns. Very generally, tables having up to six columns will fit in a single journal column; tables having up to 13 columns will fit in the double-column spread. Tables that exceed the double-column spread will be rotated 90° and set lengthwise on the page.

In books, tables having up to eight columns can fit the page width; tables having 8–12 columns will be set lengthwise on the page. Larger tables can span two pages.

In all publications, extremely wide tables can cause composition difficulties. In such cases, consider presenting the material as two or more smaller tables.

The style for the individual parts of tables (i.e., the use of capital and lowercase letters and whether the entries are centered or flushed left) varies among publications. Consult a recent edition of the journal or the instructions for authors.

Keep sections of multipart tables at similar widths. Widely divergent section widths within a table waste space and detract from general appearance.

Title

Every formal table must have a brief, informative title that describes its contents in nonsentence format. The title should be complete enough to be understood without referring to the text, and it should not contain information that is not in the text. Place details in table footnotes, not in the title.

Column Headings

Every column must have a heading that describes the material below it. Be as succinct as possible, keep headings to two lines if possible, and use abbreviations and symbols whenever practical. Be consistent with the text and with other column headings. Define nonstandard abbreviations in table footnotes. Name the variable being measured and indicate the unit of measure after a comma, after a slash, or within enclosing marks. Use the same style within and among all tables. A unit of measure alone is not an acceptable column heading. A unit of measure under a straddle rule, as in Table 1, is acceptable.

If a column heading applies to more than one column, use a rule below it that spans the columns to which it applies; this is called a straddle rule. Below the rule, give the specific headings for each column. A column heading should not apply to the entire table; information that describes all of the columns belongs in a general table footnote.

Column Entries

In many tables, the leftmost column is the stub or reading column. Usually, all other columns refer back to it. Stub entries should be consistent with the text as well as logical and grammatically parallel with each other. Main stub entries may also have subentries, which should be indented.

Material in columns can be aligned in various ways. Use only one type of alignment per column. Words are usually aligned on the left, and numbers are usually aligned on the decimals, unless they do not have the same units, in which case they are aligned on the left. Use numbers on both sides of a decimal point. Columns that are made up of numbers and words together or columns that contain a variety of sizes or types of information might call for centering or alignment on the left or right, depending on the publication's style.

Do not use ditto marks or the word "ditto" to indicate the same entry in a column; repeat the entry. Define nonstandard abbreviations in table footnotes. Try to keep all entries at similar lengths by placing any explanatory material in table footnotes. If you use a dash as a column entry, explain it in a footnote the first time you use it (e.g., "—, too low to be measured.").

Make sure that all of the columns are really necessary. If there are no entries in most of a column, it probably should be deleted and replaced with a general table footnote. Or, if the entries in the entire column are the same, the column should be replaced with a table footnote that says "In all cases, the value was *x*." or whatever is appropriate.

Footnotes

In footnotes, include explanatory material referring to the whole table and to specific entries. Examples of information that should be placed in general footnotes referring to the whole table are the following: units of measure that apply to all entries in the table, explanations of abbreviations and symbols used frequently throughout the table, details of experimental conditions if not already described in the text or if different from the text, general sources of data, and other literature citations.

Information that should be placed in specific footnotes includes units of measure that are too long to fit in the column headings, explanations of abbreviations and symbols used with only one or two entries, statistical significance of entries, experimental details that apply to specific entries, and different sources of data.

In some publications, general footnotes and sources are not cited with superscripts; they are labeled "Note" and "Source", respectively. Specific footnotes are cited with superscripts. In other publications, all footnotes are cited with superscripts. Check the directions for the publication to which you are submitting your paper.

Where superscripts are prescribed, use superscript lowercase italic letters in alphabetical order, starting from the top of the table and proceeding from left to right. Write footnotes as narrative and use standard punctuation. Short phrases such as "N.D., not determined" and "$x = 23$" are acceptable. Label each footnote with its superscript letter and group the footnotes together.

How To Cite Tables

Number the formal tables sequentially with arabic or roman numerals, depending on the publication's style, in order of discussion in the text. Like figures, every formal table must be cited in the text.

How To Submit Tables

Submit formal tables each on a separate page after the reference section of the text. Print the entire table double-spaced. If a table must contain structures or other art or special symbols, or if a table has special alignment and positioning requirements, submit it as camera-ready copy, in which case it should be single-spaced. If it is too large for the column or page, the publisher will reduce it.

Submit informal tables in place in text, double-spaced.

Lists

Sometimes you may need to give numerous examples of items, such as chemical names. In such cases, if there are too many to run into the text, they can be set as a list in some publications. Put the entries in alphabetical order, unless there is a reason to do otherwise. A list of names is not truly a figure and not really a table. Give the list an unnumbered title. In ACS journals, lists may be handled as informal tables or even as charts.

Potentially Carcinogenic Medicines

azacitidine	cyclophosphamide	methotrexate
azathioprine	cytarabine	nitrofurazone
chloramphenicol	dacarbazine	phenacetin
chlornaphazine	fluorouracil	phenoxybenzamine
cimetidine	mercaptopurine	procarbazine
cisplatin	methapyrilene	thiotepa

Reproducing Previously Published Materials

To reproduce a figure, photograph, or table that has been published elsewhere, you (the author) must obtain permission in writing from the original author and from the copyright owner (usually the publisher), and you must submit the written permissions along with your final manuscript. Even if you were the author of the previously published figure or table, you still need written permission from the copyright owner. The only exception is for a work of the U.S. government. See the section on figure captions in this chapter for examples of credit lines (pp 296–297). See also Chapter 11, Copyright and Permissions, for further information.

If you construct a figure or a table from data that were previously published as narrative text, you do not need permission, but you should reference the source of the data (e.g., "Data are from ref 7.").

If you significantly change a figure or table that was previously published, you do not need permission to publish it. Significant changes would be the addition of as much information as was in the original. The changed material must amount to a substantially new creation. If you are in doubt as to whether you have changed it enough to use it without permission, be on the safe side and obtain written permission. Even if you make significant changes, you should reference the original source and give proper credit (e.g., "Adapted from ref 60.").

If you are thinking about using a previously published figure or table, consider carefully whether citing it as a reference would be adequate.

Bibliography

Briscoe, Mary Helen. *Preparing Scientific Illustrations: A Guide to Better Posters, Presentations, and Publications,* 2nd ed.; Springer-Verlag: New York, 1996.

Fleming, J. L.; Kornacki, A. *Scanned Line Art from Authors: Avoiding Moiré and Other Rescreening Problems,* Version 2; Cadmus Journal Services: Richmond, VA, 1996.

Pocket Pal: A Graphic Arts Production Handbook, 16th ed.; International Paper: Memphis, TN, 1995.

CHAPTER 10

Peer Review

Everyone knows what peer review is and why it is done, but people have their own ideas on how to do it. This chapter presents the views of many ACS journal editors, ACS Books Advisory Board members, and other chemists who were asked these questions:

- How do you go about reviewing a manuscript? What do you look for when you review a manuscript?
- When your own manuscripts are peer-reviewed, what kinds of comments are you expecting? What kinds of comments do you consider inappropriate or useless?

By asking many scientists and presenting a broad range of opinions, we hope that this chapter, along with Appendix III, "Ethical Guidelines to Publication of Chemical Research", will provide guidance for new and experienced peer reviewers alike. We believe, though, that what you will find here is not only how to review a scientific paper, but also how to write papers that will meet reviewers' expectations and thus get published.

Here, in alphabetical order of the respondents' surnames, are their replies.

Allen J. Bard

The University of Texas at Austin
Editor, *Journal of the American Chemical Society*

The basic guidelines for reviewers are to answer the main questions: Is it new? Is it true? Is it interesting? In general, a reviewer should address the novelty of the finding and make sure that the author puts this in the context of past work in the field. The reviewer should also try to establish whether the experimental or theoretical work is done well, whether there are possible flaws or artifacts in the measurement, and whether the techniques are appropriate for the type of information sought and generally represent the state of the art. The authors should make clear what conclusions are drawn directly from the experimental or theoretical work and what are speculations. Finally, the reviewers should judge whether the work makes an interesting and significant contribution to the field as opposed to a simple and obvious extension of past work or trivial modification of an earlier study.

Both with my own manuscripts and with manuscripts that I consider as an editor, the most annoying comments are "publish without change" and no further discussion, or "this is too specialized" and "do not publish" without any qualifications or discussion of why this is so. Especially useful comments point to relevant work that is not referenced in the paper. Most authors try very hard to do a good literature search, but given the scope of the scientific literature and the fact that, for many types of papers, relevant work can appear in different fields, they may miss some important references. It is very helpful when a knowledgeable reviewer can point out other work that bears on the manuscript and has not been referenced. Of course, if reviewers note typographical errors, problems with style, or lack of clarity, that is useful, although that is not central to consideration of a manuscript.

Mary D. Barkley

Case Western Reserve University
Associate Editor, *Biochemistry*

When I review a manuscript, I read the abstract, introduction, and discussion first to find out what the point is. Then I look at the Experimental section and the data (figures and tables) to see how the point is made. Finally, I read the entire manuscript carefully from the beginning, noting questions as I go.

I look for two things: (1) whether the manuscript is worth publishing at all—I base this judgment mainly on the significance of the results and the quality of the data—and (2) whether the presentation is clear and logical so that the point is obvious.

Those who review manuscripts should give an honest opinion about the suitability of the paper for the journal. Some reviewers stick to technical criticisms, being unwilling to come out and say that a paper is weak. Comments about the significance and rigor are helpful, as are comments about the reviewer's ability to judge various aspects of the paper, so that I know how to evaluate the review.

The kind of feedback that I do not need is a list of grammatical corrections with a recommendation of minor revision. In such cases, I don't know whether the paper is great but has a few typos, or whether the reviewer does not understand the review process.

For my own manuscripts, I expect thoughtful comments from an expert who has read the manuscript carefully. Even when the reviewer is mistaken, I try to clarify that part of the text so that other readers will not make the same mistake.

Robert F. Brady, Jr.

Naval Research Laboratory
ACS Books Advisory Board

When reviewing a paper, I feel that I have some responsibility for maintaining the quality and relevance of the scientific literature. Thus, I like to see

- an introduction that states why the work was done or what the writer hoped to learn by doing the work.
- sufficient references to place the work in the context of other papers in the field, to allow the reader to understand how the work in this paper advances the field, and to allow the reader new to the field to find pertinent background information.
- an Experimental section that is clear and complete enough to allow the reader to reproduce the work. Materials must be identified by chemical name, not by generic or brand names.
- a discussion based on sound reasoning, clearly expressed, and relating the work to present knowledge. Certainly, conclusions that go beyond or against current thinking are not unwelcome, but they must be clearly placed in the context of current knowledge.
- clear figures with axes labeled and captions sufficient to understand the figure without reference to the text, but a reference to each figure given in the text.
- a summary that stands alone without reference to the rest of the paper.
- evidence that the manuscript was prepared with care, in conformance with the style of the journal and its instructions to authors.

- an engaging style of presentation that draws the reader into the paper, and some use of personal pronouns to avoid repetitious or awkward use of the passive voice.
- suitability for the journal, both in the area of technology and in the basic or applied nature of the work.

When I receive a review of my work, I like to find

- courteous and helpful comments that show respect for the time and labor invested.
- pointers to references I have missed, and alternative interpretations of the state of the art in the field in which I am writing.
- pointers to errors in reasoning.
- pointers to lapses in clarity of expression, alternative wording that would make the point clearer to more readers, suggestions on improving clarity of figures or tables.
- alternative interpretations of data and results.
- suggestions for further experiments for later papers.

Malcolm W. Chase

National Institute of Standards and Technology
Editor, *Journal of Physical and Chemical Reference Data*

When reviewing a manuscript, I first read very carefully the abstract, introduction, and conclusion (or the like) to confirm that the importance of the article is clearly stated and that the results and accomplishments suggest a new body of knowledge or understanding. Second, the list of references is very indicative of the quality of the article. I am concerned as to the proper quality of the article if either of two situations arises: (1) most of the references deal with the author's own works (too much ingrown activity), or (2) there are no recent references (say, nothing in the past 2–3 years, indicating that the author is out of date or working in a field of no current interest).

A review that simply states "good article, should be published" or the like is not useful. It suggests that the reviewer did not take time to truly examine the contents and quality of the proposed publication. I look for definitive statements as to the inclusion of pertinent data, awareness of related studies, or the lack thereof. Statements that suggest that a new approach was taken, a new interpretation was put forth, or a lack of understanding is exhibited are necessary to judge the usefulness of the article.

When my own manuscripts are reviewed, I look for statements that confirm (or not) an excellent coverage of the literature, an excellent interpretation of the available data, clues to related studies, avenues for better approach, and missed opportunities. Statements as to the clarity or lack thereof are useful.

Thomas L. Chester

Procter & Gamble Company

The reviewer has two jobs: First, the reviewer serves the editor and journal by making sure that only work of significance and of quality meeting the journal standards is accepted. Second, the reviewer serves the author by providing honest and helpful comments and suggestions for improvements, not just judgment. As a reviewer, I generally look for a few basic things in a research manuscript: significance to the field, correct level of detail, absence of technical errors, writing for the correct target audience, and so on.

A research article must be of sufficient significance to justify publication. This significance should be disclosed in the introduction (if it is not obvious). The author should explain how the current work fills a previously unmet need, answers an important question, or discloses a new capability. (The opposite, although rare, is also legitimate—some publications may state or demonstrate a need or refute previously published findings.) The editor or reviewer should reject a manuscript if it has already been published in essence by the present authors or someone else or if the work is trivial or an obvious extension of earlier work.

A research publication should usually be directed at experts in the field. It needs only enough detail for an expert reader to be able to reproduce the work. Essential parameters and techniques not obvious to someone familiar with work in the field must be disclosed, but needless detail should be avoided. Sufficient references must, of course, be included. Writing for the expert should not be construed as a license for poor explanations of new or little-known concepts. The reviewer can be very helpful to inexperienced authors in these matters.

Enough background should be given to position or contrast the current work with respect to previous publications. However, many inexperienced authors comment on every article they can find on their subject. Such detail is unnecessary and should be noted by the reviewer. Usually, even review articles should be limited to significant and useful references rather than citing everything written. Also, the reviewer may suggest including important work not cited by the authors.

Many inexperienced authors disclose everything they did in the order it was performed. Such disclosure is unnecessary, especially if the experiment was poorly designed or if unexpected results caused a change of plans. Instead, publication should focus on what was learned, backed up with sufficient and persuasive experimental evidence. The reviewer may suggest deleting superfluous experimental details and results or restructuring the manuscript to emphasize the new knowledge.

I routinely look at any equations, and particularly derivations, to make sure they are correct. I look at the units. Units on both sides of the equations must match. (It is surprising how often they do not!) I consider whether the experimental results are believable or in opposition to accepted theory and presented without convincing explanation. I also check whether the conclusions are supported by the work, or whether other conclusions can be drawn from the results presented.

I look at the figures to make sure they are necessary, on the one hand, and adequate, on the other. Figures should not duplicate information already in the text or in tables.

You might also be interested in what I do not do as a reviewer. I do not do much in the way of correcting spelling and grammar unless these errors cause confusion or precipitate technical questions. Normal copy editing belongs to the editor and his or her staff.

When my own work is reviewed, I find constructive criticism most helpful. I do not want the reviewer to become a coauthor, but I do appreciate reviewers who politely point out errors or unclear passages and offer suggestions for improvements. I really get irritated when a reviewer who is obviously ignorant of the field makes inappropriate or incorrect technical comments or inappropriately asks for more experiments. This traps the editor in the middle and causes big problems, particularly if the editor is not an expert on the particular subject. A reviewer who is unfamiliar with the subject should not be reluctant to return a manuscript to the editor without review. Also, reviewers who have a personal reason or conflict of interest that would keep them from being objective should return the manuscript immediately.

Ray A. Dickie

Ford Research Laboratory

On first receiving a manuscript for review, I scan the manuscript and ask myself a series of questions: Am I the right person to review this manuscript? Do I have the time to respond within the specified deadline? Do I know someone in my organization who could contribute to a review, or who could provide a better review?

If the answers to these questions are negative, I return the manuscript with a brief explanatory note. If I can provide a review, I give the manuscript a second and much more thorough reading, with a number of additional questions in mind. I try to develop my own statement of the intent and significance of the paper. This is really for my own benefit in completing the review, but often it will be the first paragraph of my review. This statement can form the basis for agreement between reviewer and author, or it can alert the author to potential misunderstanding of the work on my part.

Is this original work? Is it important work? Early in the review process, I try to establish whether the work is original and whether it is a major advance or a small step forward in a crowded field. Not all work that is technically correct is appropriate for publication: Careful experimental work on old materials that does not lead to new insights may be difficult to reject on strictly technical grounds, but generally does not merit publication. If I conclude that the manuscript describes insignificant work, I will return the manuscript without completing an extensive technical review.

Is the paper well-written? Is it understandable? I find it difficult to evaluate poorly written manuscripts. Although it really is beyond the normal scope of a peer review, I will sometimes edit a manuscript to make it readable enough to evaluate on technical grounds. Manuscripts submitted by non-English-speaking authors can pose special problems. If the manuscript is so poorly written that the meaning and significance of the work are obscured, I will reject it without completing the technical review.

I determine whether the subject matter of the manuscript is appropriate to the journal to which it has been submitted. A more subtle issue involves an assessment of importance. Less significant work should appear in less important journals, if at all, but I have also suggested submitting a manuscript to a journal of higher prestige or wider circulation when that has seemed warranted by the importance of the work and the quality of the manuscript.

Is the background information correct and complete, or are there important relevant references missing? Has the work been properly set in the context of prior knowledge? Too often, manuscripts are submitted with citations only to the work of the principal authors and their colleagues. I try to provide citations to related work and include brief comments describing the relevance of the references to the manuscript under review.

Are the experiments and materials described in such a way that the work can be reproduced? Experimental papers really must include good information on materials and methods. References to descriptions in theses and other hard-to-access sources are not sufficient, and I request that the authors revise their manuscript to provide additional information. For ACS journals, I may suggest inclusion of such information as Supplementary Information, but my preference is for information on techniques and methods to be included in the full published paper. Description of work based on analysis or testing of commercial materials poses a special problem. In such cases, it is essential to include both the commercial designation and a generic description.

Do the figures and tables present the data in a comprehensible, logical, and economical way? I compare the tables, figures, and discussion to be sure that the information is logically presented and consistent. When there is a well-accepted standard format for presentation of a particular type of data, I suggest that the standard format be used. I check figures to be sure that the figure caption and key (if any) provide the information necessary to understand and interpret the figure.

Does the discussion provide insight into the meaning of the results, or does it provide only a factual description of the results? In the Discussion, I expect to find a thoughtful analysis of the experimental results. Speculation should be labeled as such, but may be completely appropriate. A mere factual description of experimental results is not sufficient.

I address whether the conclusions are supported by the experimental evidence, models, and other results presented. I may suggest alternative conclusions or interpretations.

When my own manuscripts are reviewed, the comments I expect and find useful are

- New and relevant references (e.g., "Reference should be made to the work of A and B; these authors showed that changes similar to those reported in the present manuscript occur under the following conditions.")
- Alternative interpretations of data; additional structural or bonding models (e.g., "This paper would be really solid if the following 'loose ends' were addressed: bonding to the surface through oxygen lone-pair electrons, bonding of the polymer ...", accompanied by additional references and suggested models).
- Identification of statements or concepts that need further explanation (e.g., "The statement at the bottom of p 2 regarding experimental methods is not clear; were the parameters X and Y controlled independently?")

- Comments on graphical presentation of data (e.g., "Axes of Figure 2 are not well-defined: what is *f*?").
- Discussion of possible interferences or errors and suggestions for clarification (e.g., "The mixtures analyzed contain water, dimethylaminoethanol, 1-butanol, and possibly other compounds that absorb in the range 3000–4000 cm^{-1}. The authors should show that these compounds do not significantly interfere with their analysis.")

Inappropriate and useless comments include references not related to the manuscript under review; allusion to other work, or uncited work, without traceable references; demands for additional experimental work, especially when directed to secondary or follow-on questions; and suggestions that a different study would have been preferable (e.g., "The authors present results on calculation of blend modulus; from a technological viewpoint, a more important quantity is strength.").

Ernest L. Eliel

University of North Carolina at Chapel Hill

When I get a manuscript to review, I try to peruse it immediately and decide whether I shall accept the refereeing assignment or not. Possible reasons for declining are (1) serious lack of time; (2) lack of competence in the area of the work (I refuse to referee manuscripts that would force me to do extensive reading, though of course I am prepared to read or reread a few articles in the course of refereeing a manuscript that is generally in my area of expertise.); (3) lack of interest in the journal (but this would not apply to ACS journals); and (4) too many manuscripts received in a limited amount of time. (That sometimes happens with ACS journals because they are many. It is less of a problem now, because editors of a given journal communicate with each other about referees; however, I do not think editors of separate ACS journals do.)

If I decide not to referee the manuscript, I send an e-mail or fax message to that effect to the editor and return the manuscript promptly by mail. If I decide to referee the manuscript, I read it carefully once or twice and start thinking about it. As I draft my report, I reread the manuscript.

My judgment of a manuscript relates to the following factors:

1. How novel or original is the work? Will it generate interest among the readers? Or is it just a rehash of work that the senior author has published in slightly different form before?

2. How well is the experimental work done? Does the discussion properly reflect the experimental findings?
3. How well is the literature background presented? If the judgment on point 1 is high but on point 2 or 3 low, I often write extensive comments and ask for major revisions.
4. Is the manuscript suitable for that particular journal? Often very good pieces of work are so specialized as to be of interest to only a small number of readers; in that case the recommendation is to refer the manuscript to a more specialized journal.
5. Is the manuscript in acceptable English? I am prepared to make minor corrections, but not to act as a language editor for manuscripts from foreign authors.

I am very conscious of the fact that editors like to make decisions in a reasonable time frame, and so I usually observe their deadlines. I also usually fax (more rarely e-mail) my reports to the editor. Although one is supposed to discard manuscripts after review, I usually keep them for a while, in case a second review (after revision) is asked for.

Regarding my expectations when my own manuscripts are reviewed, above all, I expect a referee to be polite. I never write anything anonymously that I would not be prepared to send to the author over my name (and I sometimes sign the author's copy of my reviews). I expect others to act likewise. I would expect comments about the adequacy of the experimental work, the soundness of the conclusions, the clarity of presentation (or lack thereof), and the importance of the work, especially as to its interest to the readers of the journal in question. I am always pleased when referees point out errors (either in the writing or in the science) that I have overlooked. On several occasions, referees have thus saved me embarrassment! I consider a referee to be someone who helps me improve my paper, not someone who is there to damn it, although I am quite prepared to accept fair criticism. I am not happy when referees have axes to grind, either because of vested interests or because of preconceived notions. A referee should have an open mind.

Arthur B. Ellis

University of Wisconsin at Madison
ACS Books Advisory Board

On reviewing a manuscript, I first assess the importance of the work: Is it an incremental advance or a major breakthrough? This question relates to the journal sought for publication and the appropriate audience. I routinely recommend publication in more specialized journals, particularly for incremental advances. My advice is to submit to the more general journal, which provides more visibility if the work has the potential to have significant, broad impact.

The title should be reasonably descriptive. If an abstract is part of the paper, it should be a good summary, which helps me figure out what to focus on in the review.

The paper should have a "story line": What problem is being addressed (what is the motivation?), what was done to address the problem, and were results reasonably interpreted? The paper should have a logical flow and be clearly written. Results should be interpreted but not overly so (minimize unwarranted speculation).

The Experimental section should have enough detail to permit the work to be reproduced. There should be consistency among information in the text, tables, and figures. Figures and tables should be labeled clearly. Relevant literature should be included.

In reviews of our work, I appreciate it when reviewers identify any deficiencies relative to the aforementioned criteria. It is also helpful if reviewers can give other interpretations of our results and suggest other experiments that would help us better understand and characterize our systems.

Alan D. English

DuPont Central Research and Development
Associate Editor, *Macromolecules*

For *Macromolecules*, the associate editors examine the manuscript to make sure that it is of minimum quality, both scientifically and linguistically, so as not to waste the reviewers' time. Examples of papers that are routinely returned to authors without review or for correction prior to review are those with nomenclature that is "customized" or jargon, missing figures or figures that need original photographs to be interpreted, English language usage that is very poor, out-of-date references, and topics that are beyond the scope of the journal. My advice to authors is to read the journal's own definition of what it perceives its niche to be.

The kind of feedback we need from reviewers is specific comments about the technical merit, style, and originality of the paper. General platitudes are of limited use. It is very useful for the reviewer to indicate whether the paper can be altered and published or whether it would be better to reject it and have the authors critically examine what it is they are trying to communicate. I rely very heavily on the reviewers' comments; hence, I expect them to take the process very seriously.

For my own papers, I depend upon reviewers to give me specific technical comments. Most of all I very much want them to tell me if they think I am in error. This has happened to me twice in more than 100 publications, but I was very grateful both times—after I recovered from the shock of the intense criticism. This positive approach toward criticism comes with the passing of youth; as an associate editor, I have explained this to more than one chagrined author and have found that, once rationality sets in, most of them can accept the process as being useful, even if they do not agree with it.

Paul Hedin

U.S. Department of Agriculture, Agricultural Research Service

When I review a manuscript for the ACS Symposium Series, I take the following steps. I read the abstract to determine the conclusions, I read the introduction to determine the objectives, and I read the Results and Discussion section to determine whether the conclusions were justified. I look at the Materials and Methods section to determine the soundness of the procedures. I look at the tables and figures to check for comprehension and statistics. Tables often can be simplified; often they contain too many significant figures. I go through the text and make suggestions to improve clarity and to identify grammatical errors. I make a value judgment about the overall significance and worthiness of the manuscript. With papers in the ACS Symposium Series, which are normally invited, the focus is generally on improving the paper because it is meant to fill a niche in the book, although occasionally a symposium paper may not be suitable.

With regard to peer reviews of other manuscripts, I take the following approach. I evaluate the suitability for the journal. I judge whether the manuscript has adequate professional and scientific merit. I determine whether the manuscript provides new and unique information. I make recommendations about (1) removing erroneous information; (2) adding evidence to improve validity or adding statistical treatments; (3) shortening the manuscript, including tables; and (4) adding references.

Comments that are inappropriate or useless are

1. suggestions about doing extensive additional experiments that may not be feasible for various reasons.

2. recommendations for additional statistical treatments that may be of minimal value.

3. recommendations about using some other experimental regimen that may not necessarily be superior, but may be familiar and convenient to the reviewer.

4. a harsh review by a competitor that is subsequently not detected by the editor. In this case, the author must respond positively to the review when possible and, when not possible, rebut the reviewer's comments.

Louis S. Hegedus

Colorado State University
Associate Editor, *Journal of the American Chemical Society*

I expect reviewers to take the following steps. Carefully read the text against the experimental details to see that the data support the conclusions and that the results reported in the text are consistent with those reported in the Experimental section. Check that assigned structures are consistent with data. Carefully check that all reported compounds are fully characterized and that full, reproducible experimental procedures are provided for all key transformations. Make sure that the work is properly placed in the context of the area and that the important contributions of others are properly and thoroughly cited. Watch for inappropriate language. Make sure that the significance of the reported research is clearly stated and that the reported work does indeed have that significance. Make sure that the work is original and not previously reported. Make sure full papers are not merely summaries of previous communications without substantial new data.

Personal comments are inappropriate. Assertions that the work is not original without literature citations to substantiate the claim are useless.

William I. Higuchi

University of Utah
Editor, *Journal of Pharmaceutical Sciences*

In evaluating a manuscript for publication suitability, we ask reviewers to use "normal standards of good science" and to use these questions as a guideline: Does the paper contain an original and significant contribution to one or more of our fields of interest? Are the experimental methodology and data interpretation sound? Are there proper controls? Is the statistical analysis appropriate? Are the conclusions justified on the basis of the results obtained?

William P. Jencks

Brandeis University

When I read the literature, the first thing I look for after the title of a paper is the end of the introduction. I expect to find there the "bottom line" conclusions of the research. At least 95% of the time I stop there and look at another paper. There are more papers these days than one can examine in any depth, and I am interested in a number of different subjects.

When I review or study a paper, I start the same way. Then I read, or read again, the abstract and introduction. These sections should give a clear picture of the "bottom line" conclusions and outline the experimental basis for them. I read the Results and Discussion section and take notes on areas that might be presented more clearly or raise scientific questions that require further examination. Then I examine the Experimental section to be sure that sufficient information is presented clearly enough so that the experiments could be repeated by a first-year graduate student.

Finally, I prepare an outline of my conclusions regarding the quality and importance of the science that is presented and suggest ways to present the results and conclusions of the research more clearly. The conclusions will include my evaluation of whether I recommend the paper for publication.

When my own papers are reviewed, I hope to learn, first, whether there are scientific points that are not clear or are questionable and, second, what revisions would improve the clarity of the presentation.

Comments that do not explain clearly what the reviewer finds unclear or wrong, and value judgments that are not clearly presented or are without justification are useless.

In general, a review should be accurate and should focus first on the positive qualities of the research.

Herbert D. Kaesz

University of California, Los Angeles
Associate Editor, *Inorganic Chemistry*

A reviewer should check the abstract and contents of the work to see that it falls within his or her area of competence, and that there is no conflict of interest. Sometimes manuscripts are sent to someone whose name appears in one of the cited references. However, this can be peripheral to the work at hand, or the interests of the person could have changed and he or she is no longer current on the literature of the field. Conversely, the reviewer may be working on the same problem; if so, this should be reported to the editor.

If the paper is within the reviewer's area of competence, the reviewer should decide whether the conclusions are significant and supported by adequate experimental data and whether the relevant literature is adequately cited.

Lawrence H. Keith

Radian International LLC

I start a manuscript review by looking at the table of contents and comparing it with what I know about a subject. This gives me some idea of the completeness of the content. I also look for typos and readability in addition to technical accuracy on the basis of my knowledge of a subject. I usually do not look at tables in detail, but I am a visual person, so I look at figures for understanding.

When my own papers are reviewed, I like to receive very specific comments if something should be added or clarified. Simply to say things like "expand this section" is useless. I need to know what exactly they think I have left out.

I consider the main thing peer reviewers should be requested to do is to be very specific with questions and comments. Usually, brief corrections involving errors and things that need simple clarification are most useful if they are marked in the margins of a manuscript.

A. Douglas Kinghorn

University of Illinois at Chicago
Editor, *Journal of Natural Products*

When reviewing a natural products manuscript, I look for novelty, significance, and completeness of the study, so that a definite contribution is being made. It is important that the work of others is appropriately cited and built on as a foundation and that the authors try to answer as many questions in their own work as is feasible. The work must be carried out according to accepted standards in terms of collection and authentication of the biological material and in the rigor with which chemical structures are determined, bioassays performed, and synthetic procedures conducted.

A primary task of reviewers for the *Journal of Natural Products* is to help decide on the originality of a manuscript, as well as the thoroughness of a given study. Reviewers who are cognizant of the relevant literature are very helpful, as are those who critically evaluate the logic of an author in coming to a structural or other type of conclusion. Comments on relevance to the scope of the journal, stylistic problems, or grammatical shortcomings are of less value, although these are often somewhat useful. We must be wary of a tendency of many authors to try to fragment their work on the same species into several short manuscripts rather than produce a single comprehensive manuscript, and reviewers can often provide valuable input into this part of the evaluation process. In general, unless a paper is of outstanding merit, we expect something more than a one-sentence review from our journal reviewers.

When my own papers are reviewed, I appreciate cogent comments on the argument being presented for a compound structure proof, as well as any perceived deficiencies in the chemical or biological approaches being applied to the problem at hand. Comments that are factually incorrect, illogical, or personal in nature are not usually valuable.

Charles Kutal

University of Georgia

Upon receiving a manuscript for review, I first determine whether I possess sufficient knowledge of the subject matter to provide a competent review. Usually, the information provided by the title and the abstract allows me to make that decision. Typically, I review about 80% of the manuscripts sent to me (not counting those returned to the editor because of lack of time).

In reading a manuscript, I appreciate a concise introductory section that places the subject matter in perspective and provides relevant background information. The Experimental section should be sufficiently detailed to allow an experienced chemist to assess the quality of the data. I prefer comprehensive descriptions of experimental techniques and procedures rather than cursory accounts that leave the reader uncertain about how experiments were performed. Discretion must be exercised, however, and repetitive information such as syntheses of closely related compounds and lengthy descriptions of instrumentation already published should be minimized. The Results and Discussion section should present data in well-planned tables and figures and provide logical, well-supported, and adequately referenced interpretations. Unfounded claims and excessive speculation should be avoided.

If I am not totally comfortable with my level of knowledge of the subject, I read some of the references cited by the author. Sometimes I suggest references that the author has overlooked. If I detect errors or omissions in the presentation or interpretation of data, I note these in my review and, if possible, offer suggestions for improvement. I may also note alternative explanations overlooked by the author. I may mention other factors such as unacceptably poor grammar and improper nomenclature.

When my own papers are reviewed, I expect from other reviewers what I hope to provide them in my reviews; I suppose the Golden Rule applies here. In particular, suggestions meant to improve an interpretation of data can be very useful. In contrast, comments that reflect an inflexible bias by the reviewer are clearly inappropriate. Opposing views must be supported by fact and sound reasoning. Finally, suggestions for additional experiments can be valuable, but the reviewer must be careful to explain the relevance of any new experiment to the stated scope of the work under consideration.

Michael R. Ladisch

Purdue University

The following guidelines are useful. First, is the text of the manuscript longer than 12 double-spaced pages? If so, it is probably too long and will require some editing. If it is much longer than this, I return the manuscript to the editor with the suggestion that it be shortened and reviewed again.

I first review the abstract and the conclusions. These sections should answer the question of whether the work is significant and whether it adds significant value to the literature. If it appears from the abstract and the conclusions that the paper adds only incremental value to the literature, I look at the Materials and Methods and Discussion sections with the intent of advising the author of its shortcomings and noting where further work is needed.

If the work seems to indicate a substantial contribution to the literature and fits the journal to which it was submitted, I carry out a further detailed review. I carefully consider the overall theme of the paper and analyze the fit of the introduction and background materials with the subsequent discussion of results. The creative approaches and exciting and novel features of the work being described should be clearly stated in the abstract and introduction. Most reviewers will look for some explanation of the significance of the work and how it affects the field of research. How does it add to existing knowledge? How does the work answer an important question?

If references cited in the text of the manuscript are omitted from the References section at the end of the paper, this indicates sloppy preparation of the manuscript. Examining some of the references is often useful to confirm the accuracy of the information reported. The reader of the journal would expect prior work to be accurately represented in the manuscript.

This is particularly important in some types of engineering equations and derivations that follow from prior work. Consequently, I spot-check these journals for accuracy of representation of the data.

When a manuscript is revised several times prior to submission, there can be a mix-up between the information cited and the actual reference. Therefore, I suggest that author use an author–date format in the paper until the final version of the manuscript has been prepared. Once this process is completed, the cited references can then be changed to a numerical system if this is what the publication uses.

I compare the internal consistency of the author's results and discussion with respect to the data presented in the paper against my prior knowledge of the literature and the research area. Often authors will try to extrapolate results and propose conclusions that may not be completely warranted by the data. In this case, reviewers may present several options, including obtaining more information or data or eliminating sections that are somewhat speculative relative to the results that have been obtained.

The materials and methods are a central part of papers that present experimental (compared to theoretical) results. Most of the manuscripts that I review have both experimental and theoretical components. For experimental work, it is critical that the materials and methods are clearly described, so that a reader who is an expert in the field can reproduce the work. At the same time, certain standard procedures need not be repeated in detail. The author should cite the appropriate reference and then indicate clearly any modifications to the original procedure.

With engineering derivations, particularly those that include models with differential equations, I look for accurate representation of boundary conditions and initial conditions for these equations. Often an author will not present these, and without them these equations are meaningless. Some manuscripts, however, provide too much detail of certain types of derivations and models, particularly when these can be found in prior publications. In this case, I usually suggest a more concise representation in which the key equations, the initial and boundary conditions, and then the solution are presented. When such an equation is presented, I check that the equation is actually used in the paper. Often the derivation is presented, but the equation is never really used. This leads me to suggest that the derivation be eliminated from the manuscript and appropriate citation be made to other literature references.

The analysis of the data in the manuscript includes dimensional analysis (are the dimensions of mass, time, distance, etc., internally consistent?). When a model is compared against data, does the model actually give the results presented in the tables or figures? Are the results being modeled of sufficient significance to warrant the type of model being used? Some of these questions are subjective, but this is the same analysis that a reader would probably apply to a paper. Consequently, the author should consider these factors before the paper is submitted.

The kinds of comments that are inappropriate or useless are those making a severe criticism with no explanation. For example, a reviewer might state that the theory is inappropriate and not useful, yet give no indication why he or she came to this conclusion. Another comment that is not useful is a criticism of a literature survey or background section for omission of references without an indication of which references may have been overlooked. In other words, general criticisms need to be qualified with specific suggestions or comments.

John W. Larsen

Lehigh University
Editor, *Energy & Fuels*

I begin by reading the manuscript relatively rapidly to familiarize myself in a general way with its contents and principal arguments. My second reading is intensive, and I give particular care to the adequacy of the experimental work and the relationship between the data and conclusions. After this reading, I prepare a draft review. I then leave the paper and the review for a few days to give me some time to think about any issues that have come up. I then go over the paper a third time and, in light of this reading, make final revisions to the review.

The first thing I look for in the manuscript is whether the experiments have been done well. Next, I make sure that the data support the conclusions reached. Then I consider the overall organization of the manuscript and whether the paper accurately reflects what is in the literature and has integrated its findings and conclusions well with what has gone before. The last issue is the quality of the writing and grammatical correctness.

A good reviewer's report contains not only the evaluation of the manuscript, but also the reasons for arriving at that evaluation. It is especially important to document such assertions as "The author has ignored significant portions of the literature." This statement is of no help to an author and is unconvincing unless accompanied by a reference or two. Ad hominem comments or speculations about the author's skills or motivation have no place in a review. Reviewers' comments should be limited to an evaluation of the work and a recommendation on the suitability for publication. Often, when reviewers are sharply divided on a paper, the negative review will be more detailed and more specific than the positive one. A few sentences explaining why the work is of high quality and importance are necessary.

When my own manuscripts are reviewed, I am hoping for a constructive evaluation that helps me to improve the manuscript and that clearly points out all errors and omissions. I prefer the comments to be highly specific and to include literature references where appropriate. Comments about organization and the clarity of the presentation are always welcome. I do not appreciate ad hominem comments or speculations about my intelligence, the care with which the manuscript has been prepared, and so on.

One gray area in reviewing is suggestions by the reviewer for additional work. Sometimes reviewers request additional experimentation that would enormously expand the scope and sometimes even change the direction of the research. Other times they are asking for additional information that is absolutely necessary to confirm or validate the work. I believe authors have the right to define their own scope of work, and so I am not sympathetic with reviewers' demands for major expansions. Nevertheless, requests for necessary but omitted data are a crucial part of the reviewing process.

R. U. Lemieux

University of Alberta, Edmonton

When I receive a manuscript to review, I first read the abstract and introduction to decide whether I am competent to assess the novelty and scientific value of the paper. Should I feel seriously deficient in this regard, I will return the manuscript to the editor. Otherwise, I will state that I believe myself to be either marginally or fully competent.

I determine whether the authors have dealt fairly with the literature and how their contribution relates to the relevant previous work. It is easy nowadays to provide long lists of citations, obtained by computer assistance. Often I suspect that the authors have not read the articles because their relevance to the manuscript is not at all evident. This omission can be serious, by implying to the novice that there is little need to consult this literature. This practice, together with that of referring only to their own previous publications rather than to the key papers of the past, poses a threat to a cohesive and healthy development of chemical knowledge.

I next examine the Experimental section to decide whether it is presented in sufficient detail to enable other researchers to repeat the work reliably. Having decided that the experimental work is adequately presented, I then examine whether the conclusions drawn from the results are acceptable; that is, is reproducible fact well-separated from conjecture?

I feel obliged to document my basis for any adverse criticism and expect that editors will ignore critiques that do not do so, regardless of their origin. I have encountered too many self-serving pontifical criticisms that were largely unjustified and did considerable harm to the careers of newcomers. Such assessments not only are useless, but also may be dangerous to the welfare of chemistry.

If a paper appears to be an honest, scholarly effort, the reviewer should try to be helpful and encouraging. This I believe to be of major importance to strong intra- and interdisciplinary effort. Gratuitous recommendations for further experimental work should never be made.

Kenny B. Lipkowitz

Indiana University–Purdue University at Indianapolis
Associate Editor, *Journal of Chemical Information and Computer Sciences*

The feedback I expect from reviewers is mostly about the technical aspects of the manuscript: whether it has sound statistics, adequate or state-of-the-art techniques, and the like. I also enjoy when referees point out that the work has been published in large part in other journals or is just a redundant set of experiments from previous work. If it is a manuscript of borderline quality, I have added support for rejecting it.

What I look for in the submitted manuscript is something that is creative and new. I do not mind rehashing old work if a new way of thinking about things is presented, but "yet another calculation of such-and-so" is something I tend to avoid. So, novelty is a must, along with technical soundness. Moreover, I do not look for "timely" papers; everything is timely if it is explained or reinterpreted in a refreshing, alternative way.

For my own papers, I expect pretty much the same treatment that I expect from my group of referees. I want the reviewers of my papers to check the technical aspects of my work. I also want them to point out that I'm drawing the wrong conclusion from the results if I have done so and to offer some (unbiased) alternative suggestions. I find comments about "timeliness" of the work inappropriate, along with any kind of personal attacks on me or anyone else in the literature.

Douglas R. Lloyd

University of Texas at Austin
ACS Books Advisory Board

When I receive a manuscript, I look at the abstract and the conclusions to see whether I am qualified to review it (you might be surprised at the number of manuscripts I am asked to review for which I have no expertise). If I feel qualified and I am able to do the review by the deadline, I set aside a reasonable period of undisturbed time. At the same time, I have at least one of my graduate students review the manuscript. I then meet with my graduate students, and we discuss the manuscript. Finally we combine all of our comments, criticisms, and suggestions.

The manuscript should be clearly written, concise, logically organized, informative, original, and focused on the key topic (that is, it should not include a lot of extraneous material). The manuscript should be adequately referenced with current and historical literature. All figures and tables should be necessary, and they should stand on their own; that is, readers should be able to make sense out of the figure captions without reading the text.

When my own manuscripts are reviewed, I expect constructive criticism—suggestions on how to make the paper better. However, not all reviewers offer constructive criticism; many reviewers simply criticize. Being critical in a nonconstructive fashion (which is a growing trend) does not help improve the paper.

Comments that I consider inappropriate or useless are personal attacks on the authors. These attacks are unacceptable, unprofessional, and serve no useful function. Other comments that have no value are those that suggest that the authors do some additional research. Although this suggestion helps the authors consider future research plans, most authors are not able to delay publication until this future research is completed, especially if they do not have the money or equipment to do the work.

Royce W. Murray

University of North Carolina at Chapel Hill
Editor, *Analytical Chemistry*

When I consider reviewing a manuscript, first I try to decide by skimming through it whether I have a sufficient familiarity with the subject to do a review. Then I read the paper carefully.

I determine whether the paper has anything original to say. The level or degree of originality that should be demanded correlates with my perception of the standards of the journal. Also, I determine whether the topic is one of contemporary interest or novelty or is new, as opposed to more data on an old subject.

I assess whether the experiments seem to be carefully done, including control experiments where appropriate. I decide whether the interpretations given are supported by the data or, if they are not fully supported, whether speculation is thus labeled. I determine whether I could reproduce the experiments (or computations) from the information given or referred to. It is crucial that other workers be able to repeat an experiment, so I always look for the experimental details in the Experimental section and figure legends, and for a sensible explanation of how experiments were done (if they were uncommon ones).

I evaluate whether the paper claims undue credit for the advance made, as compared to previous workers. I check that the supporting literature is properly cited, especially key works by others. I look at the length of the paper; are there excess figures or tables of boring, repetitious discussions?

Finally, I try to cast criticism in a constructive way, especially if the paper is a good one but there are problems that need fixing.

As an editor, I expect reviewers to proceed as I have just described. Reviews that itemize problems and note pages where they occur are more helpful than rambling, general statements. However, general statements sometimes are mechanisms for saying that there is nothing technically wrong with the paper, but that the subject is just not very important in the reviewer's opinion.

Reviewers who look only at experimental details and spelling and do not respond to whether the paper has originality are the poorer reviewers.

When my own papers are reviewed, I appreciate constructive criticism. Have I explained the experiments adequately? Do the reviewers appear to understand what we did? Sometimes there are complaints that tell me that the reviewer does not understand what I've tried to say and that the writing has been deficient in explaining. If a smart reviewer cannot understand what we are doing, then how could the general reader? At other times, the reviewer may spot a misstatement or a stretched interpretation; in such cases, explicit remarks identifying the error are very helpful. I also appreciate it if a reviewer points out references that I have missed.

Statements that are personal are always inappropriate. If I see (as editor) a review that is personal and negative, I discount it and usually do not send it to the author.

Jack R. Norton

Colorado State University
Associate Editor, *Journal of the American Chemical Society*

When I review a manuscript for the *Journal of the American Chemical Society*, I look first at the importance of the issues addressed and the effect the work will have on the way all of us think about chemistry. I then look at the details: How convincing are the arguments? Do the data support the conclusions? Are there alternative interpretations of the results? Has the work been put fairly into context within the scientific literature? For *JACS* I need reviewers to compare a manuscript with others competing for the same space; short, bland, uncritical reviews are not helpful! Of course, detailed comments that answer the questions mentioned are essential.

When I review a manuscript for other journals, I am less concerned with the importance of the work and more concerned with the detailed issues mentioned.

Donald R. Paul

University of Texas at Austin
Editor, *Industrial and Engineering Chemistry Research*

An important issue is technical accuracy. As an author, I do not want to publish papers that are in error. Similarly, as an editor, I do not want to publish papers in my journal that are in error. To the extent possible, we hope reviewers will find our mistakes. In addition, we hope reviewers will provide additional insights.

As an editor, I hope reviewers will inform me about the importance of the papers we ask them to review, because we cannot publish all papers that may be regarded as technically correct. Thus, part of our selection process is to publish papers that will be interesting and useful to our readers. As an author, I naturally assume my papers are both important and useful; although occasionally reviewers do not always agree with that assumption.

As an editor, I am also interested in knowing from the reviewers whether the manuscript can be made more concise and whether it can be made more clear. Both clarity and conciseness are important issues for readers. Also, we rely on reviewers to learn whether sufficient new material is being presented to justify publication. Sometimes authors submit papers that contain information already published with little that is new. We often reject such papers.

Philip S. Portoghese

University of Minnesota
Editor, *Journal of Medicinal Chemistry*

In considering reviewers' comments, the editors of the *Journal of Medicinal Chemistry* evaluate the novelty, quality of data, and conclusions drawn from the study. For example, a manuscript describing a simple descriptive study without a discussion of the relevant target or mechanism of action normally is not acceptable. Also, a trivial extension of a series of compounds that does not provide the reader with fresh insight into the structure–activity relationship of the series would not be published.

The kind of comment that is not helpful is "This is a well-done study; publish without change." Although we often receive excellent papers, we have never seen a paper that does not require some revision. A reviewer's report that is terse and recommends no change is generally viewed as useless.

Henry Rapoport

University of California at Berkeley
Associate Editor, *The Journal of Organic Chemistry*

The most important function of peer review is to ascertain the correspondence between the narrative section of an article and its experimental foundation. Do the experimental data support and justify the claims? Are there missing data, and, if so, why? What I look for, and what I hope my reviewers will look for, is a critical and constructive evaluation. Also of importance are the usual criteria of style: brevity and precision.

Readers of articles dealing with organic chemistry often look at the graphics first and many times go no further. Therefore, the reviewer should be particularly sensitive to the inclusion of clear and informative graphics.

A major problem I have had as an editor is the problem of multiple publication, which is a cause for rejection. The reviewer has the key role in identifying papers that have been published elsewhere.

Finally, we do not want ad hominem comments from the reviewers; their critiques should be totally substantive.

Elsa Reichmanis

Bell Laboratories, Lucent Technologies
Publishing Board, *Chemical & Engineering News*

When I am asked to do a review, I read through the paper and determine how it relates to other work in the area. Are these new results that significantly advance the field, or is the work simply a variation on a given theme? All research cannot be expected to lead to significant new discoveries, but work that reiterates previously published results generally does not warrant publication. I check whether there are appropriate references (are any references missing, that is, have the authors failed to cite a body of work that was relevant?), whether there is enough detail to reproduce the work, whether the authors' interpretation of data holds together, whether logical conclusions are drawn from data (conclusions must make sense with respect to what has been presented), and whether chemical compounds that are discussed are defined or left as trademarks (chemical compounds really must be defined).

In reviews I get, I want to know that the person actually read the paper. I would rather see the reviewer give a list of items to look at than to just say "good work" or "do not publish". The reviewer must at least say why it is good work and how it contributes to the current understanding of the topic. Alternatively, if publication is not recommended, a statement of the reasons will often help in revising the paper or determining the appropriate publication medium.

John D. Roberts

California Institute of Technology

When I review a manuscript, I look for readability, clarity of presentation of difficult concepts, neatness, completeness, style, figures, and accuracy. All are extremely important! Finding inaccuracy in what I really know about scares me as to the accuracy of those things that I do not know about.

In reviews of my own manuscripts, I want comments on all of the points I mentioned, plus a clear idea of what needs to be expanded, contracted, or eliminated. Vague praise or condemnation is almost useless. Examples of problems, writing, and any other concerns should be pointed out clearly and in as much detail as possible.

Robert L. Rowell

University of Massachusetts
Associate Editor, *Langmuir*

Considerations that I look for when I review a manuscript are a certain technical style and content that fits the journal; citations from peer-reviewed, relevant, and recent journals; and new ideas and results. I ask these questions: Is it new? Is it correct? Is it of interest to our readers? Are there omissions of known ideas or important references? Can it be shortened?

When my own manuscripts are reviewed, I would like to receive the same kind of review I ask from my reviewers. Unscientific, biased comments are useless. The reviewer should respond in the same factual and reasoned manner in which I submit.

Finally, I am sure that there are a great many relevant articles that I have read and filed away somewhere, but one that pops immediately to mind is the ACS Ethical Guidelines to Publication of Chemical Research. It summarizes a lot of things that all of us should keep in mind. [See Appendix III.]

George C. Schatz

Northwestern University
Associate Editor, *Journal of Physical Chemistry*

When reviewing a manuscript, of course, the primary issue is the correctness and importance of the science being described. Often, however, poor writing and poor organization of the manuscript make it difficult to assess or appreciate the science. The most common problems are seen in manuscripts from authors whose native language is not English, where one sees many different kinds of grammar and spelling errors. These errors often make the manuscript impossible to understand.

However, even native English speakers make frequent errors in judgment with respect to organization of manuscripts, including such problems as excessive introductory material, poor integration of equations with text, tables that are too complicated, and figures that are unreadable. I would especially like to emphasize tables, as very often I see tables that are simply printed verbatim from a database program with essentially no thought about significant figures or the amount of information presented. My recommendation is that tables with more than about 200 numbers in them should be relegated to Supplementary Information, as no one is going to take the time to decipher the numbers. My primary concern with figures is that often they are prepared with little thought about what they will look like after being reduced to journal size. Font sizes for symbols are especially important, but one also needs to worry about line thickness and the amount of information presented.

In reviews of manuscripts, generally I like to see about one sentence that summarizes the reviewer's assessment of the importance of the work and one sentence that tells me the reviewer's recommendation. If the reviewer has suggestions for revision, these should be numbered, and their relative importance should be indicated. In fact, ideally the important recommendations should be separated from the trivial ones. I spend a lot of time reading reviews, and I get really upset when I am presented with a random list of comments, some with lots of substance and others very trivial. Reviews that go on for more than one page are fine, but they need to be broken up with subheadings so that the editors can decide what to spend time studying. One should never send a review that recommends rejection without some specific details. I realize that this is a lot of work, but in the absence of this the author will simply request additional review.

These comments concerning what I like to see in reviews that I receive as an editor also apply to reviews of my own manuscripts. I like to use these reviews to learn new perspectives on a problem, and often this is how the process works. But reviews with vague statements or reviews that are not well-organized are not useful.

Richard L. Schowen

University of Kansas
Associate Editor, *Journal of the American Chemical Society*

Reviewers, of course, must address the novelty and quality of the work, the clarity of the presentation, the adequacy of the references, and other such considerations, as other respondents probably have noted. But some points should not be left out of a discussion of peer review.

Reviewers need to be "impolite" enough to be absolutely clear about what needs change and what kind of change is needed. Sometimes referees are so sensitive and circumspect that the authors do not recognize their criticisms as such and thus do not take their comments seriously. The result is that mistakes get published, and reviewers feel their comments have been ignored.

Reviewers need to be polite enough that the authors will not get angry and refuse to take their comments into consideration at all. Sometimes extremely thorough and expert reviews have to be repeated by other referees because authors classify a rude reviewer as biased and self-interested and insist on jettisoning the review.

Reviews are best written as a series of numbered points, with those that relate to a specific place in the manuscript indexed to the page number and, if possible, the line number. This system makes it easy for authors to respond and for editors to check that reviews have been addressed in revisions.

Some pointers for authors on how to use and respond to reviews follow:

- Treat the review of your manuscript as the constructive response of expert colleagues, even when it appears to be otherwise. If criticisms derive from a misunderstanding of the presentation, do not attribute the misunderstanding to ignorance, carelessness, or ill will. Instead, ask yourself how to recast the presentation so that misunderstanding will not occur with other readers.

- Respond to each review in a point-by-point manner. If the review has not been written as a series of numbered points, divide it up in this way and add numbers. Then write a response addressed to the anonymous referee in which you cite changes made in the manuscript in response to each point; if you disagree with the referee on a given point, explain the referee's error. In this response, use a collegial tone and vocabulary (even if the review has not) so that the editor can send this response to the referee or other referees if need be.

- When you send your revised manuscript to the editor for reconsideration, send a copy of the old version with the changes indicated so that the editor can quickly and easily see how the reviews have been addressed. For the same purpose, also include copies of the reviews with their criticisms numbered, and a separate point-by-point response to each referee suitable for sending to the referee. Often an editor, with such a detailed response in hand, will be able to make an on-the-spot decision about accepting the paper.

- Do not substitute for these measures a cover letter to the editor that says, "All the criticisms have been adequately addressed in the revision." Busy editors may not feel they have time to research laboriously whether this has been done and may send the paper back to the referees with the cover letter and the question, "Have your criticisms been dealt with adequately?" Busy referees will be irritated that they are faced with a blind search of the manuscript to see what has been done and will hardly approach it in a sympathetic manner. In general, the less completely and forthcomingly the author responds to reviews and documents these responses, the longer and less favorable for the author the further review is likely to be.

Richard Schwenz

University of Northern Colorado

I first consider several issues only barely related to the manuscript. The first is tremendously important to the authors, namely, "Do I have the time to complete the review in an expeditious manner?" For me, this means to have the review back to the editor within 3 weeks (usually, for a journal) or whatever time the editor requests. If I cannot complete the review within that time, I respond to the editor quickly, as it is not fair to the authors or the editor to delay a response. I will then read the abstract quickly to try to decide whether I have the expertise necessary to review the manuscript. If not, then I'll inform the editor quickly.

Then and only then will I begin a careful reading of the manuscript. I look for several items in addition to the technical content of the paper, including the clarity of the English (with non-native speakers, I make an effort to clean up the grammar), the appropriateness of the manuscript for the journal to which it has been submitted (I make sure I know the readership of the journal), and the use of the figures that have been included (journal pages are expensive, and figures take a lot of space).

When considering the technical content of the paper, I look at the Background section to make sure that the literature review covers all of the appropriate literature without becoming excessively lengthy. The Experimental section must describe what was done with sufficient clarity that a graduate student with some knowledge of techniques in the field could replicate the experiment. The Data and Results sections should be complete, with no missing experiments (conducted but not included in the presentation). I spend considerable effort in determining whether the conclusions drawn from the experimental results are reasonable. Lastly, the conclusions must follow from the data and results; alternative explanations that were seriously considered should be proposed and then refuted by experimental evidence.

When I am reviewing a book, I want the writing style to be clear, pedagogically sound, and not overly repetitious. When reviewing a proposal, I want the reasoning to be clear, the background well-founded, the budget well-justified and not excessive (after all, it is my money you are playing with if it's a National Science Foundation grant), and the proposed experiments well-designed and able to answer the question asked (which must be important of itself).

When my papers are reviewed, the useful criticism is constructive, pointing out specific reasons why the manuscript or proposal is inadequate or needs correction. General comments are often confusing to me as an author and provide little to no assistance as to how to improve the manuscript.

William A. Steele

Pennsylvania State University
Editor, *Langmuir*

The most important step in the review of a manuscript is the choice of referees. The *Langmuir* editors use two and sometimes three referees chosen on the basis of their expertise in the research area of the manuscript.

Useful referees' comments deal with

1. clarity of presentation (including the English)
2. correctness of the science
3. knowledge of the current state of understanding in the field, especially regarding extensive repetition of previously published work
4. new physical insights
5. corrections of previous incorrect work
6. whether the paper is one in a long series dealing with the same subject

Clear grounds for rejection are points 2 and 5; point 3 can be grounds either for rejection or for revision. Defects in point 1 can usually be eliminated by careful revision. Points 4 and 6 are too subjective to make hard-and-fast rules about rejection.

A journal editor should expect to have his or her own manuscripts reviewed in accord with the same criteria, but of course this does not always happen.

One needs patience to handle the careless or inappropriate comments that are occasionally received—a fiery reply to a referee generally does not achieve the goal of acceptance for publication, even when it is justified.

Inappropriate referee comments include "I didn't understand a word of this; it should be rejected." "This paper is so bad that it should be rejected together with all future submissions from the senior author's laboratory."

"The paper is not interesting and thus should be rejected." "The work is okay, but it is out of date and thus should be rejected."

F. Gordon A. Stone

Baylor University

I first skim through the manuscript to gain an overview of the subject matter to determine if I am sufficiently knowledgeable to review it. However, journal editors are fairly astute in their choice of reviewers, and so unfortunately I am usually stuck with doing my duty for colleagues and thus have to review the article with some care.

I look carefully at the list of references for two reasons. First, I want to check whether the authors have given due credit to other workers who have a prior claim to similar work. It is exceedingly common for U.S. authors to be unacquainted with results published in their area in major European journals. Second, I frequently find that an author's name has been incorrectly spelled in the list of references quoted in a paper, and I am able to point this out. It is rare for me to find that the list of references is error-free.

Having amused myself with the references, I settle down to read the paper. Having written many papers myself, I tend to pay attention to the style of presentation. I find that many manuscripts can be improved and shortened by replacing a phrase or sentence with one or two words. It must be the competition for funding that has led authors to make unduly verbose statements in manuscripts. Young writers in particular should try to restrain themselves to sentences that are simply constructed. It is a good idea, having written a paper, to think how it will read 10 years on. It helps to put the article away for a day or so before rereading it.

I have to make a judgment as to whether the results reported are worthy of presentation in the journal to which they have been submitted. I regard this decision as of crucial importance. Fortunately, my rejection rate is only about 1 in 10. I believe this rate is low not because I am unduly soft, but because authors have a pretty good idea of what is acceptable.

I firmly believe that a picture is worth a thousand words and try to encourage authors where appropriate to show more structural formulas, schemes, and charts to illustrate their results.

When my manuscripts are reviewed, I do not generally have too much trouble with comments and can readily accede to suggestions for improvements, about 50% of which are helpful. For those suggested changes with which I am unable to agree, I give my reasons to the editor. I cannot recall an instance when an editor overruled my stance.

About 25% of referees' comments are unhelpful. I would divide these about equally between those that show that the reviewer has not read an earlier paper (and so is not up on the recent literature) and those that indicate a lack of empathy for the work being reported. Among the latter are comments such as "Not another paper from this group on . . . ." Fortunately, one develops a thick skin, and such comments are often accompanied by a critique from another referee that extols the work.

Donald G. Truhlar

University of Minnesota
Associate Editor, *Journal of the American Chemical Society*

A reviewer should keep in mind that his or her main responsibility is to the editor who requested the review. The editor is trying to determine whether the paper should be published in the particular journal to which it has been submitted and will have several concerns:

1. Is this paper correct?
2. Is the work new?
3. Is the work significant?
4. Are previous contributions by other workers properly referenced?
5. Is the work described clearly? Are all essential details specified?
6. Is the subject matter appropriate for this publication medium (e.g., journal, book series, or symposium proceedings)?
7. Is the style appropriate for this medium? For example, an article for a specialized journal may require a style different from that of an article in a journal directed to a general audience. Another style issue concerns length; referees can help editors by identifying unnecessary or insignificant material that should be deleted or shortened.

The editor is concerned with these issues because he or she has a responsibility to the whole community of scholars and researchers to see to it that the literature promotes the advance of the field. Clearly, the advance of the field is not promoted by publishing papers that are wrong, confusing, insignificant, or unclear or that do not credit previous work properly. In advising the editor on these issues, a referee performs a professional service to the whole community.

Referees can also fulfill a secondary role of helping authors improve their papers. Certainly it is admirable for referees to assist authors in this way, but referees should keep in mind that this is a secondary issue, and their first concern should be to fulfill their responsibility to the entire community, which is fulfilled by advising the editor on the issues listed.

Referees should not use harsh or insulting language. It is very tempting when faced with a stupid paper, especially if it reveals an intellectual arrogance on the part of the authors (which is not uncommon), to make comments that are insulting to the authors' abilities or honesty. Such reports are seldom helpful to editors. Most editors try to treat all authors with respect. It is much more helpful to provide a referee report that points out in a professional tone why the paper is wrong, insufficiently significant, or inappropriate for the medium for which it is being considered and recommends that it not be published.

Referee reports have one thing in common with other forms of business correspondence. If you feel very emotional about your response, it is probably best to put the report down and let yourself cool off overnight before deciding whether to mail it.

Most editors will be pleased if a paper can be refereed within about 3 weeks. Otherwise, you should contact the editor immediately and suggest that alternative referees be consulted. The same is true if you cannot referee a paper because the subject is outside your area of expertise. The editor will surely appreciate suggestions of alternative referees. If the editor chose you as the referee, but you are inappropriate, it probably indicates that the paper is somewhat removed from the editor's area of expertise. The editor has selected you as the most appropriate available referee, and probably you have more knowledge of the specific subject of the paper than the editor does. Therefore, you can be of great help to the editor in identifying an appropriate referee, even if you cannot do the job yourself. Because the most obvious person you think up may already be the other referee or may be unavailable for some reason, suggesting two or more alternatives is very appropriate.

Edwin Vedejs

University of Wisconsin
Associate Editor, *Journal of the American Chemical Society*

Review begins with (and often depends on) careful choice of referees. As an editor, I look for any concerns that the referees raise and evaluate them scientifically if I can. If I cannot, I will seek a third referee and will direct the referee's attention to specific topics where advice is needed. In the final stages of review, I look to see whether the scientific guidelines for the *Journal of the American Chemical Society* have been followed (rigorous proof; adequate citations; and thorough characterization of new substances). The most useful feedback gives specific reasons why the science is or is not at *JACS* level and makes a clear distinction between subjective critique (the value of the paper, its degree of novelty, and its general interest) and scientific critique (flawed arguments, experiments, etc.).

The biggest problems in review are citation practices, clear writing, and sound documentation. Problems in any one of these areas often interfere seriously with the scientific evaluation because referees focus on those problems. In the vast majority of such cases, the author has not followed *JACS* guidelines, and major delay is the result.

Useless comments are those that express personal annoyance (for example, with language, writing style, or citations) and those that do not maintain a sense of respect for the authors. Criticisms can be voiced in a positive way: "This section would be stronger if ..." "The argument would convince me if the following data were provided." "This work would be of sufficient general interest if it included the following ...."

Randolph C. Wilhoit

Texas A&M University System
Associate Editor, *Journal of Chemical and Engineering Data*

Before sending a new manuscript for review, it must fit one of the following three categories: (1) report appropriate experimental data values, (2) present a correlation or literature review of relevant data, or (3) review and evaluate literature data.

A manuscript in category 1 should then meet the following requirements: Well-defined property values (mostly numerical) on well-defined systems should be reported. At our journal, we are relaxing this requirement somewhat to accommodate important environmental data, but the systems should still be well-characterized. The test is that, in principle, other investigators could repeat and verify the reported data. In most cases, numerical values of all measured properties should be included in the manuscript. The data should be either novel or improved in some sense from what has already been published in the literature.

The paper should also clearly distinguish the newly measured property values from any literature data or calculated or estimated data given in the paper. It should clearly define the properties measured and identify the chemical components (preferably by IUPAC-approved name) of the systems investigated. (IUPAC-recommended symbols are also preferred.) If components of a mixture are identified by numbers, it should be made clear which number corresponds to which component. Each numerical property value should be accompanied by the appropriate variables of state and relevant experimental conditions. The numbers presented in tables and graphs should be clearly associated with these definitions.

The experimental technique should be identified. If the technique has been described in detail in a previous readily available publication, a reference to it can be given. However, sufficient description should be included for knowledgeable people to identify the technique. If modifications have been made to a previously described procedure, they should be described. If a new technique is used, sufficient description should be given to permit others to repeat it.

Some descriptions of the samples used for the measurements should be given. These should include the source and method of purification, or if no purification was done, this should be stated. Evidence of sample purity should be given.

Closely related previously published data should be identified. If the newly presented data differ from the previous data by an amount greater than their expected combined uncertainties, the authors should attempt to explain the difference. If they cannot, they should so state.

A limited amount of theoretical discussion or interpretation can be given, but is of secondary importance. The main justification for publishing is the experimental measurements reported.

The authors should give a numerical estimate of the uncertainty of their measurements. They should indicate whether this estimate is a measure of precision (ability to repeat the measurement) or of the overall uncertainty that results from all potential sources of error. Stating both measures is preferred.

Papers falling into category 2 should meet the following requirements: Empirical correlations should be novel and should have some advantage (better accuracy, wider scope, or simpler procedure) over published correlations of the same kind. Related published correlations should be identified.

The data set used to develop or test the correlation should be adequately described. The numerical values should be given directly, or if they are too voluminous, a reference to a well-organized compilation should be given. Directly observed values in the data set should be distinguished from selected, averaged, smoothed, estimated, or correlated data. Sources of the original measurement should be given directly or linked to a secondary source that permits a trace back to the original source. Some appropriate measure of the uncertainty in the original data set should be given.

Deviations between correlated and observed values should be shown, along with appropriate statistical measures of the agreement. Such deviations should be compared to the uncertainties in the test data set. Sufficient details should be given to allow others to repeat the calculations and use the correlation.

Papers falling into category 3 should meet the following requirements: The authors should include an exhaustive compilation of literature data along with estimates of data quality. They should supply recommended numerical values or coefficients of a fitting equation.

Peter Willett

University of Sheffield, United Kingdom
ACS Books Advisory Board

The main problem is finding time to fit in my refereeing—on average I get about one manuscript a week to referee—and the ideal is a nice, long train journey that can give me the time necessary to deal with a manuscript properly. Usually, I will do a rapid initial scan-through to see whether I have the necessary subject expertise to act as a referee. If the subject matter is appropriate, I read through it once, noting queries and comments in the margin. I look for all the normal things: whether the author knows the literature of the field, whether the material is novel, whether the paper is well-structured and carefully written, and so on. Then I read through it again, normally on a separate occasion, reviewing my comments and typing them on the referee sheet as I go.

When my own manuscripts are reviewed, I expect to be treated by my peers in the same way as I treat them. I hope that reviewers will note sections that are not clear, conclusions that appear unjustified by the evidence that has been presented, limitations in the experimental methodology that have not been discussed, additional references that might be considered, and so on. Comments based on misunderstandings are also (unintentionally) valuable because they serve to identify components of a paper that need to be rephrased for correct understanding

Comments that are a statement of how the referee would have written the paper, as against commenting on the paper that was submitted, are not helpful. Worse than this—and I guess I am lucky in that this has happened only once in more than 250 publications—is receiving abuse. We all have to write trenchant criticisms, but there are ways of doing it. Attacking the intelligence and scientific integrity of an author is not one of them.

CHAPTER *11*

Copyright and Permissions

Barbara Friedman Polansky
Copyright Administrator, American Chemical Society

One of the controversial issues affecting copyright and generating a lot of questions about copyright relates to the use of electronic media. The number of people who access the Internet continues to grow exponentially each year. Many people think that because information is freely available on the Internet, it must be in the public domain and free for the taking. This is not necessarily so. Most material that is posted on the Internet, which includes the World Wide Web (WWW), is protected by copyright. The U.S. Copyright Act (Title 17, *U.S. Code*) governs the use of copyrighted works in the United States and further protects foreign works because the United States is a signatory to the Berne Copyright Convention, which is an international agreement among nations. Similarly, U.S. works are protected abroad.

Most of the laws that apply to print apply to electronic formats "now known or later developed". Not everyone interprets the law the same way, and not everyone agrees that some portions of the law should apply to electronic formats. For example, the First Sale Doctrine (Section 109 of the U.S. Copyright Act) would apply to books in print (hard-copy format), but the First Sale Doctrine cannot apply to the electronic version. Once someone has an electronic version, they cannot "lend" it to someone else; by transmitting it, they are reproducing (copying) it. It is important to review basic copyright issues and to apply them to new technology.

This chapter is a brief overview of U.S. copyright and ACS guidelines; this information is not meant to replace legal advice. We recommend that you direct questions of a legal nature to an attorney knowledgeable about copyright law.

Definitions and General Copyright Questions

What Is Copyright?

Copyright is "the exclusive legal right to reproduce, publish, and sell the matter and form of a literary, musical, or artistic work" (*Webster's Third New International Dictionary*, Merriam Company). According to Section 102 of the 1976 U.S. Copyright Act (Title 17, U.S. Code), which became effective January 1, 1978, copyright is automatically secured when original works of authorship are "fixed in any tangible medium of expression, now known or later developed, from which they can be perceived, reproduced, or otherwise communicated, either directly or with the aid of a machine or device." So, if you write a paper, compose a symphony, paint a picture, take a photograph, or write a letter, either in print or in electronic format, you automatically own copyright, unless you did it for your employer or unless you were commissioned to do it as a "work made for hire" (discussed in the next section). You need not fill out any forms or do anything further to secure your ownership of copyright. If you coauthor a work with someone else, each of you is a co-owner of copyright, with equal rights in the work.

The creator of a work is the original owner of copyright unless he or she transfers the copyright in writing or unless the work is a work made for hire. If it is a work made for hire, the employer or other person for whom the work was prepared is considered the author and owns all the rights, unless the parties have explicitly agreed otherwise in writing.

Copyright ownership should not be confused with any agreement concerning patent rights. Both patents and copyrights are intellectual property rights, but they differ in an important way: Patents protect inventions; copyrights protect original forms of expression, not ideas. Data per se are not subject to copyright protection; however, the collection and fixation of data are protectable by copyright. Copyright does not protect titles or names; however, they may be protected under trademark.

What Is a Work Made for Hire?

According to Section 101 of the U.S. Copyright Act, a "work made for hire is—(1) a work prepared by an employee within the scope of his or her employment; or (2) a work specially ordered or commissioned … if the parties expressly agree … [in writing] … that the work shall be considered a work made for hire." Under definition 1, if you prepared a paper within the scope of your employment duties while working for the ABC Company, then the ABC Company owns the copyright to your paper. The company does not need to have a written agreement with you to claim copyright to papers that you prepare within the scope of your employment.

What Are the Exclusive Rights of a Copyright Owner?

According to Section 106 of the 1976 U.S. Copyright Act, the owner of copyright has the exclusive rights to do and to authorize any of the following, subject to Sections 107 through 120, which cover limitations on and scope of exclusive rights, including but not limited to fair use and reproduction by libraries and archives:

1. to reproduce the copyrighted work in copies or phono records;
2. to prepare derivative works based upon the copyrighted work;
3. to distribute copies or phono records of the copyrighted work to the public by sale or other transfer of ownership, or by rental, lease, or lending;
4. in the case of literary, musical, dramatic, and choreographic works; pantomimes; and motion pictures and other audiovisual works, to perform the work publicly; and
5. in the case of literary, musical, dramatic, and choreographic works; pantomimes; and pictorial, graphic, or sculptural works, including the individual images of a motion picture or other audiovisual work, to display the copyrighted work publicly.

Am I Authorized To Transfer Copyright, and How Is It Done?

Copyright transfers must be made in writing by the copyright owner. If the copyright owner is an employer, then an authorized agent of the employer must sign the transfer agreement. However, some employers give authorization for copyright transfer to each employee. Check with your employer before signing any such documents or agreements that transfer copyright to another party, such as a publisher.

Can the Copyright Owner Give Partial Rights to Another Party?

Copyright is divisible, so it is possible for the copyright owner to give someone the exclusive right to display the work, while keeping the other exclusive rights. A copyright owner can also grant nonexclusive rights. A nonexclusive right is simply a right that does not exclusively belong to any one person or organization. For instance, the copyright owner can grant a nonexclusive right to one company to make 1000 copies and then grant the same nonexclusive right to another company.

How Long Does Copyright Last?

For authored works, copyright lasts for the life of the author plus 50 years. For commissioned works and works made for hire, the term of copyright is 100 years from creation or 75 years from publication, whichever is shorter. Copyright protection in some countries is longer than the term for U.S. works.

Effective July 1, 1995, the term for copyright protection in the European Union (EU) has been extended to life plus 70 years, compared with life plus 50 years in the United States. However, the Copyright Term Extension Bill is being considered by the U.S. House of Representatives and Senate; we might see the copyright term extended in the United States to life plus 70 years. When in doubt, check the term of copyright protection for the country in which the work was first published, and try to keep abreast of changes to the laws. Two Web sites that might be useful are

1. Thomas Legislative Information on the Internet at http://thomas.loc.gov/home/thomas.html
2. The Copyright Law page of the Legal Information Institute of Cornell Law School at http://www.law.cornell.edu/topics/copyright.html

When Is a Work in the Public Domain?

A work is in the public domain in the United States (1) if *all* authors are employees of the U.S. government and have prepared the work as part of their official duties; (2) if a work was published (distributed to the public by rental, lease, sale, or lending) before 1978 without a copyright notice; or (3) when the copyright term expires.

When a work enters the public domain, people may use (reprint, republish, translate, transmit, etc.) that work without obtaining permission from another party. The work is free to use in any manner; of course, appropriate credit should be given to the original source.

What Is "Fair Use"? What Material Can I Use Without Permission?

The 1976 U.S. Copyright Act states that fair use of copyrighted material is not an infringement of copyright. Although fair use is defined in Sections 107 and 108, the definition is general and difficult to interpret; it is helpful to know the legislative history of this section and the guidelines that are included in the House Report that accompanied the Copyright Act of 1976. The guidelines are published in the U.S. Copyright Office circulars (see the section "Good Sources of Copyright Information").

The following is Section 107, "Limitations on Exclusive Rights: Fair Use":

> Notwithstanding the provisions of Section 106, the fair use of a copyrighted work, including such use by reproduction in copies or phono records or by any other means specified by that section, for purposes such as criticism, comment, news reporting, teaching (including multiple copies for classroom use), scholarship, or research is not an infringement of copyright. In determining whether the use made of a work in any particular case is a fair use, the factors to be considered shall include—(1) the purpose and character of the use, including whether such use is of a commercial nature or is for nonprofit educational purposes; (2) the nature of the copyrighted work; (3) the amount and substantiality of the portion used in relation to the copyrighted work as a whole; and (4) the effect of the use upon the potential market for or value of the copyrighted work.

So, how much of a work (a complete article, a single figure, etc.) can you reuse or adapt for the use to be considered a fair use? This question is not easy to answer because one copyright owner cannot speak for another; in some cases, 200 words or 10%, whichever is less, might be considered fair use, whereas in another situation, these amounts would not be considered fair use. For educational and other uses, refer to the U.S. Copyright Office circulars (e.g., Circular 21) for guidance, or consult an attorney who is knowledgeable about copyright law.

Regarding paraphrasing or adapting, the work that you create must be completely rewritten or redrawn so that it is considered a new work. Remember that the copyright owner has the right to make and to authorize the making of derivative works.

Use good judgment; when in doubt, it is prudent to obtain permission from a copyright owner. If the copyright owner thinks that your particular use is a fair use, then you will be informed that no permission is necessary, but credit should be given as a professional courtesy. (For example, "Courtesy of ABC Company." or "Reproduced from Quaid, D. *The ABC Book;* Los Angeles, CA, 1996. Published 1996 by ABC Company." or "Reproduced from ref 19. Copyright 1996 ABC Company.") Finally, remember that merely giving credit is not a substitute for obtaining written permission.

What Copyright Notice Should I Use?

Since 1978, member nations who are party to the Berne Convention (an international copyright agreement among nations) have not required a copyright notice on materials that are protected by copyright. Effective March 1,

1989, the United States began adhering to the Berne Convention and now does not require a copyright notice; however, there are incentives to using the notice, such as letting users know who owns the rights to a particular work.

Several publishers, including the American Chemical Society, require that a copyright notice be used on works for which you have obtained permission to reuse their copyrighted material. If a notice appears on copies, it consists of three elements: (1) the symbol © (the letter C in a circle), or the word "Copyright" or both; (2) the year of first publication of the work; and (3) the name of the copyright owner, or an abbreviation by which the name can be recognized (acronyms should not be used because several companies can share the same acronym). "Copyright © 1997 American Chemical Society" is an example of a copyright notice used and required by the American Chemical Society.

What Are the Rules for Making Photocopies?

Photocopying copyrighted material without the permission of the copyright holder is illegal, unless your copying can be considered fair use. If your use does not fall under fair use, you have a number of options: You can purchase additional originals, you can report your copying to the Copyright Clearance Center (CCC), or you can obtain permission or a photocopying license from the copyright owner.

Just because you have the capability to make copies, transmit articles, and so on does not mean that you have the rights to do so. For example, newsletter subscribers who have circulated photocopies of a newsletter to employees have been successfully sued.

Recent U.S. court cases on photocopying are discussed in the section "How Copyright Affects Us: Specific Cases", and ACS guidelines are discussed in the section "Photocopying of ACS-Copyrighted Material".

What Are the Rules for Using Software and Phonograph Records?

The Software Protection Act of 1984 and the Record Rental Amendment of 1984 require authorization from the copyright owner before one may rent, lease, or lend computer programs or phonograph records for commercial purposes. Check your licensing agreement for rights that have been granted to you. For rights not specified, contact the copyright owner for permission.

A "shrink-wrap" license is an agreement between a provider or seller of a product and a buyer or user that becomes effective when the cellophane wrapper is broken to use the product. In the past, shrink-wrap licenses were not enforceable, but a court ruled in 1996 that in certain cases, such a license may be enforceable; see the next section.

How Copyright Affects Us: Specific Cases

The Copyright Act touches us almost every day. For instance, the song "Happy Birthday" is protected by copyright until the year 2010. If this popular song is republished, sung on TV or in the movies, or performed for the public, the "infringing party", if it has not already paid royalty fees, is asked to pay a fee for each performance. If you own a digital watch that plays songs, including "Happy Birthday", you have already paid extra to cover the royalty fees.

Stores and restaurants that play music over loudspeakers are required either to obtain copyright royalty licenses to broadcast the music or to subscribe to a commercial broadcast music service. This decision was handed down by the U.S. Supreme Court on April 26, 1982, when it ruled that 525 clothing stores were infringing music copyright by playing the radio over loudspeakers as background music for their customers, unless they paid copyright royalty fees.

The Betamax case (*Sony Corp. v Universal Studios*) is another famous copyright case that went to the Supreme Court. In 1976, Universal Studios sued Sony Corporation on copyright infringement grounds and charged that Sony's Betamax videocassette recorders were being used to tape copyrighted television programs. The Supreme Court found that home videotaping of copyrighted television programs for personal use is not an infringement and that Sony was not guilty of contributory infringement.

In late 1984, the Software Protection Act of 1984 and the Record Rental Amendment of 1984 were signed into law. Both laws require authorization from the copyright owner before one may rent, lease, or lend computer programs or records for commercial purposes.

Related to photocopying for educational purposes is the case involving New York University (NYU). In April 1983, NYU reported that it had reached an agreement with nine publishers who had filed a lawsuit for copyright infringement in December 1982. The publishers charged that NYU, 10 of its faculty members, and the Unique Copy Center had violated copyright laws by photocopying textbooks and other educational material. The federal lawsuit was considered a test case to demonstrate the publishers' concerns that abuse by illegal photocopying was widespread on college campuses throughout the country and that copyright owners were suffering the financial consequences.

In an out-of-court settlement, NYU agreed to adopt specified guidelines on photocopying educational material, to request copying permission for material that is not covered by the agreement, and to pay royalty fees to the copyright owners, if requested. The guidelines do not pertain to students, who may copy articles for their own studying purposes. The guidelines permit teachers to make single copies of book chapters, articles, short stories, and other relatively short items for use in preparing their classes. Teachers may make multiple copies of copyrighted material, provided that the material is brief, that it will be needed before permission can be obtained, and that it will be used for only one semester.

A recent case involves Michigan Document Services (MDS), a copy shop that was selling course packets to students without obtaining permission from the copyright owners. A group of publishers, including Princeton University Press, Macmillan, and St. Martin's Press, sued MDS, alleging copyright infringement. On November 8, 1996, the U.S. Court of Appeals for the Sixth Circuit issued its opinion on this case, in which it reaffirmed the District Court's opinion that MDS's activities were not fair use and that copy shops need to obtain permission from copyright owners and pay license fees when substantial portions of copyrighted materials are included in customized course packets.

American Cyanamid Company and E. R. Squibb Corporation were also sued by publishers of scientific and technical journals and other reference material; out-of-court agreements were reached in these cases in 1982. The publishers had charged that the companies were infringing the publishers' copyrights by making unauthorized photocopies. Under the terms of the settlements, which were very similar, the companies agreed to pay copying fees to the CCC on (1) all copies of CCC-registered material made on central copying facilities and other attended copying equipment located on its premises; (2) copies received from outside sources, such as interlibrary loans or document-supply sources, unless notified that payment was made by the outside source; and (3) copies made on unattended copying equipment. The agreement with Squibb is different from the one with Cyanamid in that Squibb may exclude from reporting up to 6% of their total copying and claim this amount as fair use.

On October 30, 1995, Texaco and a group of 83 publishers announced that they had agreed on terms to settle their dispute over corporate photocopying of articles. In 1994, the Second Circuit Court of Appeals held that the photocopying of copyrighted articles by a Texaco employee was not a fair use. On April 28, 1995, Texaco had filed a petition for certiorari (a writ of a superior court to call up the records of an inferior court) with the Supreme Court, but later, in May 1995, Texaco and a steering committee of publisher representatives had announced that they had tentatively agreed to settle their dispute. Under the agreement, Texaco did not admit to any wrongdoing, but did agree to pay a seven-figure settlement and a retroactive licensing fee to the CCC. Texaco will also enter into a standard annual license agreement with the CCC during the next five years.

To assist large industrial and other users in reporting their copying activity, in 1983 the CCC instituted the Annual Authorizations Service, which is a photocopy authorization program that allows users to make copies of participating publications without recording or reporting internal copying activity, except during an audit period. Participating copyright owners sign an agreement with the CCC and give the CCC the right to enter into a licensing agreement with users. Basically, the users make an annual payment to the CCC for the license to make all the copies they wish of articles included in the licensing program, provided that the copies are used for internal purposes at a corporate site during a one-year period.

For users other than large industrial organizations, the new Authorizations Service does not replace the CCC's transactional reporting system, in which users record all copies made.

Regarding shrink-wrap licenses, on June 20, 1996, the U.S. Court of Appeals for the Seventh Circuit reversed a district court's decision and held that shrink-wrap licenses are enforceable unless their terms are objectionable because they violate a rule of positive law or unconscionability. The judge acknowledged that one cannot agree to hidden terms; however, the outside of the package noted that the purchase was subject to a license, which appeared inside the package. The court pointed out that many transactions involve the exchange of money before detailed terms are spelled out for the purchaser, such as with airline tickets, insurance contracts, and products for which the warranty information is included inside the box.

ACS Copyright Policy

ACS copyright policy is governed by its bylaw on copyright, Bylaw IV, Section 1(b), which states that "for any writing of an author published by the Society in any of its books, journals, or other publications, the Society shall own the copyright for the original and any renewal thereof." Therefore, authors, or their employers in the case of works made for hire (see the exception described in the next paragraph), are required to assign copyright to ACS for scholarly research material that is submitted for publication in its journals, books, and most magazines, including *Chemical & Engineering News.*

Authors of U.S. government works or works of a foreign government that reserves copyright under its national law are exceptions; however, they are required to sign the appropriate ACS form to certify this information.

Under the terms of the 1976 U.S. Copyright Act, the society is required specifically to obtain copyright transfer from authors of each paper to allow ACS to continue its past practice of holding copyright on individual articles and book chapters. This requirement is necessary for the society to carry on its normal publishing activities of disseminating information, granting reprint permissions, making and supplying reprints of articles, collecting royalty fees or other fees, repackaging and distributing its published information in print and other media, entering printed material into its databases, and authorizing rights to others, subject to certain approvals by authors and their employers. In return for copyright assignment to ACS, the society grants certain rights to authors (or to their employers in the case of works made for hire).

ACS Copyright Status Form

All authors of research papers in ACS journals and books are required to sign the ACS copyright form, which is printed every year in the first issue of each ACS research journal; it can also be found at ACS's Web site: http://pubs.acs.org/instruct/instruct.html.

Rights Returned to Authors and Their Employers

The rights that ACS returns to authors and their employers appear on the ACS Copyright Status Form. Section A of the form states that "The undersigned author and all coauthors retain the right to revise, adapt, prepare derivative works, present orally, or distribute the work in print format, provided that all such use is for the personal noncommercial benefit of the author(s) and is consistent with any prior contractual agreement between the undersigned and/or coauthors and their employer(s)."

Section B states that "In all instances where the work is prepared as a 'work made for hire' for an employer, the employer(s) of the author(s) retain(s) the right to revise, adapt, prepare derivative works, publish, reprint, reproduce, and distribute the work in print format, and to transmit it on an internal, secure network for use by its employees only, provided that all such use is for the promotion of its business enterprise and does not imply endorsement by ACS."

Sections C and D are also important for authors to note. The following statement appears under Section C: "Whenever the American Chemical Society is approached by third parties for individual permission to use,

COPYRIGHT STATUS FORM

Name of American Chemical Society Publication

Author(s)

Ms No.

Ms Title

Received

This manuscript will be considered with the understanding you have submitted it on an exclusive basis. You will be notified of a decision as soon as possible.

[THIS FORM MAY BE REPRODUCED]

Print or Type Author's Name and Address

COPYRIGHT TRANSFER

The undersigned, with the consent of **all authors**, hereby transfers, to the extent that there is copyright to be transferred, the exclusive copyright interest in the above cited manuscript, including the published version in any format (subsequently referred to as the "work"), to the **American Chemical Society** subject to the following (Note: if the manuscript is not accepted by ACS or if it is withdrawn prior to acceptance by ACS, this transfer will be null and void and the form will be returned.):

A. The undersigned author and all coauthors retain the right to revise, adapt, prepare derivative works, present orally, or distribute the work in print format, provided that all such use is for the personal noncommercial benefit of the author(s) and is consistent with any prior contractual agreement between the undersigned and/or coauthors and their employer(s).

B. In all instances where the work is prepared as a "work made for hire" for an employer, the employer(s) of the author(s) retain(s) the right to revise, adapt, prepare derivative works, publish, reprint, reproduce, and distribute the work in print format, and to transmit it on an internal, secure network for use by its employees only, provided that all such use is for the promotion of its business enterprise and does not imply endorsement by ACS.

C. Whenever the American Chemical Society is approached by third parties for individual permission to use, reprint, or republish specified articles (except for classroom use, library reserve, or to reprint in a collective work) the undersigned author's or employer's permission will also be required.

D. No proprietary right other than copyright is claimed by the American Chemical Society.

E. For works prepared under U.S. Government contract or by employees of a foreign government or its instrumentalities, the American Chemical Society recognizes that government's prior nonexclusive, royalty-free license to publish, translate, reproduce, use, or dispose of the published form of the work, or allow others to do so for noncommercial government purposes. State contract number:__________

SIGN HERE FOR COPYRIGHT TRANSFER: I hereby certify that I am authorized to sign this document either in my own right or as an agent for my employer, and have made no changes to the current valid document supplied by ACS.

Print Authorized Name(s) and Title(s)

⇒

Original Signature(s) (in Ink)

Date

CERTIFICATION AS A WORK OF THE U.S. GOVERNMENT

This is to certify that ALL authors are or were bona fide officers or employees of the U.S. Government at the time the paper was prepared, and that the work is a "work of the U.S. Government" (prepared by an officer or employee of the U.S. Government as part of official duties), and, therefore, it is not subject to U.S. copyright. (This section should NOT be signed if the work was prepared under a government contract or coauthored by a non-U.S. Government employee.)

INDIVIDUAL AUTHOR OR AGENCY REPRESENTATIVE

Print Author's Name

Print Agency Representative's Name and Title

Original Signature of Author (in Ink)

Date

Original Signature of Agency Representative (in Ink)

COPYRIGHT RESERVED BY AUSTRALIAN, CANADIAN and/or U.K. GOVERNMENTS (For other governments, refer to section E and sign this form in the top section.)

☐ If **ALL** authors are employees of the Australian, Canadian and/or U.K. Governments, which reserve copyright under national law, **DO NOT SIGN THIS FORM**; check the box as your request for the COPYRIGHT FORM FOR AUSTRALIAN, CANADIAN AND U.K. GOVERNMENTS (Blue Form) which you will be required to sign, and mail this form to: Copyright Administrator, Publications Division, at the ACS's Washington, DC address. Otherwise, mail this form to the editor's office or to the address noted above, if given. For addresses, please refer to the publication.

Control #9608197A

The ACS Copyright Status Form.

reprint, or republish specified articles (except for classroom use, library reserve, or to reprint in a collective work) the undersigned author's or employer's permission will also be required." Section D ensures authors and employers that "no proprietary right other than copyright is claimed by the American Chemical Society."

Authors Who Are Government Employees

Section E of the ACS form concerns works contracted by the U.S. government or works prepared by employees of a foreign government or its instrumentalities: "[T]he American Chemical Society recognizes that [foreign] government's prior nonexclusive, royalty-free license to publish, translate, reproduce, use, or dispose of the published form of the work, or allow others to do so for noncommercial government purposes."

The boxed section of the ACS Copyright Status Form is for certification that a work is a work of the U.S. government, which is defined as a work prepared by officers or employees of the U.S. government as part of official duties. Such works are not subject to U.S. copyright. This certification statement is to be signed only when *all* authors are or were bona fide officers or employees of the U.S. government at the time the paper was prepared. If at least one author was not an employee of the U.S. government when the paper was prepared, that author (or the authorized agent for the author or employer) should sign the top section of the ACS form, thereby assigning whatever copyright exists in portions of the paper to ACS.

ACS recognizes that some foreign governments reserve copyright under their national laws. If all authors are employees of such governments, they will be asked to sign ACS's Copyright Form for Australian, Canadian, and U.K. governments (the "blue form"), which gives ACS all the rights it needs to carry on its publishing activities. Authors may request this form by checking the appropriate space on the Copyright Status Form.

Submitting a Properly Signed Form

Substitute forms, attachments, or changes to the ACS Copyright Status Form, including the use of correction fluid, are not acceptable. Any additions or changes made to the form will delay the processing of a paper for publication. Also, an original signature or a stamped signature must be on the form; photocopied signatures are not acceptable. Likewise, forms that have been faxed are not acceptable.

An author, or the employer's authorized agent in the case of a work made for hire, must sign only one section of the form. The ACS form states that the bottom section is to be signed when *all* authors are employees of the U.S. government. However, ACS practice is to accept a copyright form if both sections are signed, but only if the signatures are those of different authors, one a U.S. government employee and one not. ACS will claim copyright ownership in those portions for which copyright exists.

These provisions are required so that ACS can efficiently handle the more than 17,000 copyright forms that are received each year. Most of the ACS publications assistants, who initially receive and process the copyright forms for the ACS's publications, are located not at ACS headquarters, but throughout the United States. Their primary concern is processing papers for publication. Although the publications assistants know that ACS requires copyright assignment from authors, they are not familiar enough with copyright to know what changes would be acceptable. Therefore, it is efficient and effective for ACS to accept only the ACS form signed as is, without changes or attachments.

If ACS publications assistants receive any unacceptable forms, they will ask the author to submit another ACS Copyright Status Form or they will ask the ACS Copyright Office staff to handle the request. Usually, production on a paper may continue; however, if a paper is ready for publication and ACS does not have a properly signed form, the paper will be delayed until an acceptable form is received. Sometimes, though, an unacceptable form does slip through and a paper is still published. But even after publication, a thorough check is made of all copyright forms, and authors are asked to submit correctly signed forms when necessary.

Liability and Rights

Authors are solely responsible for the accuracy of their contributions. ACS and the editors assume no responsibility for the statements and opinions advanced by the contributors to ACS publications.

Contributions that have appeared or been accepted for publication with essentially the same content in another journal or in some freely available published work in any format (e.g., government publications, reports, and proceedings) will not be considered for publication in an ACS journal or book. However, this restriction does not apply to results previously published as communications or letters to the editor in the same or other journals.

Manuscripts published in ACS books or journals and for which copyright has been transferred to ACS may not be reprinted elsewhere in whole or in part without written permission from both ACS and the authors. Authors who wish to reproduce their own articles for commercial use elsewhere also must have the consent of ACS. Other reproduction is permitted only after obtaining the written consent of ACS. Requests should be addressed to the appropriate ACS staff; for example, requests related to materials produced by the ACS Education Division should be directed to that division.

Seeking Permission To Reuse ACS-Copyrighted Material

Written permission and copyright credit are required. Merely giving credit is not a substitute for obtaining written permission. The copyright owner (usually the publisher) has the right to grant permission for the reuse of copyrighted material. For permission to use ACS-copyrighted material from the ACS Publications Division, send a written request to ACS Copyright Office, Publications Division, American Chemical Society, 1155 16th Street, NW, Washington, DC 20036, or fax it to (202) 872–6060, or e-mail it to copyright@acs.org. Do not send your permission request to the editor's office or to any other editorial office; this will delay the processing of your request, which will be forwarded to the address given here.

ACS will not consider oral requests. A fee may be assessed, depending on your request. Generally, ACS does not grant blanket permission. However, a royalty agreement may be negotiated in certain cases.

When submitting a request via e-mail, you must include your name, employer, employer's address, phone and fax numbers, a complete reference citation for the ACS material that you would like to use, and where and how it will be used; permission guidelines are included here for your convenience. ACS will not reply to you via e-mail, but will reply via fax or regular mail if you do not have a fax number.

When you request permission from ACS to use ACS-copyrighted material in another medium, you should specify the publication title, volume number, year of publication, and the exact information you wish to use, such as "entire article" (give the inclusive page numbers of the requested material), "Figures 3–5", or "Table II", as well as how and where you wish to reuse the material. Also, include your requests for all ACS information (even if you would like to use material from different ACS publications) on one sheet of paper, if possible. This format will speed the processing of your request.

The credit line required by ACS is "Reprinted [or Reproduced or Reprinted in part or Adapted] with permission from [FULL REFERENCE CITATION]. Copyright YEAR American Chemical Society."

You may not copy, reprint, republish, reproduce, or transmit articles or portions thereof (tables, figures, charts, schemes, photographs, excerpts, etc.) without the written permission of the copyright owner. It is the obliga-

PERMISSIONS REQUEST GUIDELINES
To Use Material from the ACS PUBLICATIONS DIVISION

All requests for permission to use ACS Publications Division's copyrighted material must be submitted in writing to the ACS Copyright Office via mail (Copyright Office, Publications Division, American Chemical Society, 1155 16th Street, N.W., Washington, DC 20036) or fax: 202-872-6060, or e-mail. When submitting a request via e-mail, you must include your name, employer, employer's address, phone, fax number, and the information listed below; we will not reply to you via e-mail, but we will reply via fax or regular mail if you do not have a fax number. No verbal requests will be considered. A fee may be assessed depending on your request. Generally, we do not grant blanket permission. However, a royalty agreement may be negotiated in certain cases.

Requesters should provide the following information:

1. Title of ACS work;
2. Full bibliographic reference information (name of ACS journal, book, or magazine, issue date, volume number, issue number, page numbers; if an ACS book, give the book editor's name(s), the chapter author's name(s), the series title and number, year of publication, and pages);
3. What you intend to use, e.g. full article (how many copies?), figures 2, 5, 6 only, a portion of the article—mention the percentage and send a marked copy of the text you plan to use, etc.; and
4. Where it will be reused
   - if making individual photocopies, where will they be distributed, such as at a conference, to customers, in-house to employees only, etc.
   - if the work will be republished, let us know if the ACS material will be published in a book, journal, or magazine, give us the title of the work, the name of the publisher of the work in which the ACS material will appear, the editor's name of the book, the title and authors of the chapter, etc.
5. The deadline by which you need this permission; We will try to accommodate your deadline, but please note that our turnaround time is 2-4 working days.

The more information that you supply to us, the faster that we can grant your request. It is helpful if you would supply photocopies of any artwork, photographs, or other materials along with your request. To expedite our handling of your request, please send a self-addressed envelope and two copies of your request. Also, please include your telephone number and your fax number in the body of your letter; we find it difficult to read phone/fax numbers on letterheads, particularly if this information is printed in a small font size.

Should permission be granted to you, we will require that you print the ACS copyright credit line on the first page of the article or book chapter. For figures and tables, the credit may appear with the figure, in a footnote, or in the reference list. For figures, tables, and journal articles, the standard credit line is "Reprinted with permission from [reference citation]. Copyright [year] American Chemical Society." Example: Reprinted with permission from *Organometallics* **1996,** *15*, 2770-2776. Copyright 1996 American Chemical Society.

For book chapters, the following credit line should be followed:

Reprinted with permission from Reed, W. In *Strategies in Size Exclusion Chromatography*; Potschka, M.; Dubin, P. L., Eds.; ACS Symposium Series 635, American Chemical Society: Washington, DC, 1996; pp 7–35. Copyright 1996 American Chemical Society.

A sample permissions request form is included at this Web site for your convenience; of course, you may write your own letter, but please include all of the requested information. Questions? Phone (202)-872–4368 or 4367 or e-mail copyright@acs.org.

The ACS Permission Request Guidelines.

PERMISSION REQUEST FORM

Date:__________

To: Copyright Office
Publications Division
American Chemical Society
1155 Sixteenth Street, N.W.
Washington, DC 20036

Fax: (202) 872–6060

From: ______________________

Your Phone No.______________
Your Fax No.______________

I am preparing a paper entitled:

____

to appear in a (**circle one**) book, magazine, journal, proceedings, other __________
entitled: ______________________________

to be published by: ______________________________

I would appreciate your permission to use the following ACS material in print format only:

From ACS journals or magazines (for ACS magazines, also include issue no.):
ACS Publication Title Issue Date Vol. No. Page(s) Material to be used*

____

____

____

From ACS books: include ACS book title, series name and number, year, page(s), book editor's name(s), chapter author's name(s), and material to be used, such as Figs. 2 & 3, full text, etc.*

____

____

____

____

* If you use more than three figures/tables from any article and/or chapter, the author's permission will also be required.

Questions? Phone (202) 872–4368 or 4367, fax (202) 872–6060, or e-mail copyright@acs.org.

This space is reserved for
ACS Copyright Office Use

The ACS Permission Request Form.

tion of the person or organization seeking to use the copyrighted material to secure this permission. Such permission must be obtained even if the material appeared in an article originally written by the requester if he or she had previously transferred copyright.

Photocopying of ACS-Copyrighted Material

Regarding ACS's primary research journals, an individual may make a single reprographic copy of an article for personal use. This permission does not include the right to put articles together to make a collective work, nor does it extend to the making of a single copy or multiple copies for use within a business unless appropriate fees are paid. Do not interpret "personal use" and "fair use" too broadly; consult a copyright attorney if you have any questions.

The ACS Publications Division has registered its titles with the Copyright Clearance Center (CCC) to provide immediate ability for users to make authorized photocopies. Registered users of the CCC can report their photocopying activities and pay the appropriate royalty fees either on a transactional basis or through one of the CCC's annual license agreements. For additional information, contact the CCC at 222 Rosewood Drive, Danvers, MA 01923; the telephone number is (508) 750–8400, and the CCC Web site is at http://www.copyright.com.

Electronic Use or Reuse of Copyrighted Material

Manuscripts submitted to ACS are considered with the understanding that they are submitted on an exclusive basis and have not been disseminated in electronic or other formats. After the paper's publication by ACS, you must obtain written permission from ACS to reproduce your paper, unless the right is already granted (see the section on the ACS Copyright Status Form on pp 352–355).

The following statement is posted on ACS's Web site (http://pubs.acs.org or http://www.chemcenter.org) for *Analytical Chemistry*:

> Submission of a manuscript to the Journal implies that the work reported therein has not received prior publication and is not under consideration for publication elsewhere in any medium, including electronic journals and computer databases of a public nature. The editors have established a policy that any material that is posted in electronic conferences or on WWW pages or in newsgroups will be considered as published in that form, in the same way as if that work had been submitted or published in a print medium.

Other ACS journals have similar restrictions. Contact the ACS Copyright Office if you have any questions.

ACS Guidelines for Classroom Use

According to the ACS guidelines for classroom use, ACS will grant royalty-free permission to make copies of journal article(s) or book chapter(s) provided that certain conditions are met. The ACS Publications Division has guidelines for classroom use for journals, books, and magazines. Presented here as an example are the guidelines that are used for journals. Contact the ACS Copyright Office, Publications Division, for a copy of the guidelines that you need.

Seeking Permission To Use Material from Other Sources for Your ACS Article

According to U.S. copyright law, material (tables, figures, charts, schemes, excerpts, etc.) that has appeared in a medium for which copyright is held by a person or organization other than ACS cannot be reprinted by ACS without the permission of the copyright holder. It is your obligation as author to secure this permission from the copyright holder as well as the approval of the author of the published material. You need to obtain permission from the copyright holder even if the material appeared in an article that you originally wrote if you transferred copyright.

Seeking Permission To Reproduce Papers Presented at ACS Meetings

When authors present papers at ACS meetings, they own copyright to their own work unless they have already transferred it or plan to transfer it to another publisher or to ACS for publication in an ACS journal or book. Authors may not grant permission to reproduce their own papers if they have already transferred copyright to a publisher. Furthermore, authors themselves may not republish their own papers in whole or in part if they have already transferred copyright to a publisher, unless the publisher allows this right. However, if an author still owns copyright to his or her work, you should contact the author for permission to reproduce the work. If the author owns copyright and grants you permission, you need not obtain further permission from ACS. You should acknowledge, however, that the work was presented at an ACS meeting and whether it was published as an ACS preprint.

If an author has already submitted or intends to submit a paper to an ACS journal, magazine, or book, and you want to reprint material from that paper, direct your permission requests to the Copyright Office of the ACS Publications Division. Such requests should also mention the title of the publication in which the work will appear.

Your reference number(s):____________________
Professor & Class name or No.:____________________
University:____________________
No. of copies/article:__________
Journal citation(s) or see the attached page(s):_______________

Here are the ACS GUIDELINES FOR CLASSROOM USE. You may have permission to make individual photocopies of our copyrighted journal article(s) for classroom use (school/university only), provided that you meet the following criteria:

1. a written request is received listing the article(s) to be copied for the current semester and the number of copies (all blank lines above must be completed);

2. the number of individual photocopies of ACS articles <u>will not exceed 25 articles</u>, with a limit of no more than 25% of any one issue (page count) or one article, whichever is greater, for the number of students taking the course;

3. the reprint/photocopy may **not** be attached to any other article copies or material;

4. the required copyright credit line appears on the first page of each ACS copyrighted article: "Reprinted with permission from [**full reference citation**]. Copyright [**year**] American Chemical Society." Please insert appropriate information;

5. no charge shall be made to the student beyond the actual cost of photocopying; and

6. use of the article(s) will not imply any endorsement by ACS.

PLEASE CHECK THE APPROPRIATE BOX AND COMPLETE ALL INFORMATION:

☐ I meet the above criteria; free permission is granted.

☐ I am the author of the requested articles; free permission is granted.

☐ If the article copy(ies) will be part of a collective work, such as an anthology (also known as a course packet, course reader, etc.), please submit your request **for all articles** to the CCC (222 Rosewood Dr., Danvers, MA 01923, Fax 508/750-4470, Phone 508/750-8400) and pay the appropriate royalty fee.

☐ I plan to photocopy more than 25 articles; each copy **MUST NOT** be attached to other materials. Beginning with the 26th article, I agree to pay $.10/page/copy, with a minimum payment of $35.00, for a total of
$__________ payable directly to ACS.
No. articles___; **Total** pages___; No. copies/article____.
Enclosed is my check__.

Please complete all information requested, sign, date, and return this letter to the ACS Copyright Office at the letterhead address or fax **before** you make any individual photocopies for classroom use.
Questions? Phone (202) 872–4368 or 4367 or e-mail copyright@acs.org

Agreed:____________________Date:__________

ACS Guidelines for Classroom Use for journal articles.

If an author's paper appears in an ACS division's preprint publication, contact the author first to determine whether he or she has transferred copyright in writing to the ACS division. Some divisions (e.g., Rubber Division) do require copyright transfer, so it is mandatory to contact either the author or the division to obtain the necessary permission to reuse an author's work. Other divisions do not require copyright transfer, so an author can present a paper at an ACS meeting, the preprint can be made available by an ACS division, and the full paper can be published by another publisher.

Seeking Permission To Use Photographs

If you have a photograph of a person and you want to use that photograph in any publication, you must obtain written permission from the person in the photograph as well as from the copyright owner of the photograph. Likewise, if you have a photograph of commercial equipment, you must obtain written permission from the manufacturer to reproduce the photograph. Just because someone has given you a photograph, you do not necessarily own any rights in that photograph. You do not own the copyright unless you have a signed agreement.

Copyright Clearance Center

ACS does not grant blanket copying permission unless you have a license-to-copy agreement with ACS or unless you report your personal use or internal use copying activities to the Copyright Clearance Center, Inc. (222 Rosewood Drive, Danvers, MA 01923); their Web site is http://www.copyright.com/.

ACS is a registered publisher with the CCC. Either in the masthead or on the first page of each ACS-copyrighted journal article and book chapter is a numeric code, referred to as the CCC code. The appearance of this code indicates ACS's consent that copies of articles or book chapters may be made for personal use or internal purposes. The consent does not extend to other kinds of copying such as copying for general distribution, for advertising or promotional purposes, for creating new collective works, for resale, or for transmission in any electronic format.

Copyright Credit

ACS grants permission to republish or reprint ACS-copyrighted material provided that requests are received in writing; that requesters agree to pay royalty fees, if any are due; and that the required ACS copyright credit line is used.

- For a journal, examples of the standard credit line are

  Reprinted with permission from Gregg, B. T.; Cutler, A. R. *Journal of the American Chemical Society* **1996**, *118*, 10069. Copyright 1996 American Chemical Society.

  Reprinted in part with permission from Rutherford, D.; Atwood, D. A. *Organometallics* **1996**, *15*, 4417. Copyright 1996 American Chemical Society.

- For a book, examples of the standard credit line are

  Reprinted with permission from Peterson, R. T.; Lamb, J. D. In *Chemical Separations with Liquid Membranes;* Bartsch, R. A., Way, J. D., Eds., ACS Symposium Series No. 642; American Chemical Society: Washington, DC, 1996; p 130. Copyright 1996 American Chemical Society.

  Reprinted with permission from Dewar, M. J. S. *A Semiempirical Life;* Profiles, Pathways, and Dreams; American Chemical Society: Washington, DC, 1992; p 14. Copyright 1992 American Chemical Society.

Sources of Copyright Information

The best source of general copyright information is the U.S. Copyright Office (Register of Copyrights, Library of Congress, Washington, DC 20559). You can phone the Copyright Office's public information number, (202) 707–3000, during business hours (8:30 a.m.–5:00 p.m. Eastern Time), or use their fax-on-demand service at (202) 707–2600.

The U.S. Copyright Office provides application forms, regulations, and information circulars at no cost. One of the most informative circulars is Circular 1, "Copyright Basics". A list of some of the circulars that you can order is given in the Bibliography; some are available at the U.S. Copyright Office's Web site at http://lcweb.loc.gov/copyright/.

ACS is also a source for copyright information. A number of symposia are cosponsored by the ACS Joint Board Council Committee on Copyrights and the Division of Chemical Information; these programs are usually held during ACS national meetings. Also, the ACS Copyright Committee holds open meetings at all ACS national meetings; check *Chemical & Engineering News* for dates, times, and locations.

You may address questions to the ACS Publications Division copyright staff in several ways:

- by mail to the Publications Division Copyright Office, American Chemical Society, 1155 16th Street, NW, Washington, DC 20036

- by phone at (202) 872–4368 or 4367
- by fax at (202) 872–6060
- by e-mail at copyright@acs.org

The ACS Copyright Status Form and other copyright-related forms and information items are on ACS's Web site at http://pubs.acs.org/instruct/instruct.html. In addition, by sending a self-addressed, stamped envelope to the Publications Division Copyright Office, you can receive a copyright information pamphlet, which you may duplicate for noncommercial purposes.

If you have questions about materials produced by other ACS divisions, please direct your questions to the appropriate staff; the main telephone number for ACS is (202) 872–4600.

Bibliography

Partial List of Circulars Available from the U.S. Copyright Office

Circular 1 Copyright Basics
Circular 1b Limitations/Information Furnished by Copyright Office
Circular 2 Publications on Copyright
Circular 2b Selected Bibliographies on Copyright
Circular 4 Copyright Fees
Circular 5 How To Open and Maintain a Deposit Account in the U.S. Copyright Office
Circular 6 Access to/Copies of Copyright Records and Deposit
Circular 7d Mandatory Deposit of Copies of Phonorecords
Circular 8 Supplementary Copyright Registration
Circular 9 Works Made-for-Hire Under the 1976 Copyright Act
Circular 12 Registration of Transfers and Other Documents
Circular 14 Copyright Registration for Derivative Works
Circular 15 Renewal of Copyright
Circular 15a Duration of Copyright
Circular 21 Reproduction of Copyrighted Works by Educators and Librarians
Circular 22 How To Investigate the Copyright Status of a Work
Circular 23 Copyright Card Catalog and the Online Files
Circular 31 Ideas, Methods, or Systems
Circular 32 Blank Forms/Other Works Not Protected by Copyright
Circular 34 Copyright Protection NOT Available for Names, Titles, Short Phrases Not Copyrightable
Circular 38a International Copyright Relations of the United States
Circular 40 Copyright Registration for Works of the Visual Arts
Circular 44 Cartoons and Comic Strips

Circular 45	Copyright Registration for Motion Pictures Including Video Recordings
Circular 50	Musical Compositions
Circular 55	Copyright Registration for Multimedia Works
Circular 56	Copyright for Sound Recordings
Circular 56a	Copyright Registration of Musical Compositions
Circular 61	Copyright Registration for Computer Programs
Circular 62	Copyright Registration for Serials on Form SE
Circular 74	How To Make Compulsory License Royalty Payments via Electronic Transfer of Funds
Circular 99	Highlights of the Current Copyright Law

Web Sites Mentioned in This Chapter

1. ACS Home Page: http://www.acs.org
2. ACS Instructions for Authors and Editors: http://pubs.acs.org/instruct/instruct.html
3. ACS Publications Home Page: http://pubs.acs.org
4. ACS's ChemCenter Home Page: http://www.chemcenter.org
5. Thomas Legislative Information on the Internet: http://thomas.loc.gov/home/thomas.html
6. The Copyright Law page of the Legal Information Institute of Cornell Law School: http://www.law.cornell.edu/topics/copyright.html
7. Copyright Clearance Center Home Page: http://www.copyright. com/
8. U.S. Copyright Office Home Page: http://lcweb.loc.gov/copyright/

CHAPTER *12*

Making Effective Oral Presentations

Larry James Winn
Western Kentucky University

Communication pervades virtually every facet of our personal and professional lives. In today's world, technical expertise is not enough; you also need the ability to communicate effectively. The good communicator has skills that form the core of all communication competencies, knows how to adapt to different communication situations, and moves comfortably from formal writing to e-mail to one-to-one communication to group discussion to public speaking.

Public speaking competence offers unique rewards, not only for the audience but also for the speaker. An effective speaker gains credibility that outlasts the occasion. You might respond, "But I have so much stage fright, I just want to make my written text my security blanket and get through the ordeal as quickly as possible." In the first place, you do not have to throw away all speaking notes to give a good speech. Second, notes used incorrectly actually make you a prisoner of your fear. This chapter will explain how you can lower your apprehension while improving what you say and how you say it.

Another person might object, "But I have neither the ability to speak effectively nor the time to learn how." Well, reading a single chapter will not catapult you to the stature of renowned orator, but applying even a few suggestions from this chapter will improve the average speech. Incorporating many of the suggestions into habits of speech can make you a more cogent, comfortable, and confident speaker who welcomes speaking engagements as opportunities.

The Oral Presentation

Oral presentations cover a wide variety of messages, for example, classroom lectures, papers at academic and scientific conferences, formal reports of group conclusions, persuasive messages to small decision-making groups, and speeches to civic clubs. Although these speaking situations offer different types of challenges to a speaker, all good oral presentations reflect a tension between an emphasis on expanded conversation and an emphasis on speaker responsibility.

We should view an oral presentation as amplified conversation. Suppose that you walk up to two friends who are talking. One of them asks the other, Jan, to tell about her recent trip to the Amazon rain forest. You both become engrossed in Jan's account and start doing more listening than talking. Another person joins the discussion, then another, then two more. A few more people see the group and join. Eventually someone asks Jan whether she has pictures from her trip. Jan steps into her office, brings out some pictures, and refers to them as she talks.

At some point in this process we can call Jan's remarks an oral presentation. She, the acknowledged expert, does most of the talking and needs to talk a little louder and more distinctly as more people join the conversation. Others face her and do more listening than talking. Jan even uses visual aids. It makes no difference that we cannot identify the precise moment at which the conversation became an oral presentation; in fact, we should not try to sharply distinguish between a conversation and a speech. A good speaker does not think in terms of "giving a speech" or "making an oral presentation". Instead the speaker who comfortably communicates during a speech thinks in terms of talking with people, looking at them, observing their reactions, asking them to offer comments or questions, and referring to some of them by name.

In other words, an oral presentation is an interaction among people, not a performance. The atmosphere should not suggest an "I the speaker/you the audience" dichotomy but rather a "we the communicators" unity, reflected in part by the fact that the speaker literally says "we" more often than "you". The audience members become "co-communicators" through their listening, their nonverbal cues, and perhaps their comments or questions. As a result, listening improves, rapport between the speaker and the audience increases, and speaker apprehension diminishes.

The speaker's responsibility is to ensure that the audience benefits in some way from the presentation. Giving a speech or presenting a paper at a scientific meeting is a privilege. A speaker should take the time to prepare a message that is focused, substantive, organized, and adapted to the audience and occasion. In addition, an oral presentation should be concise, effectively illustrated, and delivered clearly.

Proactive Preparation

Think of a future speaking engagement as a collection of question marks, each of which will turn into a plus or a minus before or during your presentation. Reactive speakers either do not know what variables the question marks represent or else leave it to chance to turn each question mark into a plus or a minus. "Chance", however, turns out to be a euphemism for Murphy's law, which turns most of the question marks into minuses to which the speaker has to react during the presentation. For example, the speaker before you speaks longer than planned, leaving you too little time to say what you need to say. Or the introducer fails to mention one of your important accomplishments that your audience deserves to hear about. Perhaps the overhead projector does not work. Maybe the audience knows less about your topic than you expected them to know.

The proactive speaker knows what the important question marks are and turns most of them into pluses before the presentation. This section presents the initial steps in speech preparation, identifies important variables, and suggests how to turn these question marks into pluses.

One variable relates to the compatibility of speaker expertise, topic, and audience. Settings such as scientific conventions tend to match speaker expertise with a rather high level of audience knowledge. Most other types of speaking situations are less predictable and thus require taking more measures to avoid problems. Your first opportunity to ask the right questions and make the right decisions comes with the invitation to speak. An invitation gives you three options: accept, decline, or negotiate.

Agreeing to speak holds out the obvious possibilities of sharing significant information, learning from others, meeting new people, enhancing your reputation, and, depending on the situation, supplementing your income. You may feel that accepting an invitation is the only graceful option, especially if you do not have a prior commitment for the date and time in question. Understandably, a program chair, having undertaken a thankless task for free, may think, "My group has the need, and you have the expertise; therefore, you have the obligation to speak." Obviously, however, saying yes to a presentation means saying no to something else, often family. The first right of a speaker, therefore, is the right to tactfully decline a speaking engagement. Some people adopt the simple solution of routinely reserving particular blocks of time for family. Such a practice enables them to say without guilt or elaboration, "I already have a commitment for that time."

In many instances, a speaking invitation calls for negotiation about such matters as the best possible alignment of audience, topic, speaker expertise, time limit, date, and physical context. Ask the general topic that the program chair has in mind. Ask specific questions about the audience.

1. What level (or levels) of expertise will the audience have on the general topic?

2. What level of interest will the audience have in the topic? Do not assume that the program chair's interest in the topic guarantees that the audience as a whole is interested in it.

3. What are relevant audience attitudes? Ask this question of more than one person to get beyond the positive answers that officers of an organization tend to give. Ideally, you should call someone whom you can comfortably ask, "What are the hidden agendas?"

4. What are relevant demographic features of the audience? Depending on the speaking occasion, the general topic, and other factors, you might need to ask about age, ethnic groups, gender, educational level, group affiliation, education, and occupations.

5. What language should I use, and what language should I avoid? As obvious as it sounds, a guest speaker needs to get the name of the hosting group exactly right. People might be insulted if you refer to their group by the wrong name, particularly if you use a rival's name.

Secure assurances about the timeline. You should know when you can expect to start speaking and when you will be expected to stop. If you must have a certain amount of time, state it orally and remind the program chair in writing. My friend Jim traveled almost a thousand miles to present a three-day workshop, scheduled from 6:00 p.m. to 8:30 p.m. on Friday, all day Saturday, and Sunday afternoon. Having carefully organized substantive and concise presentations, he needed all the time available. Despite assurances that Jim could begin his workshop by 6:10 on Friday evening, the program chair permitted one person after another to say "a few words". Furthermore, no one informed Jim that the local university basketball team was scheduled to play an important home game starting at 8:30 p.m. Jim started speaking at 7:00 and noticed that the listeners showed impatience 45 minutes into the workshop and signs of hostility after an hour. At 8:15 Jim decided he had better adjourn the session even though he had much more material that he needed to cover.

Discuss how best to align a specific topic with your expertise, the time you can invest in preparation of the presentation, the audience's needs and expectations, and the time limit of your speech. A few minutes of telephone conversation at this point can save you time later and help ensure a successful presentation. You might end up with the exact topic suggested to you, a specific topic within the suggested general topic, or a completely different topic.

I once received a call from someone whose university was hosting a conference for people from 12 states. A major speaker had suffered an automobile accident on the way to the conference and would be unable to give her speech, scheduled the next day. The caller asked me to speak and told me the intended speaker's topic. After some discussion about the audience and occasion, I suggested a different topic, one on which I had spoken previously. The caller liked my idea, so I updated my notes and adapted them to the new audience and occasion. The audience was both delightful and delighted, in significant part because of the 15-minute telephone conversation.

Take steps to ensure the proper physical and psychological setting. I gave a talk to an appreciative audience, one of whom invited me to speak on the same topic to a different group. Unfortunately, I accepted his invitation with few questions. Not only did the new audience care nothing about my topic, but also I spoke without a microphone to an audience spread out over a large area. Latecomers went through a cafeteria line during the first part of my remarks, and people ate throughout the speech.

As this example illustrates, you should ask for a description of the location or even a diagram indicating the location of the head table, the lectern, the audience, and visual aids. (A later section in this chapter addresses visual aids in more detail.)

Take charge of the speaker introduction. Heeding this advice does not imply arrogance. At least since Aristotle, experts in communication have known that speaker credibility significantly influences speech effectiveness. The audience deserves to know your credentials, and you should not have to brag about yourself. Introducers usually speak with minimum preparation and leave out important facts, so you need to keep a typed one-page list of your credentials and make sure that whoever introduces you receives a copy.

Achieving Focus

If you are presenting a paper at a scientific meeting, organize your paper to answer the following questions:

- Why was the project undertaken?

- What was done?
- What was learned?
- What does it mean?

For scientific papers and for many other kinds of presentations, four steps help to ensure a focused speech: (1) determining the general goal, (2) developing a precise objective, (3) developing a one-sentence summary, and (4) developing a title. The first three steps focus the speaker's thoughts before outlining the message. The fourth step helps to focus the audience's thoughts before the speech.

Determining the Goal and Objective

The occasion usually dictates the general goal of an oral presentation. People speak primarily to interest, inform, persuade, or motivate. Each objective subsumes the previous ones: An oral presentation can inform only if it also holds interest; it can persuade only if it also interests and informs; and it can motivate an audience to take action only if it interests, informs, and persuades.

Having determined the general goal, a speaker should develop an objective that focuses on specific audience results. A speech that does not define "success" in terms of a specific statement of objective is like an experiment that tests no particular hypothesis. A good statement of objective identifies what listeners should be able to do, what they should think, or what they should actually do after the message. The following statements illustrate such objectives:

- An oral presentation reporting research results: "After my presentation, the listeners will be able to identify my three major findings and the main implication of the research study."
- A speech to convince high school seniors to consider a career in chemistry: "Next year several of my listeners will declare a college major in chemistry."
- An oral presentation to motivate decision makers at a pharmaceutical company at which you work: "After my presentation, the listeners will be convinced that our company should begin developing a new serotonin-uptake inhibitor."

Realism as well as specificity should govern the statement of objective, particularly when you are speaking to persuade or motivate an audience on a controversial topic. In such situations, try to determine how much the audience's opinion is likely to deviate from your own. A speech whose main proposition deviates significantly from audience opinion results in a "boomerang" effect; that is, it increases the discrepancy between the audience's opinion and the speaker's opinion. For example, a speaker would only increase listener resistance by trying to persuade an audience of tobacco growers to support legislation banning smoking. Someone committed to such legislation should consider a different audience or, if speaking to the tobacco growers, change to an objective such as the following one: "After my speech, the opinion leaders in my audience will relax their objections to experimenting with alternative cash crops."

The Single-Sentence Summary

Begin to outline a speech only after you have summarized the potential presentation in a declarative sentence that captures the central idea of the message. The single-sentence summary focuses a speaker's thoughts as nothing else does and simplifies all succeeding steps in speech preparation. Ask yourself, "What is the central point that I want to make?", and then write a sentence that meets the following criteria:

◆ The sentence should make the point about the topic around which the entire presentation will revolve.

> ***Poor summary*** I will discuss the correlation between micronutrients and arthritis in women. (Such a sentence accomplishes nothing more than saying you will speak about your topic.)
>
> ***Better summary*** The media have misled the public about the effectiveness of micronutrients on arthritis in women.

◆ The sentence should reflect a realistic assessment of the audience's knowledge and interest. The better summary in the previous example points to a speech intended for a lay audience. For an audience educated in the topic area, a summary statement should focus on a narrower point.

◆ The sentence should reflect a realistic assessment of what you can accomplish within the prescribed time limit: "I will present evidence for intrachain energy transfer in polychromophores."

Developing a Title

Papers presented at scientific meetings always need titles, and sometimes other presentations need titles for inclusion in a printed agenda or in advertisements. The first purpose of a title is to attract the appropriate audience. Titles are especially helpful for people who attend conferences at which several programs run simultaneously.

The second purpose of a title is to prepare the audience mentally for the presentation. Listeners who know in advance the direction a speech will take have an advantage over those who know only the general topic. Good titles focus listeners' thoughts before a speech and increase the speed at which they grasp the central idea.

A title that attracts the right audience and focuses listeners' thoughts has the virtues of clarity, conciseness, and creativity. Strive first for a title that communicates accurately the direction of the message, and then see if you can reduce the number of words while retaining the clarity. Creative titles use alliteration, puns, questions, or other such devices to generate interest in the presentation. Although creative titles can be suitable for substantive messages, you can lower your credibility if you pay more attention to the creativity of the title than to its substance.

Organizing the Presentation

Organization enhances our ability to understand and remember what we see and hear. Beyond these values, good speech organization helps a speaker remember a presentation and thus enhances fluency and confidence. It might comfort you to know that, although research in communication indicates that good organization has positive effects, as yet no research indicates that minor disorganization has negative effects. In other words, organization needs to be good, but it does not have to be perfect. In light of this fact, a wise speaker sometimes digresses temporarily when listeners' minds tire or resistance arises.

Selecting and Arranging the Main Points

A reader can put down an article when the mind tires and review it at leisure. A listener does not have this luxury, so a speaker should choose a simple arrangement pattern that includes only a few main points. Therefore, begin organizing your presentation by identifying the three or four most important points of the central idea (the single-sentence summary). As an economist might say, an audience reaches the "point of diminishing returns" at about the third or fourth subdivision of a presentation. Divide a presentation into four points, and the listeners might remember three of them. Divide it into five points, and they might remember one or two. Divide it into six or seven points, and expect a mental rebellion.

The choice of an arrangement pattern usually dictates what the main points will be. You can select from a variety of designs, keeping your speech goal in mind. We focus here on organizational designs especially suited to the speech to inform, the speech to persuade, and the speech to motivate.

Arrangement Patterns for an Informative Presentation

◆ ***Categorical Design.*** Some topics fall naturally into categories, for example, offense and defense, advantages and disadvantages, basic research and applications, and levels within an organization.

◆ ***Sequential Design.*** A sequential design follows a time order, for example, the steps in a reaction sequence, the stages in the formation of a cave, or the main stages in the development of antibiotics.

◆ ***Spatial Design.*** A spatial design naturally fits topics that break down into physical parts. This arrangement pattern can even help listeners envision an abstract concept by tapping into common patterns of human experience. Because of the way we read, we can describe something from left to right, down, and left-to-right again. For instance, you might describe in order the top left quadrant of a model, the top right quadrant, the bottom left quadrant, and the bottom right quadrant. Or, given our experience telling time, you could move in clockwise fashion from one slice of a pie graph to another, explaining, for example, four main causes of a problem.

◆ ***Comparison–Contrast Design.*** As the name implies, this pattern starts with similarities and then describes differences. You might extend this organizational pattern to four subdivisions, for example, by (1) describing what features two compounds have in common, (2) describing how they differ, (3) emphasizing the problems that result from failure to recognize the differences, and (4) emphasizing the benefits that could result from exploiting the differences.

◆ ***Simple-to-Complex Design.*** This pattern works especially with technical topics. A speaker builds from points the audience can more easily grasp to more complex points.

Arrangement Patterns for a Persuasive Presentation

◆ ***Problem–Solution Design.*** This design works well with audiences whom you expect to show some resistance to your solution. Because people more often agree on problems than on solutions, examine the problem first. This approach will enable you to demonstrate sound logic and build credibility before moving to the more controversial part of your case.

◆ ***Residuals Design.*** This design, which uses the process of elimination, works well when the listeners already agree on a problem. The speaker can analyze possible solutions and logically eliminate solutions one at a time, leaving until last a defense of the recommended solution.

◆ ***Proposition–Proof Design.*** Using this pattern, a speaker includes a recommendation in the speech introduction and then presents reasons to support the recommendation. The proposition–proof design works best when the speaker has new evidence to present and when the listeners have an open mind.

Arrangement Patterns for a Motivational Presentation

◆ ***Monroe's Motivated Sequence.*** A variation of the problem–solution design, the motivated sequence consists of five steps designed to move the audience from interest in a problem (attention step), to concern about the problem (need step), to agreement that the speaker has identified the best solution (satisfaction step), to a vision of concrete benefits of the solution (visualization step), to action that will help bring about the solution (action step). Empirical evidence suggests the effectiveness of the motivated sequence in moving audiences to action.

◆ ***Value–Action Design.*** This design has three stages: The speaker (1) identifies a value that the audience holds, (2) inspires a deeper appreciation for the value, and (3) encourages the audience to take specific actions in accordance with the value.

Outlining the Body of the Presentation

An outline helps ensure a cohesive speech organization no less than a floor plan helps to ensure a house with a logical design. As a general rule, start with a simple outline, as you learned in elementary school. List the three or four main points (identified by roman numerals); then begin to fill in the subheadings (identified by capital letters) and the third-level headings (identified by arabic numbers). Use the following guidelines when developing the outline.

◆ If you have access to a computer outline program, by all means use it. Such a program is easy to learn and use, saves considerable time, and aids tremendously in creating and revising an outline of any type. Most word-processing software includes an outlining feature.

◆ State each main point as a simple declarative sentence. As with the speech itself, each section needs the focus that a single sentence provides.

◆ Use your audience analysis to help determine the types of information to include under each main point. For example, when talking about the synthesis of nonlinear optical polyimides for practical applications, mention the approaches that have been developed to tackle the instability problems. When talking about pollution in an area, refer to the dumping of chemicals in a nearby stream.

◆ Use your audience analysis to decide how much information to include under a particular main point or how to apportion information under a main point. Not all main points are created equal. Include as much information in a section as is necessary to achieve audience understanding of, or agreement with, the main point of a section. For example, the problem section of a problem–solution speech may need to be short or long, depending on how serious the audience perceives the problem to be.

◆ In each section, include a blend of supporting materials. People whose left brains are dominant tend to smother an audience with data, and those whose right brains are dominant rely much more on personal stories and examples. In fact, each type of supporting material has its place. There are times to be factual, times to be allegorical, times to give examples, and times to tell personal stories. Try to balance your use of the different materials. Consider also what visual aids might fit in each section. (Visual aids are discussed in more detail later in the chapter.)

Preparing Introductions, Conclusions, and Connectives

After organizing the body of a presentation, prepare an introduction that will capture the listeners' attention and orient them to the topic. An introduction is a warm-up. No one would expect a runner to drive to the track, get out of the car, and immediately go to the starting blocks and run a hundred-yard dash. Neither can we expect an audience to be ready for a speaker who has just stepped to the microphone to proceed immediately to the body of the speech. An audience that has just finished eating, talking, listening to another speaker, or rushing through heavy traffic needs at least a couple of minutes to mentally and psychologically disengage from previous thoughts and activities and adjust to the speaker and topic.

Attention Step

Begin a speech with an attention step from which you can lead into your topic. This step needs to be proactive; you need to engage the audience. Here are some ways to begin a presentation:

◆ Refer to the occasion, the location, the organization, or a recent event. Watch the news or read a newspaper the day of the presentation. Several times I have changed my introduction to incorporate a relevant event that had just occurred.

◆ Use an illustration that pertains to the topic. Most oral presentations should begin concretely, and examples and stories serve as good attention steps. Personal examples, particularly those that include humor, tend to work especially well because they illustrate a point and also "humanize" the speaker. A personal example can also compensate for a relevant experience that the introducer failed to mention. You might say, for example, "During my year at CERN, ...".

◆ Begin with an especially good quotation. Search your memory and various books to find a relevant quotation that captures a truth in a concise, unusual, and perhaps humorous way.

◆ Ask a question. It can be rhetorical, or it can elicit an answer from the listeners. It should be focused and concise and lead directly into the topic of the speech. A good question is one of the simplest ways to begin.

◆ Refer to something that has just happened on the program, perhaps something the previous speaker said: "My colleague has discussed photochemical studies of organometallic compounds. I will now discuss the photochemistry of main-group clusters, particularly the boranes."

Orientation Step

The orientation should indicate the direction that your presentation will take. An orientation step is essentially a transition from an attention step to the body of the speech; therefore, I will give a hypothetical example of an entire introduction. I used to teach on a beautiful university campus. One morning, as I walked to teach a 7:30 class, I noticed a considerable amount of litter, apparently resulting from a celebration of a football victory the previous evening. By coincidence, the first student speaker spoke on littering, citing facts taken from national magazines. The following introduction illustrates how he might have improved his speech by referring to what all the listeners had seen just 10 minutes earlier.

> ***Attention step*** As you came to class this morning, what did you see? I'm usually too sleepy to see anything when I walk to a 7:30 a.m. class, but my nervousness over this speech made me alert this morning. I saw beer cans all over the grass. I saw paper cups floating down the campus stream. I saw toilet paper hanging from the trees.

Orientation step We had good reason to celebrate last night, but even a homecoming victory is no excuse to litter. With a little effort we can preserve the beauty of our campus and our nation. Let's look at how we can begin here at Indiana University. During the next few minutes we'll focus on the extent and effects of littering on our campus and then identify what we can do about the problem.

This introduction leads to several observations. First, although a speaker should write an introduction to ensure an excellent first impression, it is best to improvise a little if the circumstances dictate. If the student who spoke about littering had thought about audience adaptation in advance, he would have planned to mention the campus littering problem. Then, after noticing the extra littering, he would have had to add only one or two sentences to the planned introduction.

Second, this introduction would have accomplished two of the main purposes of a speech opening. The question, combined with the reference to something recent and local, would have captured audience attention, and the orientation would have then led smoothly into the specific topic.

On an intangible level, the introduction would have established rapport through audience identification. The use of "we" and "let's" would have contributed to this purpose. In addition, the confession of nervousness would have reminded the other students of a feeling that most of them had experienced: the extra anxiety felt on the way to class stemming from the thought "Today, my turn will come to speak." The audience would have chuckled knowingly, thereby actually relieving the speaker's apprehension. The student would have used a little humor without the raised expectations that accompany a joke.

Concluding the Speech

Many otherwise good presentations lose their positive impact because the person behind the lectern cannot find the "exit". The audience grows progressively impatient as the speaker says "in conclusion", talks a few minutes, says "finally", makes another point, and then thinks of "one more thing". At the opposite extreme, some speeches stop rather than end. The speaker says "well, that's about it" or "thank you" and sits down.

A blunt "thank you" is certainly not a good way to conclude, but a statement thanking colleagues by name and thanking the audience for their attention is not bad. Also, inviting questions is often appropriate.

A really good way to close is with a concise, powerful line. Never say "in conclusion" unless you mean it. The conclusion should round out the speech, tie together the main points, focus the listeners' attention on the main point, and provide a sense of closure. For example, "In conclusion, contrary to what has been believed, ion–ion interactions do not affect the structure of the first hydration shell of chromium and zinc ions." Or "Clearly, the long-term opportunities for the polyhydroxyalkanoates are enormous, spanning and benefiting many industries."

Making Connections

As indicated earlier, the reader has the luxury of seeing the words of a message, reviewing as necessary to make connections, and resting when the mind tires. On the other hand, the transitory nature of spoken language necessitates that a speaker, as the saying goes, should "tell them what you're going to tell them, tell them, and tell them what you told them." In other words, a speaker, even more than a writer, should help the audience see how the parts of the speech relate and how the whole speech fits together. Beyond using a preview in the introduction and a summary in the conclusion, a speaker should introduce each main point with a transition and in some cases even summarize the point before moving to the next.

Speech transitions fall into three categories:

1. ***Questions.*** Simple questions can serve as signposts while adding to sentence variety. The problem section of a problem–solution speech can begin with "What's the problem?" and the solution section with "How can we solve this problem?"

2. ***Enumeration.*** The simplest technique of all, enumeration means using words such as "first", "second", and "finally" or statements such as "First let's consider ...".

3. ***Review–Preview.*** This technique combines a reference to the previous point with a reference to the next point, e.g., "In the same fashion that X reacts with Y, Y reacts with Z."

Using Language Effectively

The principles of grammar and word usage discussed in Chapter 3 apply to the wording of a speech. However, the unique nature of oral communication calls for some additional principles. Compared with the reader–writer situation, the speaker–audience situation has the following cluster of characteristics:

- more specific adaptation to audience and situation

- more interaction
- more immediacy
- more personal atmosphere
- more informal atmosphere

As a result of these characteristics, the following are desirable attributes of oral style.

Attributes of Informal, Conversational Style

◆ ***Personal Pronouns.*** Although writers now use more personal pronouns than they did in the past, you need to use more in a speech than you would in a written paper. Unifying pronouns, such as "we", "us", and "our", promote a sense of identification between the speaker and audience.

◆ ***References to the Audience and Local Situations.*** Refer to events, examples, and statistics that are relevant to the audience.

◆ ***Contractions.*** Use contractions when speaking, except when you want to vocally emphasize a word, such as "not" in "We will *not*!"

◆ ***Questions.*** Writers and speakers sometimes use rhetorical questions, that is, questions used for effect. Speakers can also direct questions and receive immediate responses.

◆ ***Interjections.*** Use interjections sparingly, but the speaking situation permits more latitude to reveal some emotion. Most audiences like to see a little of the human side of a speaker.

Attributes Based on Transitory Nature of Oral Communication

◆ ***Short Words.*** Listeners do not customarily bring dictionaries, so you should use the shortest and simplest words that say exactly what you mean. (I could say "avoid polysyllabic words", but of course I will not.) Obviously, when people within a specialty talk to one another, they can use words that they would not use when speaking to a lay audience. You may not have the time to define all terms, nor would it be appropriate for a technical audience. But you need to be sure that everyone in the audience is well-versed in your specialty. If your audience might include nonspecialists, you have to accommodate them. As chemists' specialties become more and more narrowly defined, it becomes harder and harder for nonspecialists to attend a talk on an unfamiliar topic and have enough definitions given to learn something.

◆ ***Short, Simple Sentences.*** Keeping sentences short is important because listeners do not have the opportunity to reread a speaker's sentence if they get lost.

◆ ***Parallelism in the Longer Sentences.*** An audience can keep up with even a long sentence if the thoughts appear in parallel grammatical units.

Attributes That Hold Listeners' Attention

◆ ***Sentence Variety.*** A succession of sentences of the same type and length sedates even the most conscientious audience. The technique of sentence combination, explained in books such as *The Writer's Options: Combining to Composing*, can be especially instrumental in helping you to vary your style.

◆ ***Concrete Language.*** A student speaker who spoke on motorcycling told his audience, "Be careful or you'll find yourself embedded in a tree." He certainly secured the listeners' attention. Compare his vivid statement with a bland one such as "You will become injured if you do not exercise caution."

◆ ***Figurative Language.*** Similes, metaphors, and other figures of speech can underscore points, add sentence variety, and give listeners little jolts, provided that such images are consistent, concise, and used with restraint. Hearing one figure of speech after another is like a constant diet of chocolate fudge.

◆ ***Active Voice.*** Active voice increases impact.

Achieving Speaking Excellence

Although I have been using the term "presentation", this word and another word used in speech textbooks, "delivery", suggest the wrong image for someone who wishes to communicate orally. Both terms imply that messages flow in one direction, from speaker to audience. Yet, as stated earlier in the chapter, communication occurs when speaker and listeners all regard themselves as co-communicators. When people talk to each other, one person says something while looking at the other person, and the second person responds, in turn initiating a response to the response.

Good speakers emulate this pattern. They look the audience in the eyes before the speech and continue this pattern to maintain an unbroken communication cycle. When they see signs of boredom, they might intensify their nonverbal communication. When they see people tense up, they might reduce the intensity of their nonverbal communication. When they observe signs of confusion, they might slow down. When they see signs of disagreement, they might introduce additional evidence or ask listeners to express their objections.

Interpreting audience reactions is not a precise science. A single nonverbal indicator, such as one person with crossed arms, may mean nothing. A speaker should look for a combination of signs and also observe the behavior of several listeners. Someone with a frown, crossed arms, and crossed legs, or any one of these nonverbal cues exhibited by several people, may indicate disagreement with the speaker, anger over something the speaker said, or tension because the speaker is "coming on too strong".

An alert speaker observes such signs quickly, interprets them in terms of what has just been said or how something was said, and adjusts quickly. For example, when reference to a particular source of evidence elicits frowns, the speaker probably needs to introduce additional evidence from a different source. Intensely negative facial expressions may indicate that the speaker needs to re-explain something that the listeners might have misinterpreted. Eyes that lose their focus and passive facial expressions usually mean the speaker needs to recapture attention by adapting content or nonverbal communication. Whatever the listeners' reactions, the speaker should generally use behaviors that have been proven to enhance audience learning by enhancing good feeling between speaker and listeners: "leaning forward, smiling, making eye contact, and speaking with a range of different pitches."

Unfortunately, the speaker who wishes to use interactive communication faces a dilemma. Conversational communication has the qualities of spontaneity, informality, and continual interaction among communicators; these qualities appear more when the speaker speaks naturally rather than reciting memorized material. On the other hand, effective language has qualities such as clarity, conciseness, correctness, variety, and vivacity; achieving these qualities usually requires forethought, revision of a speech, and adherence to what one has planned to say. Overcoming this dilemma requires the use of various styles of speaker communication to best advantage.

The best mode of communication combines a variety of styles, but most of the speech should be extemporaneous. In extemporaneous speech, the speaker outlines the speech but chooses most of the language while speaking. In addition to promoting a conversational quality, this mode of communication has the advantage of careful preparation without the rigidity of simply reading or memorizing the manuscript. Relying on an outline rather than exact language, you can contract or expand main points in response to the audience.

There may be some small parts of the speech that you would want to read, especially if you are presenting data or speaking on a sensitive issue, but keep them short. Reading precludes eye contact with the audience and can be monotonous. Also, never read the introduction or conclusion. If you have a quotation that you do not wish to memorize, type it on a note card and hold up the card and read the quotation so that the audience does not look at the top of your head.

Type the outline, using bold print and a large font. Although some speakers like to speak from note cards, typed pages can obviate the need to continually flip from one card to another. For the parts you want to read, prepare an easy-to-read manuscript. Use a large font and bold print, triple-space, and never type part of a sentence on one page and the rest on the next page.

Working with your outline, practice giving the presentation. Record part of the speech in advance and listen to yourself to make sure that you sound informal and conversational and that you are within the time limit allowed. Become as familiar with it as possible, but do not memorize it! Often memorization results in a monotonous, droning presentation. It can also be very helpful to practice your presentation in front of an audience of family, friends, or co-workers, if possible. Ask them to be honest and critical.

Become comfortable with some silence. Discomfort with silence tempts speakers to fill every second with words or sounds. This trait manifests itself in vocalized pauses such as “uh”, repeated use of “like” and “you know”, hooking sentences together with “and”, and failure to use pauses for effect. Listen to a recording of yourself practicing the speech to see if you have these tendencies. Listeners do not expect a perfectly fluent speech, but speakers should work to communicate as smoothly as possible.

Take a deep breath or two, walk slowly to the lectern, move confidently regardless of how you feel, pause briefly, and look at the audience before starting.

Make effective use of your voice. Complement the conversational quality of extemporaneous speech with inflection, sufficient volume, and careful enunciation.

Avoid rambling. Sense when to return to the next point in your notes.

Finally, seek feedback from the audience, right after the speech, or later from people in the audience whom you recognize. Ask them what worked, what did not work, what specific things you could have done better, and how else you can improve.

Visual Aids

The effective use of visual aids significantly increases speaker credibility, the persuasiveness of an oral presentation, and audience comprehension of technical material. Nevertheless, certain caveats deserve attention:

1. Some means of displaying data have disadvantages that often outweigh their advantages.
2. The visual aids that work well for writers often work miserably for speakers.
3. Speakers sometimes negate the effects of good visual aids by presenting them poorly. The remainder of this section addresses these problems and suggests ways to maximize the positive effects of visual aids.

Choosing Means of Displaying Information

The advantages and disadvantages of various visual aids are summarized as follows.

Overhead Projector

An overhead projector permits the speaker to look down at a transparency, point to parts of it, and add to it while still facing the audience. Also, a speaker can easily file transparencies for repeated use.

Using a computer projection panel combined with an overhead projector eliminates the need to create transparencies or worry about picking up the wrong one while speaking. A press of a button on the computer mouse can key the appropriate image during a presentation. A speaker can commit an entire visual presentation to a computer disk and revise it for later presentations.

Slide Projector

A slide projector is useful for projecting color pictures or combining pictures with graphs and other visual aids. It gives the speaker flexibility in updating old presentations by adding, deleting, or rearranging slides.

The disadvantages are that it demands elaborate preparation and careful execution. Speakers often must have a third party convert their artwork to 35-mm slides. Finally, a slide projector requires a semi-darkened room.

Flip Chart

A flip chart works well when the presentation requires many visual aids, when a speaker needs to create some visual aids in advance, or when a speaker wishes to tape visual aids to the wall as they are completed (a favorite approach of trainers). During the presentation, the speaker controls which pages of the flip chart the listeners see at any time and avoids the time and trouble of erasing boards. However, the flip chart can interfere with speaker–audience eye contact.

Erasable Board

A chalkboard or erasable marker board works well for highlighting a point or building from a simple display to a more complex display. The disadvantages include loss of eye contact with the audience and the fact that displays cannot be stored for future use.

Handouts

Handouts can work against or for a speaker. A handout passed from person to person guarantees that at any given time at least one member of the audience will not be listening to the speech. Handouts can be beneficial if used in one of these ways: (1) At the start of your presentation, you give everyone a set of handouts corresponding to your remarks, and your audience turns the pages as you speak. (2) At the appropriate time during your presentation, you give each listener a copy of the same page or pages and immediately explain the handout. (3) At the end of your presentation, you give everyone a handout, perhaps one with detailed information that the audience can digest later.

Television and VCR

Interspersing almost any type of presentation with appropriate video sequences can effectively illustrate points and hold audience attention. For example, video is an excellent aid to illustrate stereochemistry, to view molecular models from different angles, or to show an enzyme fitting into a binding site.

Creating Visual Aids

If graphs or text are converted directly from a printed page to a transparency, the audience will not be able to read the visual aid, will find the aid too complex to digest while listening to the speaker, or will have insufficient time to locate the parts that the speaker discusses. At worst, the listeners will forget what they have already heard the speaker say, stop listening, and become angry. At best, they will simply stop listening.

As readers, we can take the time to understand and appreciate these same visual aids. We are not trying to listen to a speaker at the same time that we study the visual aid and fear that it will be snatched from us before we can understand it. The nature of speech dictates special rules for the types of visual aids used in an oral presentation.

◆ Use the rule of sixes. If you cannot read everything on a nonprojected overhead transparency from 6 feet away, redo the transparency using larger and bolder print; otherwise, the audience will not be able to see everything on the projected image. Excluding the title, include no more than about six lines on a page of a visual aid. Include no more than about six words per line on a page of a visual aid. Figure 1 illustrates how to condense these three rules to a form suitable for a visual aid.

◆ Use color wisely. Color accelerates learning and recall by 55% or more and comprehension by 70%. However, too many colors can be distracting. As a rule, use no more than two colors unless you use a third color as an accent. Ordinarily, use bright colors as accents, not as the main colors.

◆ If you need to use a complex visual aid, build from a simpler one. The first page of a flip chart could include a simple chart, the second page could include a more detailed version of the same chart, and so on. Some speakers show a transparency with parts left out and then add the parts while speaking or else add additional transparencies as overlays.

◆ Include nothing on the visual aid that you do not intend to discuss. Of course, you need to apply this rule with some common sense. Just as background images in a painting can set a mood for more prominent features, an entire chart can set the context for specific points.

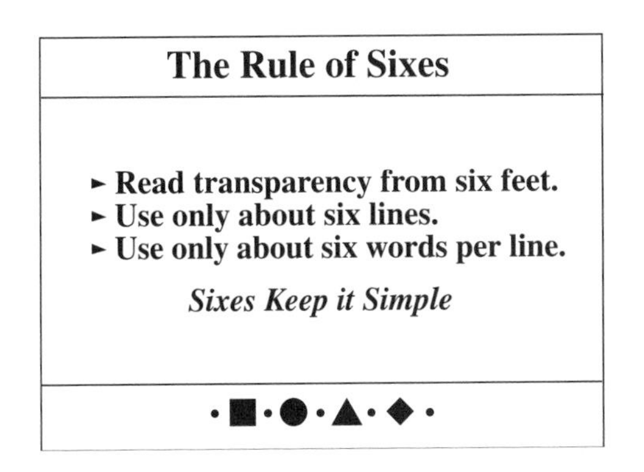

Figure 1. A bulleted list: the rule of sixes. Notice the simplicity of this visual aid and the boldness of the print.

◆ Make the visual aids look professional. A sloppy visual aid reflects poorly on a speaker. Small steps such as including titles, occasionally including borders, and using only two or three typefaces on transparencies can make a difference.

Computer-generated graphics provide the best way to develop professional visual aids. Most word-processing and spreadsheet software includes graphics capabilities. More specialized graphics programs offer the speaker a myriad of options for creating professional visual aids. I used Freelance Graphics to create Figure 2 and the other visual aids in this chapter.

Advance Planning for Visual Aids

Proactive preparation continues from the moment you accept an invitation to speak until you begin to speak. Alert the program chair as soon as you decide what equipment you will need and how it should be set up. Then assume that unless you follow up, one of the following types of problems

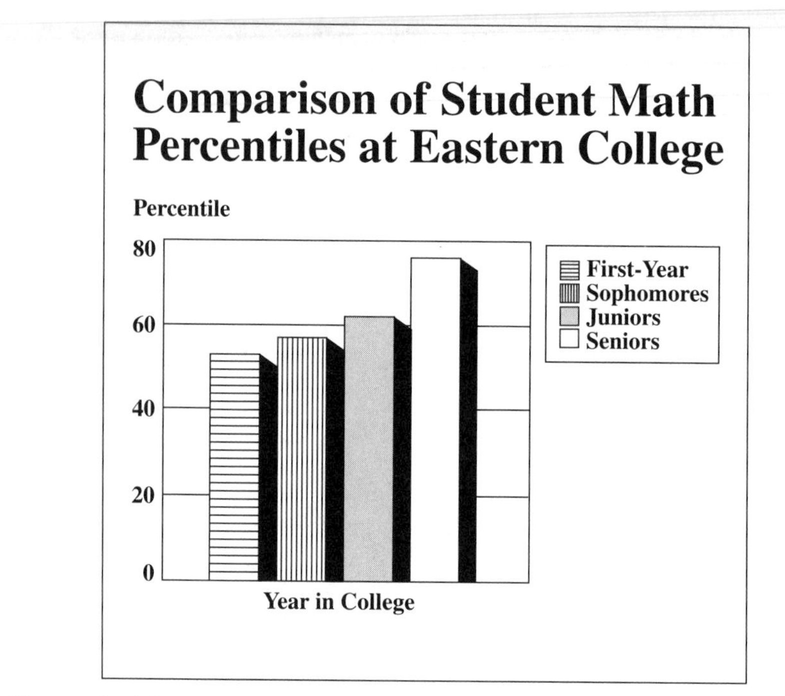

Figure 2. A bar graph: comparison of student math percentiles. Computer-generated graphics offer numerous options in terms of theme of visual aid, chart type, color selections (if color is an option), and other chart attributes, such as use of grid lines and location of caption.

will occur: no table for an overhead projector, no screen, inadequate room between the projector and the screen, no easel for a flip chart, only one marker and it is pink, and so on. Avoid these and other problems by arriving at least 30 minutes early and checking out the equipment and other physical arrangements.

Your needs may be altogether different from those of a previous speaker. Right-handed people often have to move the lectern to their left and the overhead projector to their right to be able to write on the transparencies. Left-handed people must do the opposite. Make sure that the projector is working, that it is focused, that listeners will have a clear view of the screen, and that any electrical cords are out of the way of your feet. Taking a few minutes to deal with simple matters builds your confidence and diminishes the chances of having to make mid-presentation adjustments. Using a checklist such as the one on this page might be helpful.

During the presentation, keep all visual aids out of sight until you refer to them. Make sure that all listeners can see each visual aid as you refer to it; often either the lectern or the speaker blocks the view of some listeners. Reveal only the part of the visual aid to which you are referring; cover the parts that you will show later or have already shown. The best system for revealing the right item at the right time is a computerized visual presentation with "building bullets". When you call up the next item on a list, the computer highlights this item while the previously mentioned item disappears or appears in a less bold color.

While speaking, also remember that visual aids do not have ears. Many speakers do not realize this fact and thus talk more to the visual aid than to the audience. The typical speaker stands at an angle between the visual aid and the audience, thus eliminating face-to-face communication with one-

Checklist for Physical Arrangements

- ☐ Adapter
- ☐ Chalkboard and chalk
- ☐ Chart pad and stand
- ☐ Extension cord
- ☐ Handouts
- ☐ Marking pen
- ☐ Microphone
- ☐ Note pads and pencils
- ☐ Podium
- ☐ Projector and spare lamp
- ☐ Samples or equipment
- ☐ Screen or clean white wall
- ☐ Sign-in sheet
- ☐ Table for projector
- ☐ Water and glass

third of the listeners. When referring to something on a flip chart or screen, the simple step of using a pointer can enable you to face the audience while referring to parts of a visual aid and avoid standing between the visual aid and part of the audience.

Not only do visual aids not have ears; they also do not have mouths. It is not helpful for a speaker to prepare an excellent visual aid only to do little more than nod at it during the presentation, apparently thinking that the aid explains itself. An inadequately explained visual aid does more harm than good by distracting, confusing, and frustrating the listeners. Readers can orient themselves to a visual aid and locate the relevant parts of it without distraction, but listeners perform these tasks either with the help of the speaker or while the speaker discusses the next point of the presentation. Therefore, speakers need to orally integrate their visual aids into the speech. The following example, which corresponds to Figure 3, illustrates this technique.

> ***Transition to the visual aid*** "The pie charts that we'll look at now highlight the decline in money budgeted for chemical supplies from the previous fiscal year to this fiscal year."

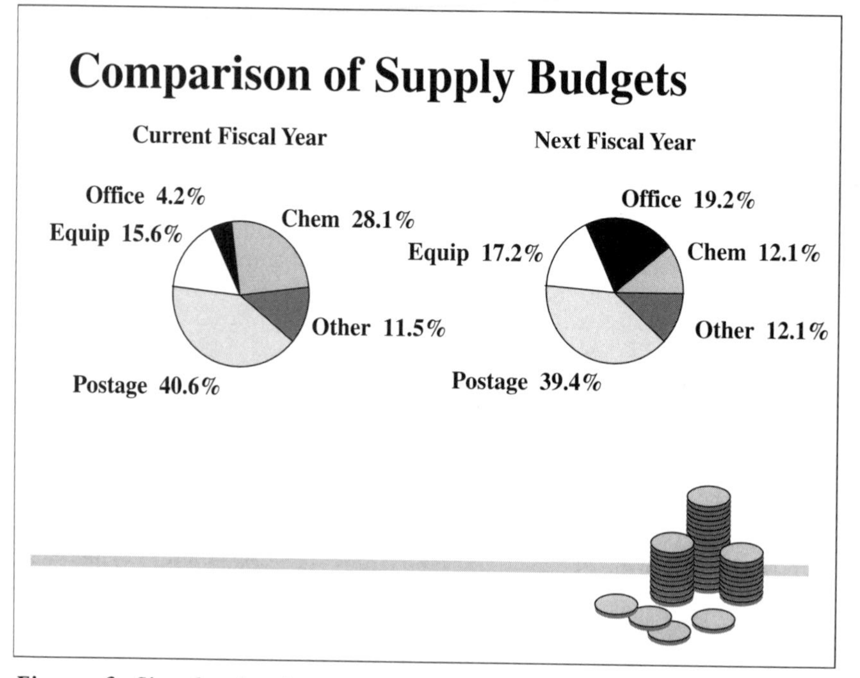

Figure 3. Simple pie charts: comparison of supply budgets. A simple chart can underscore a speaker's point.

Orientation to the visual aid "As you can see, the chart on the left focuses on our supply budget for the current fiscal year, and the one on the right focuses on our projected supply budget for the next fiscal year."

Main point "Notice that the proportion of our supply budget devoted to chemical supplies will decline from 28% to 12% unless we revise next year's budget."

Transition to the next point of the presentation "How will this reduction affect our students' education?"

Conclusion

A responsible communicator respects listeners enough to prepare a presentation that is focused, organized, and concisely worded. More than that, such a speaker chooses information relevant to the interests and needs of a particular audience and, during the presentation, makes whatever adjustments the circumstances dictate. Audiences, then, will return the respect and feel as if they have participated in a lively and illuminating conversation.

Giving oral presentations is a responsibility, but it is also an opportunity that can bring great rewards. It is a task that can be learned like any other. I hope that, by giving you the tools, this chapter will give you more confidence in your ability to make effective oral presentations.

Bibliography

Bartlett, John. *Familiar Quotations: A Collection of Passages, Phrases, and Proverbs Traced to Their Sources in Ancient and Modern Literature,* 16th ed.; Kaplan, J., Ed.; Little, Brown: Boston, MA, 1992.

Brussell, Eugene E. *Webster's New World Dictionary of Quotable Definitions,* 2nd ed.; Prentice Hall: Englewood Cliffs, NJ, 1988.

Daiker, Donald A.; Kerek, Andrew; Morenberg, Max; Sommers, Jeffrey. *The Writer's Options: Combining to Composing,* 5th ed.; HarperCollins College Publishers: New York, 1994.

Dale, Edgar. *The Educator's Quotebook;* Phi Delta Kappa Educational Foundation: Bloomington, IN, 1984.

Devine, Betsy; Cohen, Joel E. *Absolute Zero Gravity: Science Jokes, Quotes, and Anecdotes;* Simon & Schuster: New York, 1992.

Glazer, Mark. *A Dictionary of Mexican American Proverbs;* Greenwood Press: New York, 1987.

Gronbeck, Bruce E.; McKerrow, Raymie E.; Ehninger, Douglas; Monroe, Alan H. *Principles and Types of Speech Communication,* 13th ed.; HarperCollins College Publishers: New York, 1996.

Hendricks, W.; Holiday, M.; Mobley, R.; Steinbrecher, S. *Secrets of Power Presentations;* Career Press: Franklin Lakes, NJ, 1996.

Monroe, Alan H. *Principles and Types of Speech;* Scott, Foresman: Chicago, IL, 1935.

Rodríguez, José; Plax, T. G.; Kearney, P. Clarifying the Relationship between Teacher Nonverbal Immediacy and Student Cognitive Learning: Affective Learning As the Central Causal Mediator. *Communication Education* **1996,** *45* (4), 293–305.

Zuck, Roy B. *The Speaker's Quotebook: Illustrations and Quotations for All Occasions;* Kregel Resources: Grand Rapids, MI, 1996.

APPENDIX *I*

ACS Publications

Editorial Procedures

For ACS books and journals, the specifics may differ from office to office, but the general procedures for processing manuscripts from review through production are similar.

Manuscript Review

Papers submitted to ACS books and journals are considered for publication with the understanding that they have not been published or accepted for publication elsewhere and are not currently under consideration elsewhere. Appendix III, Ethical Guidelines to Publication of Chemical Research, presents ACS's guidelines for editors, authors, and reviewers.

Authors are sent pertinent reviewer comments by the journal editor or ACS Books staff editor in conjunction with the book editor, generally without identifying the reviewer. For journals, the actual sending of the manuscripts for review and receiving of reviews is done by staff in the individual journal offices. Although decisions to publish are almost always based on reviewers' comments, each editor retains the right to publish in the face of negative comments and to reject in the face of positive comments. In practice, however, editorial decisions that are contrary to reviewer recommendations are usually related to journal or book scope or involve manuscripts in the editor's particular field of expertise.

In journals, a paper that is rejected may be reconsidered if the author presents a good case for further review. In ACS books, rejections may be made on the basis of timing and scheduling if a manuscript will require extensive revision and additional review. Further review often is not possible because of time constraints.

A paper may be rejected if the author refuses to obtain the necessary permissions to reproduce previously published material (such as figures or tables) or if the author refuses to sign the ACS Copyright Status Form.

After a revised paper is accepted, it is considered to be in final form. Alterations made after acceptance may be permitted at the discretion of the editor. If alterations are extensive or if significant additions are made, additional review may be required.

Processing of Accepted Manuscripts

Manuscripts accepted for journals are sent directly from the editor to the Journals Editorial Office, where they are copyedited and prepared for the printer. Manuscripts accepted for *Chemical & Engineering News*, *CHEMTECH*, *Chemical Health & Safety*, *Today's Chemist at Work*, and the A pages of *Analytical Chemistry* and *Environmental Science & Technology* are copyedited and prepared for printing by staff in the ACS Graphics and Production Department.

Technical Editing

Technical editing of papers involves both production and copyediting. Routine production consists of scheduling, marking the copy according to publication style, selecting or specifying type, preparing page layout, designing book covers, sizing art, reviewing or proofreading galleys, checking page makeup, giving instructions to typesetters and printers, and preparing indexes. Copyediting is done to ensure consistency, clarity, and grammatical accuracy; changes are introduced to improve nomenclature, graphics presentation, and tabular format. Copy editors may contact authors or query them at the proof stage for clarification of material.

Author's Proof

One author, generally the author to whom correspondence should be addressed, receives a proof of the manuscript for final approval before publication. A paper is usually not released for printing until the author's proof or other approval has been received by the Journals Editorial Office, Graphics and Production Department, or Books Department. Hence, proofs should be checked and returned promptly according to individual journal or book instructions. To save time and expense, foreign contributors may authorize a colleague in the United States to check proofs.

Authors should check proofs thoroughly by reading them at least twice. Only corrections and necessary changes can be made in proofs. Extensive changes will require editorial approval and perhaps a revised date of receipt and will delay publication. Printer's errors are corrected at no cost to authors, but authors may be charged the cost of extensive production work made necessary by their own alterations. In books, there may not be time to do extensive alterations at all.

Instructions to Authors for Checking Proofs

- Mark corrections legibly in the margins of the proofs.
- Do not erase or obliterate type; instead, strike one line through copy to be deleted and write the change in the margin.
- Clarify complicated corrections by rewriting the entire phrase or sentence in the margin or on a separate sheet.
- Check all text, data, and references against the original manuscript. Pay particular attention to equations; formulas; tables; captions; spelling of proper names; and numbering of equations, illustrations, tables, and references.
- Answer explicitly all queries made by the technical editor.

Corrections to Published Manuscripts

Corrections for a paper that has already been published should be sent in duplicate to the editor or managing editor of the journal or to the Books Department. Most ACS journals publish corrections soon after they have been received. The Books Department will print an erratum sheet to be included in every book, and the book itself will be corrected before reprinting.

Reprints

A form for reprint orders is sent to authors with proofs of journal articles or shortly after proofs of book chapters. Authors should follow the instructions on the form itself. Payment, a purchase order, or a credit card number should be submitted with the reprint order.

Liability and Prior Publication

Authors are solely responsible for the accuracy of their contributions. The ACS and the editors assume no responsibility for the statements and opinions advanced by the contributors.

Contributions that have appeared or have been accepted for publication with essentially the same content in another journal or book or in some freely available printed or electronic work (e.g., government publications and proceedings) will not be published by the ACS. This restriction does not apply to results previously published as communications to the editor in the same or other journals.

ACS Books

ACS Symposium Series

This series provides more timely publication of original research papers presented at ACS symposia. Books in this series usually cover rapidly developing areas of chemistry, for which time of publication is especially important. However, these books are not simply proceedings; material is often added to enhance the comprehensiveness of the subject coverage. Each paper undergoes peer review. This review is conducted by the editor of the book with specific guidelines provided by the ACS Books Department. After review and revision, the authors provide camera-ready copy.

Conference Proceedings Series

This series provides a format for rapid publication of the proceedings of topical conferences relevant to the chemical sciences. These conferences are often organized by international groups and usually are not sponsored by ACS. A special review committee established by the conference organizer reviews each paper. Authors then submit the revised contributions in camera-ready form.

Profiles, Pathways, and Dreams Series

In these autobiographies, some of chemistry's most eminent and prolific scientists describe both their research and the motivation that led them to pursue this research. As they describe their personal histories, they also include stories that are touching, humorous, and even awe-inspiring.

Advances in Chemistry Series

Books in this series are collections of papers on a particular topic. They provide reviews and overviews of various aspects of mature topics and thus serve as reference works for several years. Typical volumes are multiauthored with an overall editor or editors. They may be developed from symposia, or they may be planned specifically for book publication from the outset. Papers may be reviews or reports of original research with review material included. Papers are peer-reviewed critically, and the same rigorous standards of acceptance that apply to papers for ACS journals are applied to papers for these books.

Directories

These books are reference tools for chemists and researchers.

The ACS Directory of Graduate Research (DGR) lists universities in the United States and Canada along with the names and biographical information for all faculty members, their areas of specialization, the titles of their recent papers, their telephone numbers and e-mail addresses, and departmental fax numbers. It includes listings for graduate programs in chemistry, chemical engineering, biochemistry, medicinal–pharmaceutical chemistry, clinical chemistry, polymer science, food science, forensic science, marine science, and toxicology at institutions. The DGR is published in odd-numbered years.

The *Chemical Sciences Graduate School Finder* presents department information, admissions information, degree requirements, financial information, research facilities, and faculty and research areas for chemistry and related departments of invited institutions in the United States and Canada that offer courses leading to an M.S. or Ph.D. in chemistry and allied fields. It is updated annually.

Chemical Research Faculties: An International Directory is the international complement to the DGR and covers educational institutions granting advanced degrees in chemical research fields of study. It does not include institutions in the United States and Canada that are covered in the DGR.

College Chemistry Faculties lists faculty members, their teaching fields, and their e-mail addresses for two-, three-, and four-year colleges and universities in the United States and Canada offering instruction in chemistry, biochemistry, biotechnology, chemical engineering, chemical technology, medicinal chemistry, and related disciplines.

Directories on Disc 3.0 contains the *ACS Directory of Graduate Research*, the *Chemical Sciences Graduate School Finder, College Chemistry Faculties*, and *Chemical Research Faculties, An International Directory.* The most recent edition contains information for more than 30,000 faculty members at more than 3000 departments. It is searchable using a variety of search fields, including location, name, and research specialty. It also contains e-mail addresses for individuals and allows mailing labels to be downloaded for up to 50 hits. It is available in DOS, Mac, and Windows in single-user and multiuser versions.

Other Books

ACS also publishes books that are not part of any particular series. These include reference, data, and reprint volumes as well as books that are semi-technical or nontechnical. Often these books appeal to a broader audience than the technical books. Most of these books undergo stringent peer review.

Journals and Magazines

Accounts of Chemical Research

Accounts of Chemical Research is a journal of concise descriptions of recent research developments. A typical account deals with a topic of intense interest to the author and to a considerable extent treats the author's own experimental or theoretical results. Another welcome type of article examines critically a question of current interest and draws new generalizations or new perspective from the evidence. Most manuscripts are submitted after invitation or encouragement by the editor, but unsolicited manuscripts are also considered. Comprehensive reviews do not fall within the mission of this journal.

Advance ACS Abstracts

Advance ACS Abstracts, copublished by the Publications Division and Chemical Abstracts Service, features the author abstracts of papers accepted for publication in 26 ACS journals. Each issue includes the abstracts of 600–800 wide-ranging articles that will be published in ACS journals up to 12 weeks later. *Advance ACS Abstracts* also provides the tentative publication date of the completed paper.

Analytical Chemistry

Analytical Chemistry is a hybrid research journal–magazine that covers all aspects of measurement science. The research section is devoted to all branches of analytical chemistry. Research papers either are theoretical with regard to analysis or are reports of laboratory experiments that support, argue, refute, or extend established theory. Research papers may contribute to any of the phases of analytical operations, such as sampling, preliminary chemical reactions, separations, instrumentation, measurements, and data processing. They need not necessarily refer to existing or even potential analytical methods in themselves but may be confined to the principles and methodology underlying such methods. Critical reviews of the literature, prepared by invitation, are published in June of each year in special issues that cover, in alternating years, applied and fundamental aspects of analysis.

The magazine section covers analytical chemistry from a news perspective and includes feature articles, book and software reviews, product reviews, and extensive news of interest to the analytical chemistry community. The magazine section is also available separately as *Analytical Chemistry News & Features*.

Biochemistry

This journal publishes the results of original research that contribute significantly to biochemical knowledge, particularly in understanding the structure, function, and mechanism of action of biological molecules. Preference is accorded to manuscripts that generate new concepts or experimental approaches, particularly in the advancing areas of biochemical science. Hence, the primary criterion in the acceptance of manuscripts is that they present new and germinal findings.

Bioconjugate Chemistry

Bioconjugate Chemistry reports key advances concerning the joining of two or more molecular functions. Coverage spans conjugation of antibodies and their fragments, nucleic acids and their analogs, liposomal components, and other biologically active molecules with each other or with any molecular groups that add useful properties. Topics span studies of conjugate preparation and characterization, in vivo applications of conjugate methodology, molecular biological aspects of antibodies, genetically engineered fragments and other immunochemicals, and the relationships between conjugation chemistry and the biological properties in conjugates.

Biotechnology Progress

This journal is a copublication of the American Institute of Chemical Engineers and the American Chemical Society. Its focus is the impact of the revolutions in biological sciences and their current and future impact on the practices of engineering and industry. Current papers describe the applications of discoveries in molecular and cellular biology and provide insights, examples, and new concepts for use of this knowledge. Research papers contribute to technological endeavors in the areas of applied cellular physiology and metabolic engineering, biocatalyst and bioreactor design, bioseparations and downstream processing, cell culture and tissue engineering, formulation and engineering of biomaterials, and process sensing and control. Reviews are usually written in response to invitations issued by the associate editor (for reviews), although nominations of topics and authors are often accepted. The journal also publishes topical papers on issues of current interest as well as readers' correspondence.

Chemical & Engineering News

As the official publication of the American Chemical Society and as the newsmagazine of the chemical world, *Chemical & Engineering News* is designed to keep ACS members and others informed of policies and activities of the society and to keep ACS members and others abreast of news, events, trends, and issues of importance to chemists and chemical engineers.

Chemical Health & Safety

Copublished by the American Chemical Society and the ACS Division of Chemical Health and Safety, this magazine focuses on news, information, and ideas relating to issues and advances in chemical health and safety in the laboratory. Articles provide in-depth views of safety issues ranging from OSHA and EPA regulations to the safe handling of hazardous waste, and from research or the latest innovations in effective chemical hygiene practices to the courts' most recent rulings on safety-related lawsuits.

Chemical Research in Toxicology

This journal publishes original work in the broad field of chemical approaches to the solution of toxicological problems. Papers cover structural, mechanistic, and technological advances relating to the toxicological effects of chemical agents; research on novel toxic agents and reactive intermediates; new methods for detecting adducts between toxic agents and biological macromolecules; alteration of macromolecular structure and function by interaction with chemical agents; chemical factors controlling reactivity with specific macromolecules; and the metabolism of toxic agents as it contributes to their biological effects. Cutting across the areas of biochemistry, toxicology, organic, physical, analytical, and inorganic chemistry, this journal publishes full papers, rapid communications, and invited reviews.

Chemical Reviews

Articles published in this journal are authoritative, critical, and comprehensive reviews of research in the various fields of chemistry. Preference is given to creative, timely reviews that will stimulate further research with a carefully selected subject and well-defined scope. In general, the topic should not have been reviewed in a readily available publication for about five years, although exceptions will be made if developments in the field have been particularly rapid or if new insight can be achieved through further review. Some issues of *Chemical Reviews* document progress in several areas, and others focus on a single theme or direction of emerging research. Reviews are invited by the editor in response to suggestions from the editorial advisory board or other leading scientists elsewhere in the scientific community. Unsolicited articles may also be considered, provided that the author contacts the editor and obtains preliminary approval according to the Instructions for Authors published in the journal.

Chemistry of Materials

This journal provides a molecular-level perspective of forefront research at the interface of chemistry, chemical engineering, and materials science. Its monthly issues feature communications, articles, and short reviews on all types of materials, including inorganic and organic solids, thin films, ceramics, polymers, liquid crystals, and composites. With its particular focus on chemistry relating to the preparation, processing, and analysis of materials, it is unique among the world's leading materials science journals. Special features include a continuing series of reviews, Materials Chemistry Issues in Key Technologies, and special issues on particular topics in materials chemistry.

CHEMTECH

CHEMTECH covers all the elements of the innovation process in the chemical-related industries, including the resources needed to start the process, the research and development of new technologies, and the marketplace issues integral to commercializing technology. Feature articles and regular departments explore a wide range of chemical-related technologies, commercial development ideas, environmental concerns, research and development strategies, and legal issues.

Energy & Fuels

This interdisciplinary journal contains research reports on the chemistry of nonnuclear energy sources: petroleum, coal, shale oil, tar sands, biomass, synfuels, C_1 chemistry, organic geochemistry, applied catalysis, and combustion. Papers describe research on the transformation, use, formation, and production of fossil fuels, including the molecular composition of raw fuels and refined products and the chemistry involved in processing these fuels; photochemical fuel and energy production; the analytical and instrumental techniques used in energy and fuels investigations; and research on nonfuel substances relevant to fuel chemistry.

Environmental Science & Technology

Environmental Science & Technology is both a journal and a magazine, and it features peer-reviewed research, news reports, and feature articles concerning all aspects of environmental science, technology, and policy. Research areas cover many disciplines, from fate and transport of pollution in air, water, and soil to microbiology and analytical methods. The magazine covers developments and trends in regulations, technology applications, and research. *ES&T* is also available in a *News & Research Notes* magazine edition, and research papers are available to subscribers on the World Wide Web.

Industrial & Engineering Chemistry Research

This journal reports original work in the broad fields of applied chemistry and chemical engineering with special focus on fundamentals, processes, and products. Papers may be based on work that is experimental or theoretical, mathematical or descriptive, chemical or physical. In addition to fundamental research (in such areas as thermodynamics, transport phenomena, chemical reaction kinetics and engineering, catalysis, separations, interfacial phenomena, and materials), papers may deal with process design and development (for example, synthesis and design methods, systems analysis, process control, schemes for data correlation, and modeling and scale-up procedures) and product research and development involving chemical and engineering aspects (e.g., catalysts, plastics, elastomers, fibers and fabrics, adhesives, coatings, paper, membranes, lubricants, fertilizers, ceramics, aerosols, and liquid crystals). In addition to traditional subjects, *Industrial &*

Engineering Chemistry Research encourages papers dealing with new areas of science and technology that fit its broad scope and objectives.

Inorganic Chemistry

This journal publishes fundamental studies in all phases of inorganic chemistry. Coverage includes experimental and theoretical reports on the synthesis, structures, and properties of new compounds; quantitative studies of structure and thermodynamics, kinetics, and mechanisms of inorganic reactions; bioinorganic chemistry; and some aspects of organometallic chemistry, solid-state phenomena, and chemical bonding theory. Short papers (notes), full papers, and preliminary communications of an urgent nature are published.

Journal of Agricultural and Food Chemistry

This journal is devoted to the application of chemistry in developing more efficient, economical, and safe production of foods and agricultural products. Reports of original research concern the chemical, biochemical, and nutritional aspects of foods, fibers, and feeds. Articles describe relevant analytical methods, nutrition and toxicology, flavors and aromas, chemical changes during processing and storage, the chemistry of crop and animal protection, the impact of agrochemicals on the environment, and the burgeoning application of biotechnology to agriculture.

Journal of the American Chemical Society

Original papers in all fields of chemistry are published here. Emphasis is placed on fundamental chemistry. Papers of general interest are sought, either because they appeal to readers in more than one specialty or because they disclose findings of sufficient significance to command the interest of specialists in other fields. Communications are restricted to reports (usually preliminary) of unusual urgency, significance, and interest and are limited to two journal pages (about 2000 words or the equivalent). Book reviews are also published. Specialized papers are not published.

Journal of Chemical and Engineering Data

This journal publishes high-quality experimental data on the physical, thermodynamic, and transport properties of organic and inorganic compounds and their mixtures in the gaseous, liquid, and solid states, including systems of environmental biochemical interest. *JCED* also publishes reviews, evaluations, and predictive schemes for thermophysical properties, semiempirical and theoretical correlations for predicting properties of scientific and technological importance, and the description of new experimental techniques.

Journal of Chemical Information and Computer Sciences

This journal publishes research papers in all areas of information and computer science relevant to chemistry and chemical technology. Subject fields include database search systems, the use of graph theory in chemical problems, substructure search systems, pattern recognition and clustering, analysis of chemical and physical data, molecular modeling, graphics and natural language interfaces, bibliometric and citation analysis, synthesis design and reactions databases, and chemical computation.

Journal of Medicinal Chemistry

This journal publishes articles, expedited articles, notes, and communications that contribute to an understanding of the relationship between molecular structure and biological activity. Some of the specific areas that are appropriate are analysis of structure–activity relationships by a variety of approaches; design and synthesis of novel drugs; improvement of existing drugs by molecular modification; biochemical and pharmacological studies of receptor or enzyme mechanisms; construction of mutant and chimeric receptors as a means of investigating molecular recognition of ligands; physicochemical studies on established drugs that may furnish some insight into their mechanism of action; computational, X-ray, and NMR studies that provide insight into the recognition of ligands by enzymes or receptors; and the effect of molecular structure on the metabolism, distribution, and pharmacokinetics of drugs. In addition, interpretive accounts (perspective articles) of subjects of active current interest are published at the invitation of the editors.

Journal of Natural Products

Copublished by the American Chemical Society and the American Society of Pharmacognosy, this journal publishes novel contributions covering natural products chemistry and biochemistry, including such research areas as screening methodology, isolation technology, spectroscopy, structure determination, partial and total synthesis, analysis, biosynthesis, chemotaxonomy, and pharmacology of terrestrial and marine microbial, plant, and animal natural products. Emphasis is placed on compounds with demonstrated biological activity. Contributions are accepted in the categories rapid communications, full papers, notes, reviews, and book reviews.

The Journal of Organic Chemistry

The aim of this journal is to publish original and significant contributions in all branches of the theory and practice of organic chemistry. Areas emphasized include the many facets of organic reactions, natural products, bioorganic chemistry, studies of mechanism, theoretical organic chemistry, and the various aspects of spectroscopy related to organic chemistry. This journal publishes articles, notes, and communications in all branches of organic chemistry.

Journal of Pharmaceutical Sciences

Copublished by the American Chemical Society and the American Pharmaceutical Association, the *Journal of Pharmaceutical Sciences* publishes (1) original research papers, (2) original research notes, (3) invited topical reviews, and (4) editorial commentaries and news. The area of focus is concepts in basic pharmaceutical science and such topics as processing and materials science of drugs and excipients (including crystallization and lyophilization), analysis and stability of drugs, prodrug development, biomembrane transport, drug absorption, pharmacokinetics, pharmacodynamics, metabolic disposition of bioactive agents, dosage form design, drug targeting, protein–peptide chemistry, and biotechnology, specifically as these relate to pharmaceutical technology and drug delivery. Mechanistic studies at both the macroscopic and molecular levels are most appropriate.

Journal of Physical and Chemical Reference Data

This journal is copublished by the American Chemical Society and the American Institute of Physics for the National Institute of Standards and Technology. The objective of the journal is to provide critically evaluated physical and chemical property data, fully documented as to the original sources and the criteria used for evaluation. Critical reviews of measurement techniques, which aim to assess the accuracy of available data in a given technical area, are also included. The journal is not intended as a publication outlet for original experimental measurements such as are normally reported in the primary research literature or for review articles of a descriptive or primarily theoretical nature.

The Journal of Physical Chemistry A

This journal publishes new and original experimental and theoretical basic research directed toward researchers in spectroscopy, dynamics, kinetics, and environmental physical chemistry. Coverage includes new findings and full-length studies of dynamics and relaxation; spectroscopy, gaseous clusters, and molecular beams; kinetics; atmospheric and environmental physical chemistry; and molecular structure, bonding, quantum chemistry, and general theory. Invited papers review the status of a particular topic, clarify controversies, or explore future directions. Proceedings of selected symposia and special thematic issues appear throughout the year. Rapid publications of urgent and new results appear in the Letters section.

The Journal of Physical Chemistry B

This journal publishes new and original experimental and theoretical basic research directed toward researchers in materials, interfaces, and biophysical chemistry. Coverage includes new findings and full-length studies of physical chemistry of materials from nanoparticles to macromolecules; physical chemistry of surfaces and interfaces; statistical mechanics and thermodynamics of condensed matter; and biophysical chemistry. Invited papers review the status of a particular topic, clarify controversies, or explore future directions. Proceedings of selected symposia and special thematic issues appear throughout the year. Rapid publications of urgent and new results appear in the Letters section.

Langmuir

Langmuir covers the broad area of surface and colloid chemistry. Topics include micelles, emulsions, surfactants, vesicles, wetting and interfacial films, chemisorption and catalysis, electrochemistry, and physical adsorption on liquid and solid surfaces. Original research articles and letters are accepted. There are occasional invited review-type articles. The journal is intended to provide a comprehensive coverage of the field of chemistry at interfaces.

Macromolecules

Macromolecules publishes original research on all fundamental aspects of macromolecular science, including synthesis; polymerization mechanisms and kinetics; and chemical modification, solution/melt/solid-state characteristics, and surface properties of organic, inorganic, and naturally occurring polymers. Manuscripts that present innovative concepts, experimental methods or observations, and theoretical approaches in fundamental polymer research are of primary interest. The editors welcome regular articles, notes, communications, and occasional reviews.

Organic Process Research & Development

This journal, which is copublished by the American Chemical Society and the Royal Society of Chemistry, reports original work in the broad field of process chemistry that includes aspects of organic chemistry, catalysis, analytical chemistry, and chemical engineering, with special focus on the development and optimization of chemical reactions and processes and their transfer to a larger scale, via large laboratory and pilot-plant operations, for manufacture. This journal is directed to scientists working in the areas of organic process research and organic process development as they relate to scale-up issues, safety and environmental issues, and legislation and regulatory issues. Original research papers, reviews, notes, summaries of research, and technology reports are accepted.

Organometallics

This journal publishes articles, notes, and communications concerned with all aspects of organometallic chemistry. Specifically, it covers synthesis, structure, bonding, chemical reactivity, reaction mechanisms, and applications of organometallic and organometalloidal compounds. Coverage includes organic and polymer synthesis, catalytic processes, and synthetic aspects of materials science and solid-state chemistry.

Today's Chemist at Work

Today's Chemist at Work focuses on the information needs of the non-Ph.D. chemist working in industry, with special emphasis on career development. Published 11 times a year, the magazine includes overviews of existing technologies as well as discussions of new techniques and applications in areas such as biotechnology, separation science, the environment, laboratory services, chromatography, and spectroscopy. Regular departments include computers in chemistry, regulations, health perspectives, personal financial management, industry today, and history of chemistry. Unsolicited articles are welcome and subject to review.

ACS Internet Journal Editions

All American Chemical Society journals are available as Internet editions. They are presented in two formats: as searchable text (HTML) and as complete page images (PDF). Subscribers can browse or search the entire database and print articles including the supporting information, tables of contents, abstracts, and graphics. Each journal's home page can be reached from the ACS Publications Home Page at http://pubs.acs.org.

ACS Publications on STN International

CJACS Plus Images, which is the CJACS file (Chemical Journals of the American Chemical Society), allows users to search and retrieve articles from 26 ACS journals. With full text dating back to 1982, complete page images including illustrations, halftones, graphs, and tables are now available for all articles from 1992 forward.

Chemical & Engineering News Online (STN file: cen) delivers important news, product information, market statistics, and scientific discoveries in the fields of chemical sciences, technology, chemical-related business, education, and public policy from the United States and around the world. This searchable file is updated weekly and contains the full text from all C&EN issues published since 1991.

Supporting Information in Journals: Multimedia Availability

As more and more data-gathering techniques have become automated, the amount of material being presented in scientific papers has increased significantly, and the result has been a dramatic growth in the average length of a published manuscript. Not all readers, however, need all the data used to support the arguments in a paper. As a compromise, material that may be essential to the specialized reader but does not require elaboration in the paper itself should be presented as Supporting Information. Examples of such material are extensive tables, graphs, spectra, crystallographic information files (CIF), mathematical derivations, computer algorithms, protein and nucleic acid sequence analyses, multiple regression analyses, detailed and repetitive experimental procedures, extensive experimental data, and expanded discussions of peripheral points.

In addition to providing a mechanism whereby the printed version of a paper can be effectively shortened, Supporting Information (SI) affords the opportunity to use new media for communication of ideas, such as video clips and demonstrations and "movable" molecular models. Considering the variety of media available for presenting SI, authors are able to significantly shorten their manuscripts while still publishing "larger" papers.

Authors include SI with their manuscript when they send it to the editor's office for review. SI that is submitted on paper is archived on microfiche and electronically scanned to be included in the online archives. This material is processed as received (unless the preparation requirements are not met), and no proofs are sent to authors. SI that is submitted in electronic form is made available via electronic media only. Guidelines for preparation of SI are given in the Instructions to Authors for each journal, which appear on the Internet home page and in the first printed issue of each volume.

SI is available to subscribers of the journal on the Internet on the journal's cover date of publication. Readers can also obtain SI by microfiche subscription or by ordering photocopies. Further ordering information is available on the current masthead page of each journal. SI is indexed by Chemical Abstracts Service.

Special Projects

Chemcyclopedia

Chemcyclopedia provides quick and easy access to valuable information for users and purchasers of chemicals. In addition to listing a given chemical, suppliers have provided trade names, available forms, packaging, special shipping requirements, and potential applications. Company names, mailing addresses, phone and fax numbers, and contacts are included. A searchable version is available on the Internet at http://pubs.acs.org.

Lab Guide

Lab Guide is widely recognized as the industry bible for sources of analytical instruments, lab equipment, services, and supplies. Available in print and on the Internet (http://pubs.acs.org), this comprehensive directory provides detailed information on specific products and the companies that provide them. The Internet edition contains additional product information and links to company home pages.

ChemCenter

ChemCenter is the premier electronic gathering spot for chemists, other researchers, teachers, students, and anyone else looking for essential information about chemistry. This World Wide Web site brings together critical sources of information, providing users an efficient, central point of access to the world of chemistry. From critical research information and federal R&D funding statistics to programs for educating children about chemistry, ChemCenter is an invaluable resource for everyone with an interest in chemistry. The URL is http://www.chemcenter.org.

Chemical Abstracts Service

Chemical Abstracts Service (CAS), in its mission of providing access to the world's literature related to chemistry and chemical engineering, offers a wide range of publications and services. These include the comprehensive printed *Chemical Abstracts* (CA); online chemical databases; and a broad variety of other printed, microform, CD-ROM, and World Wide Web services intended to satisfy current-awareness and retrospective searching needs. The CAS databases are made up of abstracts of documents in the scientific literature, bibliographic citations, comprehensive substance-related information, and extensive index entries. They draw upon articles that appear in 8000 scientific and technical journals published in 200 nations; patents issued by 31 patent offices around the world; and material from conference proceedings, technical reports, dissertations, and new books. All abstracts are in English, although the original literature may be in any of 50 languages.

Chemical Abstracts

Each issue of CA contains abstracts, bibliographic citations, and indexes. The abstracts are grouped into 80 subject sections. The indexes in the weekly issues include Keyword Index, Author Index, and Patent Index. The 52 weekly issues of CA are divided into two semiannual volumes. Five volume indexes (General Subject, Chemical Substance, Formula, Author, and Patent) are produced. The Index Guide is a cross-reference tool that links substance and subject terms in general use with the highly controlled CAS terminology. Collective indexes combine into a single listing the content of the individual CA volume indexes for a 5- or 10-year collective indexing period. CA is available in print, microform, CD-ROM, and online.

CA Section Groupings

CA Section Groupings divide the content of the weekly CA issues into five separate printed publications in related subject areas: Biochemistry; Organic Chemistry; Macromolecular Chemistry; Applied Chemistry and

Chemical Engineering; and Physical, Inorganic, and Analytical Chemistry. Each contains abstracts and bibliographic information reproduced exactly from specific CA Sections as well as a keyword index.

CA SELECTS

CA SELECTS is a series of about 210 current-awareness bulletins. Each *CA SELECTS* topic is a separate publication. Each topic includes CA abstracts and associated bibliographic information, selected by computer from the CA database according to a precise, special-interest profile.

CA SELECTS PLUS

CA SELECTS PLUS is a series of more than 35 printed current-awareness bulletins—each topic a separate publication. Topics include CA abstracts and associated bibliographic information selected by computer from the CA database. With *CA SELECTS PLUS*, in addition to coverage of the usual relevant journals, 1350 key journals are covered in their entirety with references including letters to the editor, meeting notes, and editorials.

CAS BioTech Updates

CAS BioTech Updates includes abstracts from *Chemical Abstracts* and business information from *Chemical Industry Notes* related to biotechnology processes, people in the industry, government activities, production, and pricing internationally. Each issue is divided into four sections: patents, papers, books and reviews, and BioTech Industry Notes. Each issue also contains keyword and author indexes.

CASurveyor

CASurveyor is a CD-ROM series of more than 15 informative summaries of worldwide science literature as published in CA. Each subject-specific topic can be searched using title, author, organization name, patent number, journal, publication year, document type, CA abstract number, keywords, and subject index terms. Abstracts, sequences, and molecular structures are displayable.

Chemical Titles

This alerting service is published every two weeks and provides titles of articles appearing in 800 leading scientific journals published worldwide. Each issue contains a keyword index (arranged in alphabetical order), bibliography (arranged alphabetically by journal CODEN with abbreviated journal titles provided), and author index (arranged in alphabetical order).

Chemical Industry Notes

Chemical Industry Notes is a current-awareness service covering activities in the chemical industry and related industrial fields. Published weekly, this publication monitors approximately 80 leading business and trade journals worldwide. Each issue covers such areas as industrial management, investment, marketing, pricing, and production with brief but informative article extracts, bibliographic citations, and keyword and corporate indexes.

CASSI Cumulative 1907–1994

Chemical Abstracts Service Source Index (CASSI, updated regularly) provides complete information on serial and nonserial publications that are monitored by CAS and held by more than 350 libraries around the world (300 in the United States). It contains complete bibliographic data for almost 70,000 publications, including information about variant titles, histories of publications, English translations of many foreign-language titles, a directory of publishers and sales agencies, and guides to the depositories of unpublished works. In addition, CASSI indicates which titles are available from the CAS Document Detective Service. CASSI is updated with quarterly supplements. A microform CASSI KWOC (keyword-out-of-context) Index is available to determine full titles for publications by use of any significant word from the title. CASSI is available in print and CD-ROM.

Ring Systems Handbook

Ring Systems Handbook (RSH) allows access to approximately 91,500 ring and cage systems formerly contained in the *Patterson Ring Index* and the *CAS Parent Compound Handbook*. The RSH consists of four volumes: a two-volume *Ring Systems File*, an index, and a cumulative supplement. In the *Ring Systems File*, entries are arranged in ring analysis order. For ring systems having a common ring analysis, ring names are listed in alphabetical order. Each entry is identified by a unique Ring File number. Structure diagrams for cage systems are grouped in ascending molecular formula order at the end of the *Ring Systems File*. The index provides two routes of access to ring information: the Name Index, where names are listed alphabetically along with a Ring File number, and the Ring Formula Index, where formulas are listed in

Hill System order and include the Ring File number. The RSH is issued once every five years, with cumulative supplements issued every six months.

Registry Handbook—Common Names

This microform listing links a common name of a substance with a CAS Registry number, as well as CA index name, molecular formula, and related names. Coverage includes 2,500,000 names and more than 2,000,000 Registry numbers.

Registry Handbook—Number Section

The *Registry Handbook*—Number Section provides the CA Index names and the molecular formulas for more than 15 million substances. The "base book" covers 1965–1971. Additions are provided in annual supplements.

International CODEN Directory

This microfiche index lists CODEN (six-character identification codes for publication titles) and full publication titles for 190,500 serial and nonserial publications. Access to these publications is available through CODEN, title, or keyword-out-of-context indexes.

STN International

STN International is the online scientific and technical information network dedicated to meeting the information needs of scientists and information professionals throughout the world. CAS represents STN in North America. The CAS databases accessible on STN are

CAplus	CHEMLIST
CA	CIN
CA OLD	REGISTRY
CHEMCATS	MARPAT
CASREACT	MARPATpreviews

STN offers information from more than 200 databases that cover a broad range of topics including chemistry, engineering, mathematics, biochemistry, physics, geology, biotechnology, medicine, energy, pharmacology, metallurgy, materials science, business, government regulations, and many more.

Some of the special advantages of STN include advanced chemical structure searching, easy searching and cross-searching of complementary databases, numeric searching capabilities, extensive online help messages, chemical reaction information, computational services, a knowledgeable Help Desk staff, and thorough documentation.

The ACS files Chemical Engineering News Online and CJACS are also available on STN.

APPENDIX *II*

ACS Divisions

Agricultural and Food Chemistry
Agrochemicals
Analytical Chemistry
Biochemical Technology
Biological Chemistry
Business Development and Management
Carbohydrate Chemistry
Cellulose, Paper, and Textile Division*
Chemical Education
Chemical Health and Safety
Chemical Information
Chemical Technicians
Chemical Toxicology
Chemistry and the Law
Colloid and Surface Chemistry
Computers in Chemistry
Environmental Chemistry
Fertilizer and Soil Chemistry
Fluorine Chemistry
Fuel Chemistry
Geochemistry
History of Chemistry
Industrial and Engineering Chemistry
Inorganic Chemistry

---

*These names are not preceded by "Division of".

Medicinal Chemistry
Nuclear Chemistry and Technology
Organic Chemistry
Petroleum Chemistry
Physical Chemistry
Polymer Chemistry
Polymeric Materials: Science and Engineering
Professional Relations
Rubber Division*
Small Chemical Businesses

Secretariats

Biotechnology Secretariat
Catalysis and Surface Science Secretariat
Macromolecular Secretariat
Materials Chemistry Secretariat

---

*These names are not preceded by “Division of”.

APPENDIX *III*

Ethical Guidelines to Publication of Chemical Research

The guidelines embodied in this document were revised by the Editors of the Publications Division of the American Chemical Society in January 1994 and endorsed by the Society Committee on Publications.

Preface

The American Chemical Society serves the chemistry profession and society at large in many ways, among them by publishing journals that present the results of scientific and engineering research. Every editor of a Society journal has the responsibility to establish and maintain guidelines for selecting and accepting papers submitted to that journal. In the main, these guidelines derive from the Society's definition of the scope of the journal and from the editor's perception of standards of quality for scientific work and its presentation.

An essential feature of a profession is the acceptance by its members of a code that outlines desirable behavior and specifies obligations of members to each other and to the public. Such a code derives from a desire to maximize perceived benefits to society and to the profession as a whole and to limit actions that might serve the narrow self-interests of individuals. The advancement of science requires the sharing of knowledge between individuals, even though doing so may sometimes entail forgoing some immediate personal advantage.

With these thoughts in mind, the editors of journals published by the American Chemical Society now present a set of ethical guidelines for persons engaged in the publication of chemical research, specifically, for editors, authors, and manuscript reviewers. These guidelines are offered not in the sense that there is any immediate crisis in ethical behavior, but rather from a conviction that the observance of high ethical standards is so vital to the whole scientific enterprise that a definition of those standards should be brought to the attention of all concerned.

We believe that most of the guidelines now offered are already understood and subscribed to by the majority of experienced research chemists. They may, however, be of substantial help to those who are relatively new to research. Even well-established scientists may appreciate an opportunity to review matters so significant to the practice of science.

Guidelines

A. Ethical Obligations of Editors of Scientific Journals

1. An editor should give unbiased consideration to all manuscripts offered for publication, judging each on its merits without regard to race, religion, nationality, sex, seniority, or institutional affiliation of the author(s). An editor may, however, take into account relationships of a manuscript immediately under consideration to others previously or concurrently offered by the same author(s).

2. An editor should consider manuscripts submitted for publication with all reasonable speed.

3. The sole responsibility for acceptance or rejection of a manuscript rests with the editor. Responsible and prudent exercise of this duty normally requires that the editor seek advice from reviewers, chosen for their expertise and good judgment, as to the quality and reliability of manuscripts submitted for publication. However, manuscripts may be rejected without review if considered inappropriate for the journal.

4. The editor and members of the editor's staff should not disclose any information about a manuscript under consideration to anyone other than those from whom professional advice is sought. (However, an editor who solicits, or otherwise arranges beforehand, the submission of manuscripts may need to disclose to a prospective author the fact that a relevant manuscript by another author has been received or is in preparation.) After a decision has been made about a manuscript, the editor and members of the editor's staff may disclose or publish manuscript titles and authors'

names of papers that have been accepted for publication, but no more than that unless the author's permission has been obtained.

5. An editor should respect the intellectual independence of authors.

6. Editorial responsibility and authority for any manuscript authored by an editor and submitted to the editor's journal should be delegated to some other qualified person, such as another editor of that journal or a member of its Editorial Advisory Board. Editorial consideration of the manuscript in any way or form by the author–editor would constitute a conflict of interest, and is therefore improper.

7. Unpublished information, arguments, or interpretations disclosed in a submitted manuscript should not be used in an editor's own research except with the consent of the author. However, if such information indicates that some of the editor's own research is unlikely to be profitable, the editor could ethically discontinue the work. When a manuscript is so closely related to the current or past research of an editor as to create a conflict of interest, the editor should arrange for some other qualified person to take editorial responsibility for that manuscript. In some cases, it may be appropriate to tell an author about the editor's research and plans in that area.

8. If an editor is presented with convincing evidence that the main substance or conclusions of a report published in an editor's journal are erroneous, the editor should facilitate publication of an appropriate report pointing out the error and, if possible, correcting it. The report may be written by the person who discovered the error or by an author of the original article.

9. An author may request that the editor not use certain reviewers in consideration of a manuscript. However, the editor may decide to use one or more of these reviewers if the editor feels their opinions are important in the fair consideration of a manuscript. This might be the case, for example, when a manuscript seriously disagrees with the previous work of a potential reviewer.

B. Ethical Obligations of Authors

1. An author's central obligation is to present an accurate account of the research performed as well as an objective discussion of its significance.

2. An author should recognize that journal space is a precious resource created at considerable cost. An author therefore has an obligation to use it wisely and economically.

3. A primary research report should contain sufficient detail and reference to public sources of information to permit the author's peers to repeat the work. When requested, the authors should make a reasonable effort to provide samples of unusual materials unavailable elsewhere, such as clones, microorganism strains, antibodies, etc., to other researchers, with appropriate material transfer agreements to restrict the field of use of the materials so as to protect the legitimate interests of the authors.

4. An author should cite those publications that have been influential in determining the nature of the reported work and that will guide the reader quickly to the earlier work that is essential for understanding the present investigation. Except in a review, citation of work that will not be referred to in the reported research should be minimized. An author is obligated to perform a literature search to find, and then cite, the original publications that describe closely related work. For critical materials used in the work, proper citation to sources should also be made when these were supplied by a nonauthor.

5. Any unusual hazards inherent in the chemicals, equipment, or procedures used in an investigation should be clearly identified in a manuscript reporting the work.

6. Fragmentation of research reports should be avoided. A scientist who has done extensive work on a system or group of related systems should organize publication so that each report gives a well-rounded account of a particular aspect of the general study. Fragmentation consumes journal space excessively and unduly complicates literature searches. The convenience of readers is served if reports on related studies are published in the same journal or in a small number of journals.

7. In submitting a manuscript for publication, an author should inform the editor of related manuscripts that the author has under editorial consideration or in press. Copies of those manuscripts should be supplied to the editor, and the relationships of such manuscripts to the one submitted should be indicated.

8. It is improper for an author to submit manuscripts describing essentially the same research to more than one journal of primary publication unless it is a resubmission of a manuscript rejected for or withdrawn from publication. It is generally permissible to submit a manuscript for a full paper expanding on a previously published brief preliminary account (a "communication" or "letter") of the same work. However, at the time of submission, the editor should be made aware of the earlier communication, and the preliminary communication should be cited in the manuscript.

9. An author should identify the source of all information quoted or offered, except that which is common knowledge. Information obtained privately, as in conversation, correspondence, or discussion with third parties, should not be used or reported in the author's work without explicit permission from the investigator with whom the information originated. Information obtained in the course of confidential services, such as refereeing manuscripts or grant applications, should be treated similarly.

10. An experimental or theoretical study may sometimes justify criticism, even severe criticism, of the work of another scientist. When appropriate, such criticism may be offered in published papers. However, in no case is personal criticism considered to be appropriate.

11. The co-authors of a paper should be all those persons who have made significant scientific contributions to the work reported and who share responsibility and accountability for the results. Other contributions should be indicated in a footnote or an "Acknowledgments" section. An administrative relationship to the investigation does not of itself qualify a person for co-authorship (but occasionally it may be appropriate to acknowledge major administrative assistance). Deceased persons who meet the criterion for inclusion as co-authors should be so included, with a footnote reporting date of death. No fictitious name should be listed as an author or co-author. The author who submits a manuscript for publication accepts the responsibility of having included as co-authors all persons appropriate and none inappropriate. The submitting author should have sent each living co-author a draft copy of the manuscript and have obtained the co-author's assent to co-authorship of it.

12. The authors should reveal to the editor any potential conflict of interest, e.g., a consulting or financial interest in a company, that might be affected by publication of the results contained in a manuscript. The authors should ensure that no contractual relations or proprietary considerations exist that would affect the publication of information in a submitted manuscript.

C. Ethical Obligations of Reviewers of Manuscripts

1. Inasmuch as the reviewing of manuscripts is an essential step in the publication process, and therefore in the operation of the scientific method, every scientist has an obligation to do a fair share of reviewing.

2. A chosen reviewer who feels inadequately qualified to judge the research reported in a manuscript should return it promptly to the editor.

3. A reviewer (or referee) of a manuscript should judge objectively the quality of the manuscript, of its experimental and theoretical work, of its interpretations and its exposition, with due regard to the maintenance of high scientific and literary standards. A reviewer should respect the intellectual independence of the authors.

4. A reviewer should be sensitive to the appearance of a conflict of interest when the manuscript under review is closely related to the reviewer's work in progress or published. If in doubt, the reviewer should return the manuscript promptly without review, advising the editor of the conflict of interest or bias. Alternatively, the reviewer may wish to furnish a signed review stating the reviewer's interest in the work, with the understanding that it may, at the editor's discretion, be transmitted to the author.

5. A reviewer should not evaluate a manuscript authored or co-authored by a person with whom the reviewer has a personal or professional connection if the relationship would bias judgment of the manuscript.

6. A reviewer should treat a manuscript sent for review as a confidential document. It should neither be shown to nor discussed with others except, in special cases, to persons from whom specific advice may be sought; in that event, the identities of those consulted should be disclosed to the editor.

7. Reviewers should explain and support their judgments adequately so that editors and authors may understand the basis of their comments. Any statement that an observation, derivation, or argument had been previously reported should be accompanied by the relevant citation. Unsupported assertions by reviewers (or by authors in rebuttal) are of little value and should be avoided.

8. A reviewer should be alert to failure of authors to cite relevant work by other scientists, bearing in mind that complaints that the reviewer's own research was insufficiently cited may seem self-serving. A reviewer should call to the editor's attention any substantial similarity between the manuscript under consideration and any published paper or any manuscript submitted concurrently to another journal.

9. A reviewer should act promptly, submitting a report in a timely manner. Should a reviewer receive a manuscript at a time when circumstances preclude prompt attention to it, the unreviewed manuscript should be returned immediately to the editor. Alternatively, the reviewer might notify the editor of probable delays and propose a revised review date.

10. Reviewers should not use or disclose unpublished information, arguments, or interpretations contained in a manuscript under consideration, except with the consent of the author. If this information indicates that some of the reviewer's work is unlikely to be profitable, the reviewer, however, could ethically discontinue the work. In some cases, it may be appropriate for the reviewer to write the author, with copy to the editor, about the reviewer's research and plans in that area.

D. Ethical Obligations of Scientists Publishing Outside the Scientific Literature

1. A scientist publishing in the popular literature has the same basic obligation to be accurate in reporting observations and unbiased in interpreting them as when publishing in a scientific journal.

2. Inasmuch as lay people may not understand scientific terminology, the scientist may find it necessary to use common words of lesser precision to increase public comprehension. In view of the importance of scientists' communicating with the general public, some loss of accuracy in that sense can be condoned. The scientist should, however, strive to keep public writing, remarks, and interviews as accurate as possible, consistent with effective communication.

3. A scientist should not proclaim a discovery to the public unless the experimental, statistical, or theoretical support for it is of strength sufficient to warrant publication in the scientific literature. An account of the experimental work and results that support a public pronouncement should be submitted as quickly as possible for publication in a scientific journal. Scientists should, however, be aware that disclosure of research results in the public press or in an electronic database or bulletin board might be considered by a journal editor as equivalent to a preliminary communication in the scientific literature.

APPENDIX *IV*

The Chemist's Code of Conduct

The American Chemical Society expects its members to adhere to the highest ethical standards. Indeed, the federal charter of the society (1937) explicitly lists among its objectives "the improvement of the qualifications and usefulness of chemists through high standards of professional ethics, education, and attainments".

Chemists have professional obligations to the public, to colleagues, and to science. One expression of these obligations is embodied in "The Chemist's Creed", approved by the ACS Council in 1965. The principles of conduct enumerated herein are intended to replace "The Chemist's Creed". They were prepared by the Council Committee on Professional Relations, approved by the Council (March 16, 1994), and adopted by the Board of Directors (June 3, 1994) for the guidance of ACS members in various professional dealings, especially those involving conflicts of interest.

Chemists acknowledge responsibilities to

◆ ***The Public*** Chemists have a professional responsibility to serve the public interest and welfare and to further knowledge of science. Chemists should actively be concerned with the health and welfare of co-workers, consumers, and the community. Public comments on scientific matters should be made with care and precision, without unsubstantiated, exaggerated, or premature statements.

◆ ***The Science of Chemistry*** Chemists should seek to advance chemical science, understand the limitations of their knowledge, and respect the truth. Chemists should ensure that their scientific contributions, and those of their collaborators, are thorough, accurate, and unbiased in design, implementation, and presentation.

◆ ***The Profession*** Chemists should remain current with developments in their field, share ideas and information, keep accurate and complete laboratory records, maintain integrity in all conduct and publications, and give due credit to the contributions of others. Conflicts of interest and scientific misconduct, such as fabrication, falsification, and plagiarism, are incompatible with this Code.

◆ ***The Employer*** Chemists should promote and protect the legitimate interests of their employers, perform work honestly and competently, fulfill obligations, and safeguard proprietary information.

◆ ***Employees*** Chemists, as employers, should treat subordinates with respect for their professionalism and concern for their well-being, and provide them with a safe, congenial working environment, fair compensation, and proper acknowledgment of their scientific contributions.

◆ ***Students*** Chemists should regard the tutelage of students as a trust conferred by society for the promotion of the students' learning and professional development. Each student should be treated respectfully and without exploitation.

◆ ***Associates*** Chemists should treat associates with respect, regardless of the level of their formal education, encourage them, learn with them, share ideas honestly, and give credit for their contributions.

◆ ***Clients*** Chemists should serve clients faithfully and incorruptibly, respect confidentiality, advise honestly, and charge fairly.

◆ ***The Environment*** Chemists should understand and anticipate the environmental consequences of their work. Chemists have responsibility to avoid pollution and to protect the environment.

APPENDIX V

Proofreaders' Marks

On galleys, mark in the margins and use these symbols. For Greek letters, mathematical symbols, or other special symbols, draw them in the margin but also spell out the word and circle it.

Mark	**Operational Signs**
	Delete
	Close up; delete space
	Delete and close up
#	Insert space
¶	Begin new paragraph
no ¶	Run paragraphs together
	One em space
	Move right
	Move left
	Center
	Move up
	Move down
	Align horizontally
	Align vertically
tr	Transpose
sp	Spell out
stet	Let it stand
fl	Flush left
fl rt	Flush right
ctr	Center

Mark	**Typographical Signs**
lc	Lowercase a capital letter
cap	Capitalize a lowercase letter
sc	Set in small capitals
ital	Set in italic type
rom	Set in roman type
bf	Set in boldface type
wf	Wrong font; set in correct type
	Superscript
	Subscript

Mark	**Punctuation Marks**
	Insert comma
	Insert apostrophe (or single quotation mark)
	Insert quotation marks
	Insert period
?	Insert question mark
;	Insert semicolon
:	Insert colon
/=/	Insert hyphen
	Insert em dash
	Insert en dash

The photochemistry of α,β unsaturated ketones has attracted much attention and is still a field ~~field~~ of current interest. 1 Numerous examples of such photochemical transformations are well-documented for cyclic enones and dienones, including both cycloaddition re actions and rearrangements. For example, cyclopentenones 1 and 2 readily rearrange to cyclopropyl ketenes upon irrandiation. Recently, the related cyclohexadienone/butadienyl ketene rearrangement has been shown to be a highly useful tool in the synthesis of natural products and macrocyclic lactones 2

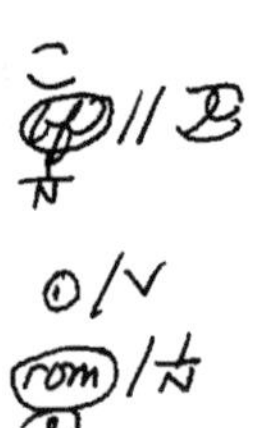

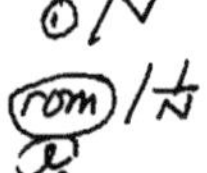

Whereas cis/trans isomerization, photodimerization, and [2 + 2] cycloadditions of acyclic α,β-undsaturated ketones are well-investigated photochemical transformations, comparatively little is documented concerning the photochemistry of such enones involving photodissociation, rearrangement or both. Clearly, the absence of ring strain lowers the reactivity twoard bond cleavage and renders an initial Norrish type I reaction unlikely. Introduction of radical stabilizing groups in the α-position of the enone may, however, be expected to change the reactivity of the enone in facor ofthe photochemical α-cleavage and subsequent reactions derived from the resulting radical pairs.

Sample of a proofread galley.

On galleys, corrections on a line are separated by slashes. Two slashes in a row indicate that the first type of correction should be repeated at the second insertion point.

On manuscripts, you can mark in place, not in the margins. Some of the symbols differ, too. When you use these marks, you need not explain them in the margin of the manuscript. They are standard.

~~word~~	Strike through to delete a word or words.
word	Triple underline to capitalize the "w".
Word	Slash to make the "w" lowercase.
wrod	Transpose two letters.
words two	Transpose two words.
Word	Double underline to make "ord" small capitals.
word	Draw a wavy line to indicate bold face.
word	Underline to indicate italic type.
word1	Draw an inverted carat to indicate superscript.
word2	Draw a carat to indicate subscript.
keep ~~word~~	Put dots or short dashes under copy that you wish to retain as it originally appeared

The photochemistry of α,β-unsaturated ketones has attracted much attention and is still a field ~~field~~ of current interest.[1] Numerous examples of such photochemical transformations are well-documented for cyclic enones and dienones, including both cycloaddition re actions and rearrangements. For example, cyclopentenones 1 and 2 readily rearrange to cyclopropyl ketenes upon irrandiation. Recently, the related cyclohexadienone-butadienyl ketene rearrangement has been shown to be a highly useful tool in the synthesis of natural products and macrocyclic lactones.[2]

Whereas cis-trans isomerization, photodimerization, and [2 + 2] cycloadditions of acyclic α,β-undsaturated ketones are well-investigated photochemical transformations, comparatively little is documented concerning the photochemistry of such enones involving photodissociation, rearrangement, or both. Clearly, the absence of ring strain lowers the reactivity twoard bond cleavage and renders an initial Norrish type I reaction unlikely. Introduction of radical stabilizing groups in the α-position of the enone may, however, be expected to change the reactivity of the enone in favor of the photochemical α-cleavage and subsequent reactions derived from the resulting radical pairs.

Sample of a corrected manuscript.

Index

A

B

C

D

E

F

G

J

M

Q

R

T

U

X

Y

Z

Production editor: Paula M. Bérard
Indexer: Zeki Erim
◆ ◆
Text design and typesetting by Betsy Kulamer, Washington, DC
Cover design by Auras Design, Inc., Washington, DC
Printing and binding by R. R. Donnelley & Sons Company, Harrisonburg, VA